Flood Scour for Bridges and Highways

About the Author

Mohiuddin Ali Khan, Ph.D., P.E., M.A.S.C.E., M.I.C.E., D.I.C. (London), is a consulting engineer and an adjunct professor at Temple University in Philadelphia, Pennsylvania. He is the author of several textbooks on bridge engineering, including *Bridge and Highway Structure Rehabilitation and Repair* (McGraw-Hill Education), and of the chapter on bridge engineering in *Handbook of Transportation Engineering* (McGraw-Hill Education). Dr. Khan helped to develop the New Jersey Department of Transportation Design Manual for Bridges and Structures section on scour at bridges. He is licensed as a Professional Engineer in five states and is a Chartered Engineer in England. Dr. Khan has received professional awards from ASCE and the Institution of Civil Engineers (England), and is an officeholder of ASCE Southern New Jersey.

Flood Scour for Bridges and Highways

Prevention and Control of Soil Erosion

Mohiuddin Ali Khan, Ph.D., P.E.

Mc Graw Hill Education

New York Chicago San Francisco Athens
London Madrid Mexico City Milan
New Delhi Singapore Sydney Toronto

McGraw-Hill Education books are available at special quantity discounts to use as premiums and sales promotions or for use in corporate training programs. To contact a representative, please visit the Contact Us page at www.mhprofessional.com.

Flood Scour for Bridges and Highways: Prevention and Control of Soil Erosion

1 2 3 4 5 6 7 8 9 10 QVS/QVS 20 19 18 17 16 15

ISBN 978-0-07-182507-8
MHID 0-07-182507-X

This book is printed on acid-free paper.

Sponsoring Editor
Lauren Poplawski

Editorial Supervisor
Stephen M. Smith

Production Supervisor
Lynn M. Messina

Project Manager
Raghavi Khullar,
Cenveo® Publisher Services

Copy Editor
Patti Scott

Proofreader
Amy Rodriguez

Art Director, Cover
Jeff Weeks

Composition
Cenveo Publisher Services

Contents

Preface

In addition to my books on bridges with other publishers, my earlier books on bridges with McGraw-Hill are *Bridge and Highway Structure Rehabilitation and Repair* and the bridge engineering chapter in their widely used *Handbook of Transportation Engineering*. During my career as a consulting engineer and teacher of structural engineering subjects, I observed that few textbooks were available on bridge design or inspection and rating. Bridge engineering background information needs to be presented beyond AASHTO (American Association of State Highway and Transportation Officials) LRFD specifications and the manual for condition evaluation and LRFR for highway bridges. What is especially needed is coverage of recent state-of-the-art developments.

A bridge engineer has to face many aspects, structural and administrative. This book is about problem solving and providing insights into important issues. I find that there is a need for a book to address new technology on planning, detailed design, and safety of river bridges. Topics such as scour countermeasures need to be summed up in the form of a book for the benefit of students and practicing engineers.

Although AASHTO LRFD and state codes have been updated to include some of the effects of floods and erosion, it takes time for them to catch up with innovations and incorporate new procedures. These codes may serve as guidelines but are not aimed at educating the reader in fundamentals. Some state codes are more enterprising in developing the specific scour-resistant technology compared to federal codes. The subject matter covered by these voluminous specifications is extensive, diverse, in-depth, and mathematical. This book will serve basically as a companion reference manual to the codes and specifications, with emphasis on new topics, and it will focus on both traditional and nontraditional design problems.

Practicing engineers continuously find the need for a book that can simplify the presentation and make it more palatable for office use. The book should address the state of the art of bridge engineering, highlight major issues, offer necessary explanations, provide sample solved examples for day-to-day design issues, and use case studies of practical problems. If the developments in the subject are reported by this book and understood by teachers and students at the university level, then future practicing engineers will have a head start and the purpose of the book will be well served.

Objectives

The issues basically present an overall perspective of the multibillion-dollar transportation industry, the resulting advantages of which to society appear to surpass those offered by other disciplines, except perhaps the medical profession. Indeed, transportation is a complement even to the medical profession, by facilitating an ambulance trip to the hospital in the shortest time, without being stuck in a traffic jam or delayed by a roadway closed for repairs.

This book's theme and scope are about water engineering, intense rainfall, and collapse of bridges from soil erosion. Its special features are about facing "ground realities" or "river realities" when fixing a bridge. No more mathematics should be required than is actually needed, since mathematical technique is a means to an end and is not an end in itself. Innovative ideas for structural planning and precast connection details developed by design and sometimes construction teams over the years will be presented, making this book unique.

Need to Incorporate Growing Technologies

There is a need to update the growing volume of knowledge about the important and practical subjects of hydrology, hydraulics, and soil erosion. More important is to speak the engineer's language and take a series of small steps rather than one big leap. Also recommending one or more outdated books to the students does not seem to solve the problem.

For bridges located on waterways, use of deeper foundations and adequate bridge openings is recommended. An earlier handbook on protection against flood scour (a major cause of bridge failures in the United States), which I coauthored for the New Jersey DOT jointly with the City University of New York, has now been approved by the Federal Highway Administration (FHWA) for use by consultants. I also developed Sections 45 and 46 of the 2002 *NJDOT LRFD Bridge Design Manual* pertaining to seismic design and bridge scour.

I find that it is the right time to address the latest changes in the code requirements (such as important changes in the Structural Inventory and Assessment Datasheet), and the method of rating needs to be based on the new LRFR method. New FHWA Hydraulic Engineering Circulars for scour analysis and countermeasures design are other examples.

Track Record of Successful Bridge Design and Construction

The bridge engineer needs to be groomed for a technical decision-making position, starting from a junior engineer, then to a senior engineer, and then to a team leader's sharing of greater responsibilities. In the design environment, there is a need to develop an engineering sense of issues related to flood estimates and to resolve them by countermeasures, in the time available.

The audience for whom the book is intended includes practicing structural and bridge engineers in the United States and abroad, government agency engineers, planners, highway and traffic engineers, geotechnical engineers, hydraulic engineers, transportation specialists, contractors, countermeasures product vendors, senior engineering students, and university professors. Also, the book is intended for those with the following job titles: engineering managers, project managers, design engineers, construction supervisors, instructors, and final-year students. In addition, the book is intended for those who might be affiliated with the following professional associations and organizations: American Society of Civil Engineers (ASCE), American Society of Highway Engineers, American Concrete Institute, American Institute of Steel Construction, AASHTO, FHWA, Transportation Research Board, and state-board, professional engineering licensing agencies.

Key Benefits

Personal experience in teaching the subject and in the design and construction supervision of bridges located in the United States, England, and Asia can be utilized in developing a book that can be recommended to students and bridge engineers alike. Some of the prestigious bridge projects in which I have successfully participated as a designer include curved steel bridges at Washington National Airport; continuous-span curved bridges currently under design at Washington Dulles Airport; the elevated Monfayette Expressway project in western Pennsylvania with high piers; prestressed concrete box bridges for the Jeddah-Mecca Expressway; numerous rehabilitations of highway transit bridges in New York City, Boston, and Philadelphia; and one of the first integral abutment bridges on Route 46 over the Peckman River in New Jersey.

In addition, the book will address the following:

1. Recent advancements in bridge inspection, rating, and hydrology software such as StreamStats and HEC-RAS, and the availability of other specialized software.
2. Changes in AASHTO design codes for resistance to soil scour from floods, the design manuals, and typical guidelines of state design manuals.
3. Modern materials technology as well as developments in new types of concrete, steel, and other construction and repair materials.
4. Developments in construction methodology such as manufacture of modular bridges, the use of long-span cranes, hauling of long-span girders, and accelerated bridge construction.
5. Revisions to design criteria based on recognition of recurring failures of bridges from floods.
6. Updates on important developments in river bridges and disaster management techniques.

Applications to Analysis, Design, Inspection, and Drawings Preparation

Professional engineers will learn to create a more-efficient design process, and construction managers will learn how to save time and money with a better bridge management system for river bridges.

Recent innovations in the telecommunications, computer software, automobile, aircraft, and telephone industries seem to have influenced the overall format of engineering disciplines. It is important for any book to generate sufficient interest in the subject matter being utilized by university students. A realistic approach is likely to help students so that they are mentally prepared for the design tasks ahead, rather than focused only on the design formulas. The book may be used as both a textbook and a desk reference.

There is a wealth of information available in the research and technical publications listed at the end of each chapter. Publications include the latest proceedings of international conferences, technical journals, and periodicals. Listed papers are mainly by learned authors and speakers from all over the world. Notable among the conferences are the SEI Proceedings, the New York City Bridge Conference, the Pittsburgh Bridge Conference, and specialty conferences and seminars. Conferences I chaired in the past 15 years on structural engineering themes are those organized by the ASCE Structural

Engineering Institute; Temple University, Philadelphia; and the OMICS hydrology conferences in Raleigh, North Carolina, and Hyderabad, India.

Journals, which provide the latest articles on related issues including vendor products from the world of construction, include *Civil Engineering, New Civil Engineer, Roads & Bridges, Structural Engineer,* and *Structure*; Elsevier's *Journal of Hydrology*; and ASCE journals *Bridge Engineering, Journal of the Structural Division,* and *Modern Steel Construction.*

This book contains the much-needed information on innovations presented in selected technical resources. A section is included on disaster management, following the recent collapse of bridges on waterways such as the Skagit River in Washington State and the Mississippi River (I-35W Bridge). Many chapters and the solved examples contain references to the applicable codes and specifications. This is also a legal requirement that the engineer needs to be well-versed with the relevant codes of practice.

Mohiuddin Ali Khan, Ph.D., P.E.

Symbols

C	Coefficient
D_{50}	Median stone diameter, ft
Fr	Upstream Froude number
K	Coefficient for pier shape
K	Velocity magnification factor
O_{95}	Opening size where 95 percent of pores are smaller, ft
Q	Discharge, ft^3/s or fps
Ss	Specific gravity of riprap
V	Flow velocity, ft/s
W	Width of abutment at base, ft
W	Width of pier at base, ft
W_R	Width of riprap layer adjacent to footing, ft
X	Width of abutment footing, ft
Y	Design depth of riprap, ft
d	Thickness of riprap
d	Water depth
y	Scour depth, ft
y_0	Flow depth, ft
α	Location factor
β	Flood factor
ζ	Scaling factor for application to geology of soil
η	Scaling factor for application to width of bridge opening
ξ	Scaling factor for application to river training measures
χ	Scaling factor for application to remaining bridge life assessed
ψ	Scaling factor for application to underwater inspection
ΨCR	Stability factor
θ	Angle between impinging flow direction and vertical wall
θ	Side slope angle with respect to horizontal plane
ρ	Density of water, lb/ft^3
ρcob	Density of block material, lb/ft^3

Other symbols are defined and described next to equations in the text.

Flood Scour for Bridges and Highways

PART 1

Erosion Problems, Review of Failures, and Progress Made

CHAPTER 1

Water May Quench Thirst But May Also Drown

1.1 Generation of Floods and Disruptions to Life

There is a need to control the recurring floods in our rivers and coastal areas. Floods result in the loss of lives, damage to crops and agriculture, soil erosion, and damage to property and the infrastructure, while at the same time lowering the standard of living and increasing the cost of living. The causes are both natural and introduced by humans. Floods are caused by intense rainfall, monsoons, snowmelt from mountaintops and glaciers, hurricanes, tsunamis, and occasional dam, reservoir, and levee breaks.

Scope of Book on Flood Prevention and Remediation

In the nine chapters of the book and the two appendices, the essential multiple aspects of promoting scour resistance technology for transportation structures are discussed. The textbook comprises research results by the author, notable design publications by federal and state transportation and environmental agencies, university research results, the American Society of Civil Engineers (ASCE) and international conferences, and useful data provided by field visits. A logical and scientific approach is used in compiling the format of the book.

The focus of this book is the protection of infrastructure such as bridges, highways, and highway structures. This book reviews the adverse effects of floods on the transportation network, bridges, and highway structures. The objectives of the book include the provision of comprehensive and detailed guidelines on bridge scour countermeasures, as a means to mitigate the scour resulting from riverine floods and coastal tides both in the United States and abroad.

This book can help in planning a bridge on a river to reduce the scour risk—by designing deep foundations and countermeasures for foundations of abutments, wing walls, piers, and retaining walls—and minimize failure through added protection.

The guidelines recommended in this book can be supplemented with the available countermeasure design data of any manufacturer's development work. Detailed specifications for several proprietary countermeasures may be obtained directly from the vendor's website, which may develop new products after research and experimentation.

This introductory chapter covers floods and the latest technology in the mitigation of erosion and scour. It addresses the issues of recurring floods that lead to failure of bridges by erosion and scour or the highway approaches connecting the bridges with highways, and in addition the methods of their mitigation. The disciplines include asset management by owner, the options of repair versus replacement, and application of various types of countermeasures and sometimes forensic engineering as a result of designing unsafe structures. Stream instability and obstruction to flow address the influence of the location and setback of abutments, wing walls, and piers in the rivers, which may cause obstruction to fast-flowing floodwater and to vortex formations; a scientific approach may be taken to achieve stream stability during floods. A brief overview of the organization of these chapters follows. Relevant citations are included in the Bibliography at the end of each chapter.

Chapter 1 presents an introduction to floods and a definition of scour problems and mitigation. In addition, it discusses

- The hydrologic cycle of surface water and groundwater and multiple sources of floods
- A checkup on infrastructure health and a survey of recent floods that affect infrastructure
- The need to maintain the transportation network at reasonable cost as well as asset management to include flood mitigation measures
- Global problems of flood hazards, drought, and water conservation and the development of an environmental policy for resisting floods

A literature review of the subject and publications by the Federal Highway Administration (FHWA) Hydraulic Engineering Circulars, the U.S. Army Corps of Engineers (USACE), the American Association of State Highway and Transportation Officials (AASHTO), the National Cooperative Highway Research Program (NCHRP), the Environmental Protection Agency (EPA), the Federal Emergency Management Association (FEMA), and state design manuals for scour at bridges has been carried out.

See Figure 1.1*a* and *b*. In Figure 1.1*a*, many towns are located near riverbanks and are inundated with floods. Figure 1.1*b* shows an example of flooding of a low-lying town, situated close to a river floodplain and thereby disrupting traffic, commerce, and industry.

Chapter 2 addresses a wide range of flood-related issues including these:

- The description of river instability caused by flow obstruction and floodplain formation
- The erosive effects of floods and coastal tides on transportation
- Multiple types of riverine flow processes and a review of the basic concepts of open-channel hydraulics and the resulting hydraulic issues of bridges on waterways
- The role of movable bridges in facilitating navigation in rivers
- Debris accumulation upstream of bridges
- Current bridge scour control practice; planning and design issues to restrict future scour at scour-critical bridges; AASHTO superflood provisions and extreme-load combinations
- Funding provisions for repairs and replacement of bridges damaged during floods

(a)

(b)

FIGURE 1.1 (a) Floods in summer 2013—rescue staffers repair the damaged canal levees broken due to flash floods at Jhang-Faisalabad Bypass Road in Northern Pakistan. (b) Process of flood formation in streams results from the combined effect of surface flow from precipitation and the rise of the groundwater table due to infiltration.

Chapter 3 introduces various types of scour countermeasures, including river diversion and the addition of auxiliary openings:

- Modifications to the Federal Highway Administration Hydraulic Engineering Circular (HEC) No. 23 (HEC-23) countermeasures matrix
- Applications of HEC-18 and HEC-23 procedures
- Case studies of scour-prone and scour-critical rivers
- Methods for investigating conditions of existing footings
- Planning considerations for safer bridges
- Selection of flood-resistant countermeasures
- Factors affecting the detailed design of countermeasures
- Use of rock riprap as temporary armoring
- Application of modern techniques to contain erosion
- Applications of geotextile and granular filters
- Design guidelines for alternate armoring type of countermeasures

Chapter 4 describes these innovative tools of flood disaster management:

- An overview of scour repairs
- Need to adopt precautions to prevent failures
- Introduction of a fresh approach in countermeasure design and construction that uses advances in construction technology
- Applications of modern construction materials and use of recyclable materials technology
- Advances in concrete technology, use of fiber-reinforced polymer (FRP) concrete and products, and developments in steel materials and metallurgy
- Accelerated bridge construction over rivers and replacement methods developing design codes for river bridges
- Analysis of performance of steel bridges over rivers
- New techniques for monitoring and reconstruction
- Postflood disaster management

Chapter 5 traces the history of failures of bridges from scour, failure analysis, and the lessons to be learned. It covers the following:

- How river bridges become structurally deficient
- Study of bridge failures including the vulnerability to failure of bridges on waterways; analysis of river bridge failures; typical bridge failures from scour; examples of bridge collapse over rivers; lessons learned from European bridge failures; bridge failures from ship impact on the substructure or superstructure; and an additional list of failures of river bridges; studies of well-known bridge failures
- Forensic engineering application to bridge failures and recommendations for avoiding failures

- Rehabilitation and repair of existing structures; underwater repairs
- Preventive actions for vessel collision or floating ice impacts

Chapter 6 discusses the influence of soil conditions on the flow of water and open-channel hydraulics including

- Standard procedures for hydrologic and hydraulic analysis and design as well as applications of the AASHTO specifications
- Modeling for hydraulic analysis; performance of scour analysis; scour of piles, pile groups, and caissons
- Contents of a hydraulic report including guidelines for preparing geotechnical reports and plans, specifications, and estimates
- Foundation modification by structural countermeasures
- Investigation of unknown foundations
- Types of foundation selection for new bridges

Chapter 7 deals with hydrology and the identification and selection of scour-critical rivers and bridges, including

- Causes and effects of floods
- Methods of hydrologic analysis with practical StreamStats examples
- Scour practice survey conducted by the author's team
- Identification of scour-critical bridges on flood-prone rivers
- Planning of substructures in relation to hydrology data
- Planning considerations for superstructures based on hydrology and hydraulics

Chapter 8 focuses on computing scour depths and the case study of practical design of scour countermeasures for protection of new bridges and rehabilitation of existing deficient highway structures. It includes the following:

- Need for scour studies and remediation
- Use of a hydraulic assessment checklist
- Rating of scour-critical bridges using the Pennsylvania's Scour Critical Bridge Indicator (SCBI) method
- Summary of factors affecting scour
- Vulnerability index for scour-critical bridges
- Scour case studies by the author

Chapter 9 explains the selection, planning, monitoring, and design of countermeasures in detail, including

- Monitoring of scour-critical bridges
- An essential scour monitoring program
- Methods for investigating conditions for vulnerable foundations
- Adoption of a flood watch list for scour-critical rivers

- Description of monitoring countermeasures
- Case study of the Pennsylvania Department of Transportation (PennDOT) bridge scour evaluation program
- Studies of inspection reports based on monitoring inventories
- Formation of scour holes at bridge sites
- Maintenance of the environment through monitoring
- Meeting the permit requirements for reconstruction
- Emergency bridge protection measures
- Interference from underground utilities
- Permitting procedures to be followed (according to Department of Environmental Protection [DEP] regulations and acts of Congress)

Appendices A and B are applicable as supporting information to Chapters 1 to 9.

Flash Floods The term *flash flood* refers to peak flooding that occurs within 6 h of heavy rain. Flash floods are caused by slow-moving thunderstorms and the intensity and duration of rainfall. They may occur within minutes or hours and without warning. Flood watch is required. Flooding of streams and low-lying areas occurs. Flash floods are generally unexpected and can cause overtopping of the deck, disrupting traffic flow.

Management of Flood Disasters and Severe Droughts

Water supplied by nature in the rivers and streams may be regarded more as a friend and less as a foe. Floodwater generally exceeds the customary demand. The excess water needs to be disposed of before it causes several types of damage to property, agriculture, and cattle. Although floods are considered a natural disaster, the large quantity of water from nature if controlled can be beneficial. The floodplain that is created helps the farmlands after the water recedes. It provides crops and vegetation, and in some countries where the water supply is scarce or there is drought, it can be most beneficial and is considered a gift from God. By creating reservoirs or by river diversion, the flood water can be utilized.

The past 15 years of drought conditions in the West constitute a megadrought. The drought that has been experienced in California in recent years is part of a larger pattern throughout the whole western United States that hasn't been seen in more than 850 years.

Ocean water is the most abundant substance on earth. It is the only substance that exists in all three states: solid, liquid, and gas (i.e., ice, water, and vapors). Its destructive properties are seen in the creation of kinetic forces, crushing effects, freezing, erosion of soil, and carving out of valleys.

Navigation in rivers, sea transport of ships, and the megamovers have been the backbone of the development of human civilization. The coastal areas are likely to be affected by the recurring floods, owing to the rise in seawater elevation.

Issues of economics and health arise out of the effects on the chemistry of water due to occasional floods. Modern engineering efforts will be required to control the forces of water and to make the demand and supply or distribution of water more equitable. The sources of the water supply, including floods and the disposal and consumption of water, are linked. Transportation networks and bridge foundations are also affected.

Recurring Floods Much of the rainfall falls on the oceans. The rainfall that falls on land is likely to cause floods. Flooding is a natural and essential feature of many ecosystems. Along rivers and streams, floodwaters inundate floodplains, transporting nutrients and sediments to the floodplain ecosystem. Soil erosion by fast-flowing water leads to the collapse of highway embankments and settlement of bridge foundations and other highway structures.

Also the floodplains are habitats for many plants and animals. Upstream urbanization tends to increase downstream flooding—as an unnatural event—thus putting humans and other ecosystems in jeopardy.

Floods and Drought Are an Act of God But Can Be Contained

Environment and natural elements cannot be ignored. Old Native American concepts of the four natural elements are always applicable; i.e., "water, fire, wind, and earth are gods unto themselves." Too much water or less water but with greater frequency are God sent. Other disasters come in different forms and cause havoc.

1. The fury of major floods can lead to loss of many lives.
2. The impact of recurring small floods can also be detrimental.
3. Collapse, near failure, and damage or potential damage are all undesirable and difficult to fix.

The study of bridge failures are like the study of open books. The experience gained from failure analysis can be applied to develop existing design code and improve construction practices. The failure mechanism and a lot more can be learned from studying the reasons for failure. Through failure case studies we learn the technical issues, e.g., why regular inspection of transportation facilities has been critical to public safety over the past 20 years. Technical issues may include

- Implementing regulatory changes
- Fixing identified problems
- Continued growth of the population and commerce, which has resulted in a greater need to perform regular structural inspections of transportation facilities
- A problem of growing damage to bridges identified over time which has resulted in an increase in the minimum design stress

The study of bridge failures includes

1. Common structural damages
   a. Material failure
   b. Abutment cracking, movement, and settlement of the pier
   c. Skew effects
   d. Expansive soils
   e. Unknown foundations
   f. Defects in concrete arches
2. Modification of design criteria

Flooding may occur due to usual hydrologic cycle, global warming, and climate changes. It refers to a long-term event that may last a week or more. The Weather Forecast Office of the National Weather Service usually issues warnings of incoming rain and floods. Floods are accompanied by soil erosion. Floods are an environmental phenomenon caused by natural causes, such as rainfall, tsunamis, snowmelt, and melting of glaciers, or human-created causes. Damages incurred by peak floods can be measured by

- Duration of floods
- Area inundated
- Number of lost lives or displaced persons
- Crop and property damage
- Number of record river levels

Variables Controlling the Hydrology of a Region These comprise

- Precipitation (source, type, rate, and amount)
- Vegetation
- Temperature
- Soil
- Geology
- Topography
- Stream gradient
- Changes in demography
- Addition of reservoirs and other human-made structures

The operation of more than 7500 active stream gauges nationwide enables the U.S. Geological Survey (USGS) to provide data for a variety of needs such as flood prediction and characterization. The stream gauges provide useful data for

- National Weather Service flood forecasting
- USACE activities for water control, fighting floods, and mitigation
- FEMA management and mitigation of emergencies
- Numerous state and local agencies
- Accurate rating curves for calibration

Misuses of Water and Environmental Effects of Floods

Engineering efforts to supply purified water to the people and dispose of used sewage water have reduced the dangers of ill health and increased the average human life span by 30 years. Moreover, the shortage of good-quality water can become a serious health problem and may even put the survival of the human race at risk. Also, livestock and marine life will be adversely affected.

Water may be seen as a friendly monster, which needs to be tamed. Examples of the fury of water can be seen in coastal tides, tsunamis, intense rainfall, and river floods.

1. An effective drainage system of the watershed is required.
2. Protection of the coastline, highway embankments, and bridge footings from continued erosion is among the foremost duties of civil, structural, and hydraulic engineers.
3. Oceans are slowly becoming contaminated due to acid rains that carry factory smoke and impure gases in the air. Poor-quality seawater can damage marine life, reducing the essential fish supply and the whale population. The seawater is not suitable for domestic use. Desalination is an expensive process. A line in the poem "The Rime of the Ancient Mariner" by Samuel Coleridge describes it well: "water, water everywhere, nor any drop to drink."
4. It has come to the point where our daily drinking water is being manufactured or processed in billions of plastic bottles (with 0-Calorie labels) in factories, to make water fit for drinking, even when piped water supply from the township or the water department is available. Some households need to use expensive filters in the kitchens for safety. Our forefathers did not experience such outside dependence or even interference, for what used to be the free and pure water supply came from nature. Human existence depends upon a pure and adequate water supply.

Preventing scour and soil erosion is a relatively new subject linked to demography. With the increase in population, the river water supply, evaporation, and precipitation have increased. Bridges designed and built over 50 years ago did not have to satisfy the same criteria for the protection of bridge foundations. Many older bridges do not conform to the requirements of new technology. Hence suitable types of countermeasures are required.

Flood Mitigation by Using Various Countermeasures

Engineering can be employed to help contain floods. The engineering disciplines required for mitigation are the sciences related essentially to

- Water resources
- Hydrology
- Hydraulics

as well as the structural, geotechnical, and environmental sciences.

1.2 The Hydrologic Cycle and Multiple Sources of Floods

A literature review of the hydrologic cycle and water technology is carried out in the chapters of the book.

A Hjelmfelt and Cassidy publication by Iowa State University Press in 1975 defined the *hydrologic cycle* as follows:

In the hydrologic cycle, transformation of water occurs from land surface to the atmosphere by evaporation and from atmosphere to the land surface, oceans and rivers by precipitation and to the subsurface by percolation.

The world and the continents we live on are surrounded by water. There is no dearth of water resources owing to the subsurface water, wells, ocean tides, rainfall, snow, glacier melts, floods, and the overtopping of the riverbanks, breach of levees, or occasional dam bursts, sometimes washing away townships. The daily hydrologic cycle comprises evaporation, precipitation, subsurface flow, and flow in rivers.

Urban runoff can be highly polluted with grease, oil, solvents, heavy metals, and trash. Daily demand is mainly for the domestic water supply, farming, and agriculture and industrial production. If there is a drought, the water supply is less than the demand, and food shortages will result. On the other hand, when there are floods, the water supply exceeds the demand. The excess water will damage the infrastructure. It is also known as the water cycle, and it comprises two processes.

Evaporation and precipitation: Driven by solar energy, evaporation from plants, waterways, and lakes occurs. Changes in temperatures cause precipitation.

1. Evaporation of water is followed by precipitation or rainfall: Evaporation from leaves is called *transpiration*. The total loss from leaves and soil is called *evapotranspiration*. The water drops formed in the clouds are relatively pure because when water molecules evaporate, the heavier impurities are left behind. The amount of moisture in the air depends on the temperature. Warmer air contains higher amounts of moisture.
2. Precipitation is followed by surface runoff: Less than 10 percent of available freshwater runoff is collected mainly for crops and livestock consumption. Industrial use accounts for 7 percent of the runoff, infiltration in ground, and percolation. Relative humidity is a measure of the amount of moisture present in the air compared to how much the air can hold at a particular temperature, if fully saturated.

When the moisture exceeds saturation, a cloud or fog forms. Air contains small particles known as condensation nuclei. A single drop of rain contains a very large number of droplets. If the temperature falls below freezing, snowflakes form. The daily solar energy drives the wind and deposits rain or drizzle, hail, or snow by the force of gravity. The hydrologic cycle is completed when water drops fall on land, lakes, and streams from where it originated.

Effects of Climate Changes The world's water supply is maintained by the hydrologic cycle.

However, the distribution of water is not uniform across the globe. Climate changes can also cause drought conditions or heavy rains and floods. Water shortages are common in some countries in South America, parts of Africa, Australia, the Indian subcontinent, the Far East, Spain, southern Italy or Spain, and the desert in the Southwest states of the United States. Worldwide, over 150 river basins are shared by two or more countries.

Water demand: Water quantity or water supply pertains to the delivery and volume of water. It is expected that the demand for domestic water supply, irrigation, and industrial use will increase manyfold, causing further shortages. It is important to conserve water. In the United States, agriculture is the biggest water user. Average domestic use is about 90 gal per person per day while growing food may take 800 gal. With industrial consumption for power supply and nuclear plants, etc., an average of 1500 gal per person per day is required. These are some other causes of water shortages:

- Overdraft of groundwater. In coastal areas it may increase saltwater intrusion in soil. Thousands of hectares of farmland were eventually taken out of production after several years, due to saltwater content increase. One example is loss of corn in the midwestern United States.
- Water pollution.
- Flooding. As mentioned earlier, snowmelt and rainfall cause floods. Soil erosion occurs due to fast-flowing water that damages highways and bridge foundations. Vegetation slows down runoff and increases infiltration.
- River sedimentation.
- Drainage of wetlands.
- Degradation of aquifers, bays, and estuaries (e.g., demand of water and overuse due to scarcity or drought, in cities are likely to dry up the resources of water).

The *Elsevier Journal of Hydrology* published a paper by the University of Kansas in May 2000 entitled "From Safe Yield to Sustainable Development of Water Resources—the Kansas Experience." In general, the sustainable yield of an aquifer must be considerably less than the recharge if adequate amounts of water are to be available to sustain both the quantity and quality of streams, springs, wetlands, and groundwater-dependent ecosystems. To ensure sustainability, it is imperative that water limits be established based on the hydrologic principles of mass balance.

Meeting the Daily Demand of Pure Water Supply for Domestic Needs

- The domestic water supply can be saved on rooftops for domestic use.
- In some rural areas, the common practice is to fetch water in containers from distant ponds.
- Water can be saved on a large scale by constructing dams and creating reservoirs.
- However, habitats of sheep and bears may be destroyed (e.g., the case of the Waterton Canyon Dam near Denver, Colorado).
- According to World Health Organization, up to 80 percent of diseases in some countries are caused by contamination of domestic water. Floodwater can carry impurities from the soil surface.

Watershed Management

Population centers sometimes do not have sufficient water supply sources nearby, so these areas must rely on distant water sources, necessitating the transportation of good-quality water.

During Drought Conditions, Both Surface and Ground Water Supplies Get Curtailed

1. Water quality is a measure of the physical, chemical, and biological conditions of a body of water. Pollution causes the water quality to deviate from normal conditions.

2. Prevention of the dumping of hazardous waste and contaminated waste into rivers will keep the rivers and lakes clean. Estuaries and oceans also get contaminated by floodwater runoff and overtopping of riverbanks.
3. Erosion of banks and bridge footings damages the essential infrastructure and results in failures of bridges and losses to commerce and industry.

Effects on Groundwater Groundwater is the primary source of drinking water for about one-half of the U.S. population. Groundwater is the water found beneath the earth's surface within the zone of saturation, below the water table. According to an estimate, of the average 40 in of annual precipitation in Pennsylvania, 20 in is lost through evaporation and transpiration, 10 in becomes surface runoff, and 10 in recharges the groundwater. Most of the water used by humans comes from groundwater aquifers or surface water. For example, 50 percent of the average annual precipitation in Pennsylvania evaporates or is transpired by plants; nationally, it's closer to 66 percent, and this is water that never enters our water supplies.

But evaporation and transpiration rates usually vary by

1. The season and local temperatures
2. Types of vegetation grown
3. Type of soils
4. Land use
5. Local conditions
6. Increase of ground water during floods

The nature's recurring hydrologic cycle usually will allow only about 34 percent of the precipitation to recharge groundwater or to enter the surface water supplies. This will result in water supply shortages in areas of low precipitation and runoff. In some high-water-demand areas, water usage patterns exceed the sustainable water supply.

Wetlands Wetlands are areas that are saturated with water for a portion of the year and have a specific type of soil and vegetation. Various ecosystems that exist adjacent to rivers become submerged after the river is dammed, severely damaging or destroying the ecosystems. This includes areas known as wetlands. Wetlands can be freshwater or saltwater. Wetlands are considered "waters of the United States" and are regulated by the government.

Overdraft Whether the water supplies are sustainable will depend on the water usage and precipitation frequency and intensity. It is estimated that water usage in the United States will exceed the water supply by 13 percent by the year 2020. Even though groundwater in the United States is plentiful, it is unevenly distributed, and in some areas, withdrawals exceed recharging. This condition is called *groundwater overdraft*. Although water seems to be everywhere, it's not where it's needed. Adequate water supplies can be retained by the construction of dams and reservoirs. Construction of dams can have significant effects on the environment. Damming a river forms a reservoir behind the dam. The stored water then may be used for

- Municipal or agricultural water supply
- Hydroelectric power generation

- Flood control
- Improved navigation
- Recreation

The Hoover Dam, located on the Colorado River, is a major source of water supply to several states. In fact, so much water is stored or taken out of the Colorado River that only occasionally does it flow into the Gulf of California in Mexico. Water conservation efforts are required for agricultural water by reducing water from canals and for domestic water by preventing water wastage. Conservation of huge volumes of water by dams comes at a price. There are several environmental effects, and it affects the water table.

1.3 Need for Nationwide Checkup on Infrastructure Health

An attempt has been made to assess the nation's highway infrastructure health in general and for bridges located on rivers in particular. A detailed review of the published literature listed in the bibliography was made. Results from a study of infrastructure health, conducted by the FHWA in coordination with AASHTO, are now available online in a series of four reports. The study's goal was to define a consistent and reliable method to document infrastructure health, focusing on bridges and pavements on the Interstate Highway System. A related goal was to develop tools to provide the FHWA and state transportation agencies with key data that will produce better and more complete assessments of infrastructure health nationally.

NBI Bridge Ratings Performance measures for bridges include

Tier 1 ratings for *structural deficiency* (SD)

Tier 2 ratings for structural adequacy based on the National Bridge Inventory (NBI) ratings

These measures were evaluated on I-90 in Wisconsin, Minnesota, and South Dakota. The I-90 corridor runs for 1406 km (874 mi), with average annual daily traffic ranging from approximately 5000 to 90,000 vehicles. Evaluations were done using the Highway Performance Monitoring System (HPMS) and NBI data, as well as data collected by the FHWA project team and provided by the participating state highway agencies. State information covered

- Documentation of the state's systems
- Processes
- Corridor inventory and pavement management system data

The good/fair/poor rating for bridges proved to be a viable approach, with NBI data sufficient for the performance management assessment. However, a bridge's SD status was not as easily incorporated into the analysis. The study report notes that a measure of structural adequacy based on NBI ratings would be a viable supplement to SD status, as a national measure of bridge condition.

As part of the study, a sample health report was prepared for the pilot corridor. This report uses several metrics to assess the overall health of a corridor, including the good/fair/poor measures, age, remaining service life for pavements, and traffic volumes.

Infrastructure Health Reports To improve infrastructure, a rating of the infrastructure health is required. A prototype tool that would automate the creation of infrastructure health reports is described in the pilot study report *Improving FHWA's Ability to Assess Highway Infrastructure Health* (Pub. No. FHWA-HIF-12-049).

As part of the pilot study, a section of Interstate 90 through South Dakota, Minnesota, and Wisconsin was evaluated in order to test approaches for categorizing bridge and pavement condition as good, fair, or poor that potentially could be used across the country. As a result of the pilot study, it was found that a bridge good-fair-poor methodology can be implemented nationwide today.

Visit www.fhwa.dot.gov/asset/pubs/hif12049/hif12049.pdf.

The FHWA and AASHTO presented the study results to senior-level state transportation agency representatives at a national meeting held in October 2011, in Detroit, Michigan. To view the national meeting report, which summarizes discussion about the recommended condition ratings and health reporting, visit www.fhwa.dot.gov/asset/health/workshopreport.pdf. A table in this report shows the varying condition of bridges in each state in the United States. This study helps in funding by the owner agency and the federal government. Bridges are prioritized for repairs or replacement based on their structural condition.

1.4 A Survey of Recent Floods That Affected Infrastructure

A study of recent historic data available in United States and abroad was carried out. River bridges, which are subjected to floods, require the following:

1. Construction of deep foundations in flowing water and maintenance of the bridge substructure underwater
2. Regular painting of corroded girders
3. Recognition of any regional differences in the types of rivers
4. Funding limitations

It is prudent to reexamine predictive scour models for analysis and estimated scour depths. In the United States, detailed floodplain maps are prepared by FEMA. NASA's *Terra* satellite acquires images of the distribution of floods.

Overcoming Widespread Floods in the United States Major rivers such as the Ohio and Mississippi Rivers have recurring floods. Both northern and southern California were subjected to heavy rain and floods along the Pacific Coast. Major floods occurred in 1938 and 1958. The floods were followed by landslides, resulting in casualties and road closures. The following steps were taken to overcome the major disasters:

- Hundreds of miles of tall retaining walls had to be constructed to protect the highways, costing billions of dollars.
- For the Pacific Highway, three states financed the construction costs: California, Oregon, and Washington.

- Drainage of the cliffs during and after the rains was introduced by a network of PVC pipes. Many nearby towns lying at lower elevations than the overflowing banks were inundated with floodwater, incurring losses of property and life.
- Also highways were closed for long periods, adversely affecting commerce and industries and causing many problems for commuters, who were forced to make detours through narrow roads.
- Prevention of bridge scour has now been a national priority for two full decades.

Colorado Floods Resilience should be the focus of flood recovery (Erica Meltzer, meltzere@dailycamera.com, posted December 3, 2013). "Boulder needs to rebuild and recover from September's historic flooding in ways that leave the community more resilient than before. It is likely to cost roughly $400 million to build a wastewater and storm water infrastructure capable of withstanding an event like this year's flooding."

Record Floods in Kansas Channeling and levee construction altered how the floods affected various areas along the Missouri River. A comparison of Kansas City data for the three big floods since the early nineteenth century follows.

Great Flood of 1844: This was the biggest flood of the three in terms of the rate of discharge at Westport Landing in Kansas City. It is estimated that 625,000 ft^3/s was discharged in the flood.

Great Flood of 1951: The 1951 flood was the second largest, in terms of the rate of discharge, at a rate of 573,000 ft^3/s. The flood was the most devastating of all modern floods for Kansas City because the city's levee system was not built to withstand it. It destroyed the Kansas City stockyards and caused Kansas City to build Kansas City International Airport away from the Missouri River bottoms, to replace the heavily damaged Fairfax Airport in Kansas City.

Great Flood of 1993: The 1993 flood was the highest of the three but had the lowest discharge at 541,000 ft^3/s. Although the 1993 flood had devastating impacts elsewhere, Kansas City survived it relatively well because of levee improvements after the 1951 flood. This 1993 flood was also used to model other natural disasters and simulate potential flood impacts in areas awaiting more extensive flood control efforts.

Kentucky Floods If the $2.5 billion Brent Spence Bridge project gets funded, engineers say it would also take care of another northern Kentucky problem—Covington's flooding. A 12-ft-diameter storm pipe along I-71/75 would handle 377 million gal of storm water runoff that courses through this corridor each year. Neighborhoods along the way from Fort Mitchell to Covington could channel rainwater into the pipe that would discharge it into the Ohio River. The combined storm and sanitary sewers underneath Covington overflow during heavy rains, spilling a mix of sewage and water into some neighborhoods.

The Great Mississippi and Missouri Rivers Flood of 1993 These floods occurred in the Midwest, along the Mississippi and Missouri Rivers and their tributaries, from April to October 1993. The flood was among the most costly and devastating ever in the United States, with $15 billion in damages. Here is what happened:

- A rainy autumn in 1992 resulted in above-normal soil moisture and reservoir levels in the Missouri and Upper Mississippi River basins.
- During the winter of 1992–1993, the region experienced heavy snowfall. These conditions were followed by persistent spring weather patterns that produced storms over the same locations.
- Soils across much of the affected area were saturated by June 1, with additional rainfall all running off into streams and rivers, instead of soaking into the ground.
- These wet-weather conditions contrasted sharply with the droughts and heat waves experienced in the southeastern United States.

Torrential Rains in 2011 There were record flash floods resulting from these rains. In the Mississippi River, the damage to property from water overflowing the banks was very high compared to the 1927 record floods. Similarly, the fast-flowing Ohio River was subjected to floods, and bridges needed to be shut down. The Mississippi River floods in April and May of 2011 were among the largest and most damaging recorded along the U.S. waterway in the past century, comparable in extent to the major floods of 1927 and 1993.

Flood Damages: In April 2011, two major storm systems deposited record levels of rainfall on the Mississippi River watershed. When that additional water combined with the springtime snowmelt, the river and many of its tributaries began to swell to record levels by the beginning of May. Areas along the Mississippi River itself that experienced flooding included Illinois, Missouri, Kentucky, Tennessee, Arkansas, Mississippi, and Louisiana. The U.S. President had declared the western counties of Kentucky, Tennessee, and Mississippi to be federal disaster areas. For the first time in 37 years, the Morganza Spillway was opened on May 14, deliberately flooding 4600 mi^2 of rural Louisiana to save most of Baton Rouge and New Orleans.

Floods in 1993 Storms, persistent and repetitive in nature bombarded the upper Midwest with voluminous rainfall. Portions of eastern central Iowa received as much as 48 in of rain between April 1 and August 31, 1993, and many areas across the central northern plains had precipitation levels 400 to 750 percent above normal.

The hydrographic basin affected cover around 745 mi long and 435 mi wide, totaling about 320,000 mi^2. Within this zone, the flooded area totaled about 30,000 mi^2 and was the worst such U.S. disaster since the Great Mississippi Flood of 1927. In some categories, the 1993 flood even surpassed the 1927 flood, which at the time was the largest flood ever recorded on the Mississippi River. There are recurring examples of floods in the Midwest.

Floods in 2008 Floods were some of the worst in the history of the United States. The U.S. Geological Survey published a report entitled "Flooding in the United States Midwest, 2008." Record precipitation amounts coupled with saturated soils created the worst outcome. Indiana had the most recurrent flooding during 2008. Peak stream flows were recorded at 147 USGS stream gauges.

Arkansas Floods Fourteen people were killed in Arkansas, with 392 killed across seven states in the preceding storms in 2010.

Louisiana Floods The Mississippi River watershed floods in April and May 2011 were among the largest and most damaging recorded along the U.S. waterway. When that additional water combined with the springtime snowmelt, the river and many of its tributaries began to swell to record levels by the beginning of May. The U.S. President declared the western counties of Kentucky, Tennessee, and Mississippi federal disaster areas. The Morganza Spillway was opened to deliberately flood 4600 mi^2 of rural Louisiana to save most of Baton Rouge and New Orleans. The USACE stated that an area in Louisiana between the cities of Simmesport and Baton Rouge was expected to be inundated with 20 to 30 ft of water. The unprecedented extensive rainfall from four storms, combined with springtime snowmelt from the upper Midwest, created a 500-year flood along the Mississippi.

Minnesota Floods Following the fatal collapse of an interstate bridge in Minnesota the spotlight was thrown on the poor condition of many major structures. Multiple floods occurred in Minnesota in the past such as in September 2010 and June 2012. The ensuing investigations showed a large number of bridges in the United States to be functionally obsolete or structurally deficient.

Nevada Floods Urban flooding in Nevada City from Truckee River, with its origin at the outlet of Lake Tahoe, will require rerouting of the waterway. It will help protect the industrial areas in Sparks, Nevada. The river has caused extensive flood damage in both Reno and Sparks. Recent floods in Nevada were in September, 2014. (Also, refer to the website NevadaFloods.org).

New Hampshire Floods Scott A. Olson's investigations of New Hampshire floods were published by USGS as Scientific Investigations Report 2008-5206, "Estimation of Flood Discharges at Selected Recurrence Intervals for Streams in New Hampshire" (performing organization: U.S. Geological Survey). The report provides (1) estimates of flood discharges at selected recurrence intervals for stream gauges in and adjacent to New Hampshire and (2) equations for estimating flood discharges at recurrence intervals of 2, 5, 10, 25, 50, 100, and 500 years for ungauged, unregulated, rural streams in New Hampshire. The equations were developed using generalized least-squares regression. Flood frequency and drainage basin characteristics from 117 stream gauges were used in developing the equations.

A technique for adjusting a flood discharge frequency curve computed from a stream gauge record with results from the regression equations is described in this report. Also a technique is described for estimating the flood discharge at a selected recurrence interval for an ungauged site upstream or downstream from a stream gauge using a drainage area adjustment.

Oklahoma Floods in 2013 Oklahoma was hit on June 2, 2013, this time with flash floods. Eight people lost their lives and many houses collapsed. This was like adding insult to injury, following the tornadoes that had hit only a few weeks earlier.

The USGS investigated the *annual exceedance probability* (AEP), using flood probability analysis based on selected parameters and peak flows by the large number of installed stream gauges, thereby computing the trend magnitudes for the 1908 floods in the Midwest. The percentage changes in median annual peak stream flow values indicated clustering of increasing and decreasing trends. However, it was observed that none of the trends were consistent in the Midwest for reliable future predictions. This uncertain trend is somewhat similar to that in predicting other natural disasters such as windstorms, tsunamis, and earthquakes. However, the level of risk can be estimated from the historical data by analyzing the various factors affecting hydrologic cycles, described above.

New Jersey Floods Snowmelt combined with heavy rain can cause floods. Hurricane Floyd in September 1999 turned Interstate Route 46 into lakes, causing travel disruption.

Texas Floods After flash floods occurred across Texas in May 2013, another peak flood occurred in May 2014, which resulted in a record rise in rivers in Texas. Another example of flood is the record rise in river level in the Southern states, such as in San Antonio, Texas, in which a woman drowned on May 24, 2014.

Washington State Floods The most recent interstate bridge collapse in Washington State occurred on May 23, 2013, when the bridge carrying Interstate 5 over the Skagit River collapsed, dumping a number of vehicles into the river. The four-lane highway bridge is about 62 mi (100 km) north of Seattle in the Pacific Northwest state of Washington and provides a key link to neighboring Canada. It is one of over 78,000 bridges in Washington State highlighted as not being up to standard. But the incident has reopened the issue of the safety of bridges in the United States. Thousands of dry streams and rivers are subjected to both seasonal and out-of-season floods. Prevention of bridge scour has been a national priority for two full decades now. Because of federal EPA requirements and tight budgets, Sanitation District 1 (SD1) must spend most of its capital budget on projects other than flood abatement.

A 21st-Century Solution to a Century-Old Urban Flooding Problem

The $2.5 billion Brent Spence Bridge project in Kentucky, if completed, could address flooding for the thousands who live along I-75 from Fort Mitchell to Covington. Many of Covington's storm pipes were designed to handle the storm water 100 years ago when Covington was surrounded mostly by woods and farmland.

Rhode Island Floods

As reported by Alex Kuffner in the *Providence Journal* of Rhode Island (December 3, 2013), "The Narragansett Bay Commission is building a key part of a system of pipelines and tunnels to contain contaminated storm water that inundates Providence's century-old sewer system during rains. It could take nearly two decades to complete and could cost more than $1 billion. The overflows were once estimated at 2 billion gal/yr. Following passage of the federal Clean Water Act in 1972, the EPA ordered Providence to fix the problem. The first phase of the project—a 3-mi-long, 300-ft-deep tunnel beneath the city—was completed in 2008."

In the second phase of the project pipes will be installed under neighborhoods along the Woonasquatucket and Seekonk rivers, which both empty into the bay.

The commission has planned a third series of upgrades that will take place farther up the Seekonk River and will divert runoff from Pawtucket and East Providence to a treatment facility at Bucklin Point. The first phase is estimated to have captured 40 percent of storm water overflows and the second phase an additional 20 percent. The third phase will bring the total to 98 percent.

1.5 Need to Maintain Transportation Network at Reasonable Cost

Soil Erosion due to Recurring Floods and Need for Mitigation Measures

After floods, agricultural produce may be destroyed, houses washed away, and roads inundated. Floods cause casualties and some victims to require medical treatment, food, and clean water.

Soil conditions at rivers and hydraulics are based on geotechnical considerations, the flood velocity, and water surface elevations. The estimation of discharge requires the computations of selected software for 50-, 100-, and 500-year floods giving rise to the magnitude of scour with no scour on rocks. Following the collapse of Schoharie Creek, a multispan bridge in New York State, the criteria of flood discharge have now been upgraded, for the design of new bridges, from 100-year to 500-year probability. The peak 500-year flood discharge is approximately 1.7 times the 100-year peak flood discharge. The selection of scour-critical bridges and hydrology deals with those bridges which are scour-critical and hydrology of 50-, 100-, and 500-year floods. The estimated scour depth and description of countermeasures are important.

1. For using erosion and scour theories and estimated scour depths, Hydraulic Engineering Circulars are used. The scour analysis method applies AASHTO procedures to compute scour depths upstream and downstream of the bridge. HEC-18 presents detailed guidelines on the planning of new bridges to reduce or minimize the effects of scour.
2. The important aspects of repairs and retrofit of existing bridges are discussed in Hydraulic Engineering Circulars. These pertain to the selection of countermeasures for existing bridges. Information on the constructability of scour countermeasures for existing bridges is given.
3. Countermeasures are described in detail in Chapters 3 and 9. Design guidelines are given for filters—which are essential to the effectiveness of countermeasures—such as riprap and gabions. They present detailed information of countermeasures based on scour type and costs, recommended countermeasures for piers and abutments, scour countermeasures for culverts, selection of countermeasures, and monitoring countermeasures.

Bridge scour has been a major threat to the structural integrity of bridges over rivers. It has been estimated that 60 percent of all bridge failures result from scour; scour was the most common cause of highway bridge failure in the United States during the 1961–1976 period. Field inspections for scour have been a regular feature. The USGS and other federal agencies have installed monitoring gauges on selected rivers, which are providing data for flood risk assessment studies. Postflood disaster management needs to be introduced.

Maintenance of Highway Structures

The United States is a vehicle-focused nation, which has become highly dependent upon the transportation network for travel, trade, and commerce. According to a report published by U.S. Department of Transportation (DOT) in 2013 on the status of the nation's highways, bridges, and transit conditions and performance, there are 600,000 bridges located over 4 million mi of public roadways and 163,000 mi forms part of the national highway system. Nearly $120 billion is spent on maintenance of the transportation network each year.

Constructing Water Tanks, Dams and Reservoirs

1. Fast-moving streams and rivers stay cooler than the relatively still waters of reservoirs, so dams will increase the water temperatures, affecting the spawning behavior of aquatic species.
2. Dams trap sediment behind them in the reservoirs that would normally travel downstream, thus causing sedimentation starvation downstream.
3. Dams will interfere with or block the routes of migratory fish or cause damage to fish attempting to move through hydropower turbines or sluiceways.
4. Dams and reservoirs will alter the water table as well as groundwater recharge and discharge patterns.

1.6 Global Problems of Flood Hazards, Drought, and Water Conservation

Floods occur across the globe wherever there are fast-flowing rivers. Examples of major floods in some of the flood-devastated countries are presented below. Both floods and droughts are dangerous for our survival. Floods cause over abundance of water but a decrease in water quality for daily use and irrigation. Water is precious and misuse of water needs to be avoided.

Flooding Issues in Australia and New Zealand

Guidelines for Ensuring Specified Quality Performance in Bridge Construction According to the Austroads publication "Road networks represent a significant asset to governments in Australia and New Zealand. In Queensland the state-controlled road network consists of 33,000 km of road, which contains about 2850 bridges and many thousands of culverts. The replacement value of these bridges and culverts is around $3 billion. Analysis of life cycle costs, including the costs of replacement, shows that long-life structures (such as 300 years) are the most economical in almost all cases." Investigations by Carse and Bell (2004) have found that concrete structures are acid sulfate soils and sodic soils which cover 45 percent of Queensland. Sodic soils may cover a wide range of pH values from 3.5 to 11.5, and they have associated salinity issues depending on the attached anion species. Concrete can be designed to withstand these harsh conditions, and in some cases extra steps are taken such as coating piles and using polythene sheeting to protect concrete from very low-pH ground conditions.

The majority of Queensland's concrete bridges typically require maintenance at a life of 50 years, with probably only about 1 percent of these structures reaching a

100-year design life without significant maintenance. This report discusses the importance of quality and durability in the construction of bridges and other transport infrastructure and the basis for a set of guidelines for achieving specified quality in construction is presented.

Bangladesh Floods

Each year in Bangladesh about 26,000 km^2 (around 18 percent) of the country is flooded, killing many thousands of people and destroying millions of homes. The floods have caused devastation in Bangladesh throughout history, especially during the years 1966, 1987, 1988, and 1998. The 2007 South Asian floods also affected a large portion of Bangladesh. Recent floods occurred in 2010.

Bolivia Floods

Torrential rain and floods in Bolivia killed 38 people and left many homeless in the first week of February 2014. The Ministry of Defense said it flew in food aid and the army evacuated families from the worst affected regions, such as the Amazonian and cattle-rearing department of Beni, northeast of the capital, La Paz.

Around 100,000 head of cattle were threatened and 6000 hectares of crops were destroyed, around 0.2 percent of Bolivia's total, the government said. More than 40,000 households were affected by the heavy seasonal rains since October 2013, about five times more than the previous year, with the central and northern regions of the country being hardest hit.

China Floods

In August 2010 China had two major floods which killed as many as 1500 people.

Indian Floods

In June 2013, flash floods and landslides in northern India killed over 1000 people in the Himalayan state of Uttar Akhand, and with thousands missing or stranded the toll was expected to rise.

The National Disaster Management Authority of the government of India provided practical guidelines. With the continuous growth in the country's population, the demand for freshwater is going up continuously.

Floods contribute to many environmental issues in India, and the quality of the water supply is adversely affected. Far-reaching effects are listed below, and adequate flood control measures need to be implemented:

1. Per capita availability of utilizable water is going down.
2. There is an ever-rising standard of living, so more gallons are needed per person per day.
3. Rapid industrialization and urbanization have further increased demand.
4. Unabated discharge of industrial effluents into water bodies is further aggravating the scarcity of water of acceptable quality.
5. For farming needs, at the global level, about 60 to 70 percent of total annual water consumption is for irrigation. In India water use for irrigation is about 83 percent of the current level of total water utilization.

6. Water audit is an effective management tool for minimizing losses, optimizing various uses, and thus enabling considerable conservation of water in both the irrigation sector and other sectors such as domestic, power, and industrial water use. Thus apparently there is ample scope of water saving in the irrigation sector.

"General Guidelines for Water Audit and Water Conservation," published by the Central Water Commission, Ministry of Water Resources, serves as a useful reference for undertaking water-saving measures in all sectors of water use and helps state governments to formulate their own region-specific, project-specific, system-specific, or service-specific guidelines.

Benefits of Water Audits in India A water audit improves the knowledge and documentation of both distribution system problems and risk areas and enables a better understanding of what is happening to the water after it leaves the source point. Leak detection programs help in minimizing leakages and tackling small problems before they become major ones. These programs lead to

1. Reduced water losses
2. Improved financial performance
3. Improved reliability of the supply system
4. Enhanced knowledge of the distribution system
5. Better safeguards of public health and property
6. Reduced legal liability
7. Reduced disruption, thereby the improving level of service to customers

Water conservation is a key challenge that requires public participation. The simple information on the typical use of water for domestic purposes and how to save water may help in creating awareness.

Water in every country is essential for survival of the people. Planning of reservoirs to conserve water and the distribution system through pipes in each area will require an estimate of the daily demand and the projected supply. Wastage of water at the individual level needs to be discouraged in drought conditions and during floods. An estimate of average demand and supply of water is shown in Table 1.1.

Drinking, 4 percent
Cooking and other kitchen uses, 8 percent
Personal hygiene, 29 percent
Washing clothes, 10 percent
Toilet flushing, 39 percent
House cleaning, gardening, etc., 10 percent
Total 100%

TABLE 1.1 Typical Uses of Water Household in India

Table 1.1 shows typical percentages of daily water use in India. Savings can be made for daily water use such as:

1. Use of a bucket: Bathing or washing clothes with a bucket with limited quantity of water compared to using shower or running tap water
2. Use of a large mug: Brushing teeth, washing hands/face or shaving shaving compared to use of running water
3. Use of a wet cloth: Washing the floor or washing the car with repeated wet cloth
4. Using a new style water closetand latrines for flushing in place of old style flushing. Total daily saving may exceed 900 litres per family per day.

Total Saving per Typical Household

For a family of six, as much as one-half of water consumption can be saved daily. During floods or in drought conditions, water rationing may be imposed in urban areas. Examples are restrictions on washing cars and the use of swimming pools for certain months in the year. In the rural areas, however, where water is collected manually from wells or lakes, water rationing may be difficult to implement as the daily consumption cannot be monitired on a daily basis and impose penalties.

Common sense savings of water help greatly at the times of drought and when water supply is disrupted for days during recurring floods.

ASCE Task Committee on Flood Risk Management

In April 2013, a comprehensive summit was held in Herndon, Virginia, called "Building a Framework for Flood Risk Management: Goals, Roles, Responsibilities, Resources, and Systems." The need for people to have a common understanding of flood risk management was emphasized. Some progress in this direction has been made by the ASCE Task Committee on Flood Safety Policies and Practices (TCFSPP), and Congress created the Unified National Program for Floodplain Management (UNPFM).

ASCE chapters in the United States and their structural engineering institutes (SEIs) abroad have been active in holding continuing education seminars on scour-critical bridges for the award of professional development hours (PDHs) required for renewal of licenses to practice engineering. The 2014 winner of the most active chapter was awarded to the Philadelphia chapter of SEI. Figure 1.2 shows the author of this book receiving the award as chairman and founder of the Philadelphia chapter.

Advantages of Flood Risk Management The following are some of the advantages of flood risk management:

- Mitigation of the consequences of flooding
- Use of structural and nonstructural measures for a resilient infrastructure
- Assessment of risks
- Effective and sustainable management of risks to life, health, economic activity, cultural heritage, and the environment

FIGURE 1.2 President of SEI of ASCE (left) giving the best chapter award to Mohiuddin A. Khan in Boston in April 2014.

- Risk sharing by all stakeholders
- Funding allocation on priority basis

In 2007, Congress established a new federal flood policy. It gave equal attention to economic considerations, public safety, and environment objectives. In 2013, the U.S. Water Resources Council issued new principles to allow consideration of nonstructural measures:

1. Risk reduction measures by federal, state, and local agencies. A high level of initial risk faced by a community can be mitigated by cumulative risks (expressed in descending order of magnitude) such as from
   - Zoning
   - Building codes
   - Risk communication
   - Evacuation plans
   - Natural storage
   - Levees and flood walls
2. Individual homeowner or business risks:
   - Flood proofing
   - Insurance: The National Flood Insurance Program (NFIP) in many states of the United States helps to protect life and property from the recurring natural disaster of floods.

3. Residual risk (that remains after all measures are taken):
   - There is a need for national flood risk management legislation that will define clearly the roles of the federal, state, and local governments.

Drought in Kenya The government of Kenya has made a commitment to end the worst of the suffering caused by drought by 2022. The actions needed to achieve this are set out in the Drought Risk Management and Ending Drought Emergencies Medium Term Plan (MTP) for 2013 to 2017, which is part of the Kenya Vision 2030 MTP2.

Water conservation and water quality during low-rainfall seasons are of paramount importance. The MTP has six pillars: peace and security, infrastructure, human capital, sustainable livelihoods, drought risk management and coordination, and institutional development and knowledge management. Each of these pillars is being operationalized through a common programming framework in order to harmonize the actions and resources of the government of Kenya and its development partners.

Floods in Pakistan In July 2010 the heavy monsoon rains caused floods in northwest Pakistan. Nearly 2000 people lost their lives, and floodwater destroyed the agricultural crops. Jacobabad received the heaviest monsoon downpour of over 12 in. Flash floods triggered by heavy rains caused the following widespread destruction across vast swathes of the country, breaking a 24-year rainfall record:

- There were 100 dead in upper Sindh.
- Punjab's Dera Ghazi Khan and Rajanpur areas experienced devastation.
- Five districts of eastern Baluchistan were cut off from the rest of the country.
- In the Sindh province, more than 100 people were reported dead and scores of others injured. Flooding destroyed thousands of houses, leaving a large number of families homeless.
- Flooding also hit urban areas, bringing life to a standstill and submerging roads and streets in all cities and towns of upper Sindh.
- Widespread damage to standing crops and agricultural land and residential settlements in rural areas was reported from seven districts of the province.
- The Suleman Range hill torrents caused havoc in the Dera Ghazi Khan and Rajanpur districts.
- At least five districts of eastern Baluchistan were cut off from the rest of the country after floodwater submerged railway tracks and highways in Nasirabad and Jaffarabad. Hundreds of houses were washed away, and thousands of families were left homeless in the province.
- Train service between Baluchistan and other parts of the country was suspended as many railway tracks were inundated. Figure 1.3 shows Frontier Corps personnel distributing relief goods among flood-affected people of Baluchistan Province in Pakistan. The engineering wing of the army was plugging the canal breaches. Meanwhile, rain kept lashing other towns of central and south Punjab. This subject is discussed in Chapter 7.

Figure 1.3 Frontier Corps personnel distributing relief goods among flood-affected people of Baluchistan Province in Pakistan.

A joint paper on the computation of flood volumes in recurring Indus Valley floods by two professors (Professor Irfan Ph.D. and Professor Gabriel Ph.D.) at the National University of Science and Technology, Islamabad, and the author was presented in the Second International Conference on Hydrology & Ground Water in 2013 OMICS Second International in Raleigh, North Carolina.

Scour Mitigation by Applying FHWA Scour Technology Of the more than 575,000 bridges in the U.S. national bridge inventory, approximately 84 percent are over water. Each year in the United States, highway bridge failures cost millions of dollars as a result of both direct costs necessary to replace or restore bridges and indirect costs related to disruption of transportation facilities.

Of even greater consequence is the loss of life from bridge failures. Hydraulic factors such as stream instability, degradation, contraction scour, and local scour account for more of our bridge failures (approximately 60 percent) than all other factors combined. Ongoing screening and evaluation of the vulnerability of the nation's highway bridges to scour by state departments of transportation (DOTs) have identified the following:

1. More than 18,000 bridges are considered scour-critical and in need of repairs or replacement.
2. Almost 100,000 bridges with unknown foundations have been identified.
3. As state DOTs develop plans of action to address the effects of scour and stream instability, there is a need for innovative, effective, and economical countermeasures to be considered for the design of new bridges and for repairing bridges.

1.7 Asset Management to Include Flood Mitigation Measures

The major issue is combating the floods by effective asset management and maintenance of bridge substructure, which is subjected to floodwater. Effects of floods are both aggradation and degradation, which result from sediment transport and continued erosion of soils supporting the foundations. With the current economy and the need to spend public dollars wisely, asset management policy and practice are a high priority for highway agencies. It is important to ensure accountability, performance measurement, and transparency. Asset management practice helps to achieve these goals.

The FHWA defines *asset management* as a "systematic cost-effective process of maintaining, upgrading and operating physical assets." The approach supports and complements the New Jersey Department of Transportation's (NJDOT's) federal and state-mandated investment planning documents. The following additional programs are included:

- The 10-Year Capital Investment Strategy
- The 10-Year Capital Transportation Improvement Program
- The Annual Transportation Capital Program
- The Annual Study and Development Program

Asset management is being adopted by other agencies that manage large infrastructure assets, including water and sewage authorities and public transit agencies. Nationally, the FHWA maintains an Office of Asset Management, and the AASHTO considers asset management critical for making sound and practical capital spending decisions.

Initiatives by NJDOT. The NJDOT adopted asset management as the official institutional approach to managing its infrastructure assets and making capital investment decisions in January 2008. With the promulgation of NJDOT's asset management policy, an Asset Management Steering Committee was established. The NJDOT has taken a multistep approach to implement and advance asset management:

- Develop an asset management plan for NJDOT assets, through a variety of investments.
- Prepare an inventory of specific assets, their condition, performance targets, and a plan to achieve these targets.
- Advance an asset management improvement strategy that examines NJDOT's proficiency and maturity in asset management practice, identifies strengths and weaknesses, looks at methodologies and practices, and sets goals and objectives for improvement.

Through asset management, flood control and mitigation will save lives, property, and damages to bridges and highway structures that inevitably result in the loss of hundreds of million dollars from each recurring flood. It is important that these losses be kept to an absolute minimum. Examples of mitigation measures follow:

- Create effective drainage by swales and by grading the landscape.
- Build pumping stations at the waterlogged and submerged highways.

- Construct reservoirs and dams.
- Minimize erosion of foundations of bridges and highway structures by shielding.
- Through mitigation measures, the lakes of water formed by floods can be stored for future irrigation use and for growing bumper crops to increase agricultural production.
- In some countries, the higher elevation can be taken advantage of by diverting floodwater across the border into neighboring countries, although this is not the best solution. Practical solutions need to be tailored on a case-by-case basis. The designer often is challenged by steep slopes, unsuitable soil, and limited site area. Standard countermeasure details need to be modified at many sites.

In Pennsylvania, for example, the design engineer is required to inspect and certify that erosion controls such as swales are installed and function as designed, significantly elevating the design of easily constructible and maintainable facilities.

European Practice for Inspection and Monitoring According to the NCHRP report "1998 Scanning Review of European Practice for Bridge Scour and Stream Instability Countermeasures, July 1999," *Research Results Digest*, no. 241, most of the European countries have initiated efforts to develop a bridge inspection or scour evaluation program comparable to the National Bridge Inspection Standards (NBIS) in the United States.

In Switzerland, some specific guidance on the stability of structures in water has been published (1998) as a multiagency guideline by the Federal Office for Highways (ASTRA), the Federal Office for Transport (BAV), the Federal Office for Water Management (BWW), and the Swiss Federal Railways (SBB). This document is presented as a "recommendation for the preservation and maintenance of existing structures and hints for the construction of new structures."

In Germany, the Federal Highway Research Institute (BAST) is investigating the use of the FHWA PONTIS bridge management system.

The *Railtrack Handbook 47* on scour prepared in the United Kingdom apparently contains some guidance on inspection, but a copy of this document was not available for review. In response to the Glanrhyd Bridge failure in 1987, high-water marks were painted on most railroad bridges in the United Kingdom to guide inspectors' decisions on caution or closure, but this approach apparently met with only mixed success, as it resulted in unnecessary bridge closures.

Only in the United Kingdom was an effort made comparable to the U.S. National Cooperative Highway Research Program Project 21-3, to develop fixed instrumentation for measuring and monitoring scour at bridge piers and abutments. This resulted in the patented Wallingford "tell-tale" device which was installed on several railroad bridges following the Glanrhyd bridge failure. In none of the above European countries was technology available to determine the characteristics of unknown bridge foundations, that is, foundations for which design or as-built drawings do not exist. In the United Kingdom there are numerous unknown foundation bridges, and the problem is considered as serious as it is in the United States.

Bioengineering Applications Europe is considered a leader in the use of bioengineering techniques for river instability problems. While bioengineering techniques are integrated with traditional engineering countermeasures for river system management in Europe, hydraulic engineers would not recommend reliance on bioengineering countermeasures as the only countermeasure technique when there is a risk of damage to property or a structure, or where there is potential loss of life. The primary concern was a lack of knowledge about the properties of the materials being used in relation to the force and stress generated by flowing water, as well as the difficulties in obtaining consistent performance from countermeasures relying on living materials.

Risk Analysis Some form of risk analysis is used to determine the level of effort and investment in countermeasure design and installation in all countries visited by a selected ASCE Team of engineers.

Switzerland uses a matrix to code flood hazard zones into increasing degrees of hazard (yellow, blue, red), and infrastructure design and maintenance decisions are guided by the degree of risk involved.

The Netherlands uses both a regional risk analysis, considering such factors as distance from the coast and vulnerability to storm surge, and a detailed analysis of the risk of failure of individual components of a project. These factors influence the decision on investment in scour countermeasures.

In European countries, both a design flood and a larger "check" flood are used in bridge and scour countermeasure design, comparable to the 100-year frequency design flood and the 500-year frequency (or superflood) check flood used for scour assessment in the United States. In the United Kingdom the superflood has a 200-year return period.

1.8 Developing an Environmental Policy for Floods

Environmental impacts are considered for scour and stream instability countermeasure selection, design, and installation in all countries visited. In general, the approach is to emphasize environmental enhancement and sustainability without creating an undue risk to lives and property in applying environmental policy to structures in a riverine system.

Floods can become a major environmental issue. They adversely affect townships, disrupt highway travel, and damage bridges by foundation scour. Any disruption to the interstate highway network, for example, will cause enormous financial problems and discomfort. The major damage to transportation structures in the United States is from excessive flooding in local rivers. Unlike earthquakes, floods seem to hit at regular intervals in many parts of the Northeast and other parts of the United States. Environmental impacts are considered for scour and stream instability countermeasure selection, design, and installation in many countries.

In the United Kingdom, the Environmental Agency through the National Centre for Risk Analysis and Options Appraisal has prepared a Guidance Note (no. 18) on River Geomorphology: A Practical Guide, which outlines a geomorphological approach to river management. In general, their approach is to emphasize environmental enhancement and sustainability without creating undue risk to lives and property in applying environmental policy to structures in a riverine system.

Table 1.2 shows the pollutants generated by industries as reported by Central Water Commission, India.

Nature of Pollutants	Type of Industries
Organic pollution	Distillery, sugar, tannery, pulp and paper, dairy, coke-oven, refinery, and food industry
Dissolved solids	Chemical industry, fertilizer, pharmaceutical, pesticides
Toxic chemicals	Electroplating, coke-oven, tannery, chemical industry, pesticides, pharmaceuticals, H-acid, G-acid plants, dye and dye intermediates
Cooling water	Thermal power plants, cable, rolling mills, PVC and plastic mounting

*Classification of industrial discharges according to nature of pollutants.
Source: Central Water Commission, India.

TABLE 1.2 Adverse Environmental Effects in India*

Design Philosophy for Scour Resistance

1. The general design approach in Switzerland, Germany, and the Netherlands is to prevent scour from occurring or to move scour away from the structure by including scour and stream instability countermeasures in the initial design and construction of their bridges. In general, these countries feel that they do not have a significant bridge scour problem, largely because of this design approach. The fact that their major navigable waterways and canals have been stabilized by extensive river training works obviously contributes to the success of this approach.
2. The bridge scour problem in the United Kingdom is similar to that faced in the United States. Both countries are applying the latest scour technology to the design of new bridges, but both face significant problems with scour and stream instability at thousands of bridges on an aging infrastructure. As a result of a catastrophic railroad bridge failure with fatalities at Glanrhyd in 1987, rail track has taken the lead in developing scour assessment techniques and countermeasures in the United Kingdom.

Role of Movable Bridges during Floods

During floods, the river water surface elevation is elevated. On navigable rivers, fixed bridges can be an obstruction, and accidents are likely. However, vertical-lift bridges or drawbridges can be raised to allow ships to pass through, provided the safe speed of ships is not exceeded. Regular maintenance of the drawbridge and lift mechanism will ensure improved performance.

Some movable bridges require one or more piers in the river, depending upon the width of the river. During floods, the operation of a movable bridge becomes unsafe. Further, the pier foundation is subjected to increased scour. Suitable countermeasures are required. The following types of movable bridges are in use in the United States:

- Bascule bridges
- Drawbridges

- Swing bridges
- Vertical-lift bridges

Many of the movable bridges are historic. For example, the Cortland Street Drawbridge in Chicago was built over the Chicago River in 1902. A comparative study on Maryland's Susquehannah River Swing Bridge (horizontal sway) Replacement Project with a bascule (vertical lift) is in progress at Temple University under the supervision of the author. In floods, the water level may be higher than the bridge deck elevation, making the navigation of ships difficult. However, greater clearance is provided by the vertical-lift bridge.

Environment Policy to Include Effective Flood Risk Management and Legislation Much water had flowed both under and over the bridge before it was recognized that the magnitude of risk posed by flooding needed to be tackled through engineering, by finding new ways to reduce the vulnerability to flooding.

- It has been nearly 10 years since Hurricane Katrina devastated the U.S. Gulf Coast in August 2005.
- Hurricane Rita struck in 2005 in the Gulf Coast States.
- Hurricane Ike came in 2008 in Gulf of Mexico and Texas.
- The Midwest floods occurred in 2011.
- In New Jersey, Hurricane Floyd struck in 1999 and Hurricane Sandy in 2011.

The above listed hurricanes led to over 2000 lives lost and the loss of billions of dollars to commerce and damages to property. The ASCE Report Card has taken the lead in waking up the owners of the bridges and the transportation agencies in many states. The author served on a panel for the grade allocation for Pennsylvania bridges for 2014. The massive effort for flood risk management involves

- Identifying scour-critical rivers and bridges
- Making emergency preparations in the short term
- Strengthening the flood protection systems
- Actually obtaining the huge funds required for minimum funding for the long term due to unexpected damages.

Similar to the ASCE Report Card, both government and nongovernment organizations have echoed the ASCE approach to the problem. Both the public and the decision makers need to be fully motivated. Due to population growth and climate changes, floods are more frequent. As a result, levees continue to become damaged during peak floods, along the Mississippi and Ohio Rivers in United States. According to FEMA, the 100-year floodplains in the contiguous states will have expanded by 45 percent by the end of twenty-first century. After the 2005 Hurricane Katrina disaster, the USACE set up a task force called the Interagency Performance Evaluation Task Force (IPETF) to conduct an inquiry into the failure of the important New Orleans Hurricane Protection System. At the request of the USACE, the ASCE External Review Panel prepared a report, which gave with following main recommendations:

- Communicate the quantified risks to the public and decide how much risk is acceptable.
- Place safety first.
- Correct the deficiencies and upgrade design procedures.
- Improve interagency coordination.

1.9 Current Bridge Scour Control Practice

The Nation's Bridges Desperately Need Repair

There are more than 200 scour-critical bridges in New Jersey and a similar number for yearly review in Pennsylvania. All over the United States, the number of bridges vulnerable to floods is increasing after each flood, and runs in the thousands. The maintenance cost of such bridges is also increasing. Of the more than 575,000 bridges in the National Bridge Inventory, approximately 84 percent are over water.

1. Each year in the United States, highway bridge failures cost millions of dollars as a result of both direct costs necessary to replace or restore bridges and indirect costs related to disruption of transportation facilities.
2. Of even greater consequence is the loss of life due to bridge failures.
3. Hydraulic factors such as stream instability, degradation, contraction scour, and local scour account for bridge failures (approximately 60 percent) more than all other factors combined.
4. Ongoing screening and evaluation of the vulnerability of the nations' highway bridges to scour by state DOTs have identified more than 18,000 bridges that are considered scour-critical and in need of repair or replacement.
5. Almost 100,000 bridges with unknown foundations have been identified.
6. As state DOTs develop plans of action to address the effects of scour and stream instability, there is a need for innovative, effective, and economical countermeasures to be considered for the design of new bridges and for repairing existing bridges.
7. For years bridge designers in the United States have used FHWA Publication HEC-18 as a principal tool to determine scour depths. Increasingly, however, practitioners recognize that some of the circular's standard equations overpredict the scour depth for certain geologic and hydraulic conditions.

NJDOT Survey of Scour Practice in the United States. In an effort to improve scour design and evaluation methods, NJDOT recently conducted a survey of the current scour practices of DOTs across the United States. The 10-question survey queried agencies about their design standards, experiences with failures, monitoring programs, and countermeasure preferences, among other things.

A paper by the author and Professor Schuring's team of New Jersey Institute of Technology (NJIT) presented at the San Francisco International Conference of Scour and Erosion summarizes the results of the nationwide scour survey. Highlighted are the creative and diverse approaches taken by some states to either modify HEC-18 procedures or develop alternative scour prediction methods. The paper also discusses critical geologic, hydraulic, and hydrologic parameters for rational evaluation of scour depth, gleaned from both the survey and local experience with New Jersey bridges. The situation could worsen. Bridges located on scour-critical rivers present special threats. While water is a daily indispensable need for survival, too much water during floods can be a threat to our existence. For example, a 50- or 100-year flood is a natural disaster and cannot be easily controlled, even by modern technology. Its effects are seen in many forms and are not just restricted to the disruption of our vast transportation system. The design issues for bridge and highway structure foundation in rivers need to be addressed.

Magnitude of the Problem Nearly one-tenth of the 607,380 bridges in the National Bridge Inventory, a database of information on bridges and tunnels, are rated as structurally deficient by the Federal Highway Administration (FHWA); the average age of those bridges is 42 years.

FHWA budget forecast: The FHWA has estimated that the nation would need to spend about \$20.5 billion per year to eliminate the backlog by 2028; the United States spends about \$12.8 billion annually. To remain in service, these deficient bridges are often posted with weight limits restricting the gross weight of vehicles permitted on the bridge.

Potential for failure: In the United States, over 36,000 bridges are either scour-critical or scour-susceptible. Examples of recent bridge failures due to scour such as Wilson Bridge in Virginia and Peckman's Bridge in New Jersey, are given in this chapter. A new study says that over 63,000 bridges are in need of repair. *USA NOW* host Carly Mallenbaum talks about the busiest bridges on the list. (*USA TODAY, USA NOW*, Larry Copeland, April 24, 2014). Figure 1.4 shows an example of a river bridge serving both passing ships for navigation and motor vehicles. The substructure needs to be protected from impact of navigation of ships.

Planning of Shallow and Deep Foundations to Minimize Erosion Foundations require guidelines on river training measures that may be used without armoring countermeasure guidelines on structural countermeasures for existing bridges. In some states, riprap is considered to be a temporary countermeasure. There is general flooding over vast areas

FIGURE 1.4 A long-span river bridge on a wide river allows navigation under the bridge deck, but navigation is not possible when the water level rises during the flood season.

which is not just confined to the rivers or the highway structures, as well as flooding specific to the foundations of bridges. Floods cause the following major problems:

1. Soil erosion at bridge foundations and riverbanks
2. Soil erosion at bridge approaches and retaining walls
3. Sediment deposit and aggradation at rivers
4. Settlement of highway pavements in the flood zone
5. Damage to the drinking water supply and other utilities

The above issues pose serious challenges to the methods used for computing hydrology, hydraulics, and structural engineering solutions. The engineering objectives are to keep the transportation system functional both during flood duration and after the flood has subsided.

1.10 Precautions and Solutions

In the United States, the well-known hurricanes in the recent past such as Floyd, Katrina, Irene, and Sandy were responsible for billions of dollars in losses in addition to the loss of lives. It was seen that during floods the supply of water exceeds the demand for water. The excess water causes damage to the transportation network, including bridges. Bridges on waterways have the following problems:

1. There are restrictions from environmental agencies, in placing piers in riverbeds, resulting in much longer spans using abutments only.
2. It is difficult to design and construct deep foundations in flowing water.
3. It is also difficult to maintain bridge substructure underwater and during floods.
4. Regular painting of the girders that are corroded due to evaporation and condensation is expensive and time consuming.

This chapter reviewed the impacts of recurring floods leading to scour and damage to life and property in the United States and a survey of recent floods and a nationwide check of infrastructure health. Flood hazards are a global problem, hence the review of European practices and a brief review of floods and flood control in countries such as Australia, New Zealand, and Pakistan. Review of European inspection and monitoring programs and manuals in relation to the NBIS has helped to improve the U.S. practices. Recognition of the importance of geomorphic reconnaissance and analysis in the selection and design of countermeasures is required.

1.11 Conclusions

The discussion in this chapter may be summarized as follows:

1. A preliminary discussion of the observation of design philosophy, scour prediction techniques, and inspection and monitoring practices is presented.
2. Detailed descriptions of the findings and the final recommendations are presented in later chapters.

3. Additional information is given in several appendices. Forensic engineering procedures are usually required following failures that adversely affect public lives. Hence, asset management should include timely monitoring and scour prediction to avoid any failure. Repair versus replacement options need to be considered for mitigation of the issues.
4. The following items provide recommendations for implementation or further evaluation:
   - The economics of including scour and stream instability countermeasures in the initial planning and construction of a bridge needs to be taken into consideration.
   - Risk analysis can be employed in bridge and countermeasure design, particularly for existing scour-critical or unknown-foundation bridges. Consideration of risk to the structure, lives, or property in relation to environmental policy should be weighed against the additional costs. Legislation may be required both in United States and abroad.
   - Use of improved techniques as in Europe to analyze and predict scour, particularly for complex flow situations such as wide piers, pressure flow, tidal scour, debris, and the interaction of general and local scour components, will help develop the new subject of controlling the effects of soil erosion from floods and providing countermeasures.
   - The applicability of sediment management as a means to counteract long-term riverbed degradation problems should be evaluated.
   - "Lessons learned" can be developed from case studies.
5. There has been a significant increase in scour resulting from limited bridge opening sizes and from settlement of footings due to geotechnical conditions. Many river bridges that are more than 40 years old are subjected to overtopping during floods and to sudden increases in contraction and local scour. HEC-23 matrix methods for river training, armoring, and monitoring need to be adopted. In particular, the strategy should be as follows:
   - At the time of river bridge planning, there were no design criteria for scour. Asset management must include mitigation and environmental policies for flood control. Opening sizes in many cases are smaller than required (to allow unrestricted flow of water). For existing bridges with high rate of scour, adjacent auxiliary bridges for increasing the opening size between the abutments and alternative river diversion may be considered, in addition to providing armoring countermeasures.
   - Comprehensive hydrology methods for analysis such as the scientific StreamStats software for computing flood discharge should be used.
   - The shallow footings can cause settlement compared to deep foundations.
   - Average daily truck traffic has increased due to demography and an increase in commerce. Unprecedented global weather changes caused heavy rains and melting of glaciers with recurring floods.
   - Deck slab elevations are not high enough to allow minimum free board of 1.0 ft for a 100-year flood to prevent overtopping. For example, during the

1990s, floods in the Ohio River caused closure of important highways located on the river due to overtopping. It is not easy to replace these bridges, but they can be placed on a priority list for shielding the footings, using alternative river training measures and biotechnology.

Recommendations and Prevention The problems arising from flood scour need to be identified, and economical solutions need to be implemented. The following is a list of the issues to be addressed:

1. Bridges are subject to displacement, movement, and sway of the substructure due to floods.
2. If design and reconstruction criteria are correctly implemented, many failures could be prevented.
3. Some failures may occur because of experimentation with new types of materials or new systems, such as undefined and unpredictable material properties in cast-in-place or precast construction.
4. Collapse can be reduced by using standard details and, in addition, flexible and ductile moment resistance connections.
5. The expected life of 75 years or more for modern bridges and their components may not be achieved without effective inspection, structural evaluation, and timely rehabilitation. The trends in peak stream flows are important to investigate, as trends indicate emergencies arising out of the resulting floods. This generally leads the infrastructure managers to reexamine the levels of risk to the public. For the Midwest floods, the USGS has computed the trend magnitudes based on Sen's (1968) slope estimator.

1.12 Bibliography

AASHTO LRFD Bridge Design Specifications, Washington, D.C., 1998, with periodic revisions.

ASCE, *2013 Report Card for America's Infrastructure* (www.infrastructurereportcard.org).

Durrant, J., and Whitten, B. "Report by the Task Committee on Flood Safety Policies and Practices," *Magazine of Civil Engineering*, 2014.

ASCE, "Chicago Style: The Cortland Street Drawbridge, a History Lesson by Jeff Brown," *Magazine of Civil Engineering*, April 2014.

Landers, J., "Nevada City to Reroute Waterway," *Magazine of Civil Engineering*, April 2014.

Austroads, *Waterway Design—A Guide to the Hydraulic Design of Bridges, Culverts and Floodways*, Sydney, Australia, 1994.

Carse, A., and Bell, A., "Design of Queensland Road Infrastructure for High Risk Environment," 13th International Soil Conservation Organisation (ISCO) Conference, Brisbane, Australia, July 2004.

Chiras, D. D., *Environmental Science*, Benjamin Cummings, San Francisco, 1991.

Elsevier Journal of Hydrology, "From Safe Yield to Sustainable Development of Water Resources—the Kansas Experience, M. Sophocleous," *Kansas Geological Survey, The University of Kansas*, May 2000.

FHWA Pilot Study Report no. FHWA-HIF-12-049, Government of India, Ministry of Water Resources, *General Guidelines for Water Audit and Water Conservation* Central Water Commission, Evaluation of Water Utilization Directorate, New Delhi, December 2005.

HEC-18, *Evaluating Scour at Bridges*, DOT, FHWA, Washington, D.C., 2001.

HEC-18, *Evaluating Scour at Bridges*, 5th ed., DOT, FHWA, Washington, D.C., 2012.

Hjelmfelt, A. T., Jr., and Cassidy, J. J. *Hydrology for Engineers and Planners*, Iowa State University Press, Ames, 1975.

Khan, M. A., Baig, S., and Zarriello, J. "Factors Affecting Stream Stability at Scour Critical Bridges in New Jersey," *First International Conference on Bridge Scour*, Texas A&M Kenya Vision 2030 Sector Plan for Drought Risk Management and Ending Drought Emergencies, 2013-17 (izzy.birch@ndma.go.ke) University, College Station, 2002.

Melville, B. W., and Coleman, S. E. *Bridge Scour*, Water Resource Publications, Highlands Ranch, CO, 2000.

Richardson, E. V., and Lagasse, P. F. NCHRP 21-3, "Instrumentation for Measuring Scour at Bridge Piers and Abutments," Final Report, Phase III, Transportation Research Board, Washington, D.C., 1994.

National Cooperative Highway Research Program (NCHRP) Research Results Digest—241, Hydraulics and Hydrology, "1998 Scanning Review of European Practice for Bridge Scour and Stream Instability Countermeasures," July 1999.

Parker, G., Toro-Escobar, C., Voigt, R. L., et al., NCHRP 24-7, *Countermeasures to Protect Bridge Piers from Scour, Users Guide and Final Report*, National Cooperative Highway Research Program, Transportation Research Board, National Research Council, 1998.

New Jersey Department of Transportation, Bridge Scour Evaluation Program, "Stage II—In-Depth Scour Evaluation," Phase 3, June 2000.

New Jersey Department of Transportation, *Soil Erosion and Sediment Control Standards*, 1989.

Parker, G., Toro-Escobar, C., Voigt, R. L., et al. "Countermeasures to Protect Bridge," 1999.

Pennsylvania Department of Transportation, Publication 100A, *Coding Manual*, office version, Bureau of Design, Bridge Management System (BMS), 1993.

Pennsylvania Department of Transportation, *Bridge Design Manual*, No. 4, Chapter 7, "Scour," 2006.

Fotherby, L. M., "Alternatives to Riprap for Protection against Local Scour at Bridge.

Piers," *Transportation Research Record*, vol. 1420, TRB, National Research Council, Washington, D.C., 1993.

Tyler Miller, G., and Spoolman S., *Living in the Environment*, 18th ed., National Geographic Learning, Independence, KY, 2012.

U.S. Army Corps of Engineers, Hydraulic Engineering Center, "HEC-6: Status and New Capabilities of Computer Program HEC-6: Scour and Deposition in Rivers and Reservoirs," CA, June 1990.

U.S. Army Corps of Engineers, EM1110-2-1601, *Hydraulic Design of Flood Control Channels*, Washington, D.C., 1994.

U.S. Army Corps of Engineers, "Wire Mesh Gabions," Report CW-02541, 1991.

U.S. Bureau of Reclamation, *Guide for Computing Degradation*, Denver, CO, 1963.

U.S. Department of Agriculture (USDA), "Streambank and Shoreline Protection," Natural Resources Conservation Service, *Engineering Field Handbook*, Washington, D.C., 1996.

U.S. Department of the Interior, U.S. Geological Survey, "Procedures for Bridge Scour Assessments at Bridges in Pennsylvania," Open File Report 00-64 by P. J. Cinotto and K. E. White (in cooperation with Pennsylvania Department of Transportation), Lemoyne, PA, 2000.

U.S. Department of the Interior, U.S. Geological Survey, *Software User's Guide for Determining the Pennsylvania Scour Critical Indicator Code and Streambed Scour Assessment Rating for Roadway Bridges*, by M. F. Henneberg and J. L. Strause, Open-File Report 01-446 (in cooperation with Pennsylvania Department of Transportation), New Cumberland, PA, 2002.

U.S. Department of the Interior, "Guidelines for Bridge Maintenance and Rehabilitation," by Secretary, Washington, D.C.

U.S. Geological Survey, "Flooding in the United States Midwest," Professional Paper 1775, by R. Holmes, Jr., T. Koenig, and K. Kaestensen, Reston, VA, 2008.

U.S. Geological Survey, Scientific Investigations Report 2008-5206, Prepared in cooperation with the New Hampshire Department of Transportation, "Estimation of Flood Discharges at Selected Recurrence Intervals for Streams in New Hampshire," By S. A. Olson, 2006.

U.S. Department of Transportation, FHWA, *2013 Status of the Nation's Highways, Bridges and Transit: Conditions and Performance*, Washington D.C., January, 2014.

CHAPTER 2
River Instability Caused by Flow Obstructions

2.1 Floodplain Formation

After a flood event, which may last for several days, it may be weeks before the water recedes and the land is dry once again. The floodplain water will infiltrate the soil, and groundwater will give rise to wetlands depending on the type of soil. The sizes of floodplains vary for each bridge and depend upon

- Intensity of rainfall and its duration
- Topography and slopes
- Natural drainage
- Provisions of pumping stations
- Debris accumulation after the flood

The abutments, piers, and superstructure of the bridge, when placed in the path of moving water, create a combined obstruction to the natural free flow of water. The velocity of water is increased considerably, and the banks of the river or channel become unstable. The continued flow of water, the fallen trees, or debris accumulation at the upstream face of the bridge further submerge a wider area around the bridge and cause a floodplain to form. The increased volume of water will cause overtopping of banks and place the highway approaches and the bridge deck out of service; the entire highway needs to be closed down. The undesirable effects of floods are twofold:

1. Overtopping of riverbanks and breach of levees
2. Soil erosion around bridge foundations

In this book, these hydraulic effects are investigated and engineering solutions are presented to combat them. In this chapter, the basic concepts of water flow, scour, and erosion arising from the floods are explained.

Applications of the modern methods for scours, hydrologic, hydraulic and geotechnical analyses will benefit the engineering design tasks in terms of safety, economical design of foundation, and solution of underwater construction issues. In Chapters 3 and 4, there is a relevant literature review of publications on this relatively new subject, including applications of the Federal Highway Association (FHWA) Circulars and the American Association of State Highway and Transportation Officials (AASHTO) provisions.

The Hydrologic Cycle Causing Rain and Intense Monsoons

The earth's hydrologic cycle causes water flow and floods. Nature's hydrologic cycle is repeated at regular intervals and comprises the following processes:

- Evaporation due to the sun's thermal energy
- Condensation due to cloud formation in the atmosphere
- Precipitation to the earth surface due to gravity
- Surface flow of rainwater, causing floods

The kinetic and potential energy of rapidly flowing water causes aggradation (sediment and debris deposit) and degradation (erosion of soil) and infiltration and percolation through the soil (the sponge effect), which causes wetland and groundwater flow. Figure 2.1*a* and *b* shows examples of cold weather constructions on river bridges.

(*a*)

FIGURE 2.1 (*a*) Cold weather construction on a river bridge was an author's bridge replacement project. (*b*) Suspended rail traffic during the repairs of a railway bridge located on river.

(*b*)

Figure 2.1 (*Continued*)

Soil resistance: The geological conditions for rock and noncohesive and cohesive soils will govern the rate of erosion and the resulting structural damage by foundation scour. Rain is caused as a result of the processes of evaporation, condensation, and precipitation. The hydrologic cycle describes the distribution of water through earth's biogeochemical system.

Rainwater flows across the land surface into rivers and streams. The rate of gravity flow or drainage depends upon the runoff. The runoff factors are longitudinal and cross slopes, type of soil to allow absorption and percolation (for subsurface groundwater), and the intensity of rainfall per hour, depending on the rise in temperature from solar energy.

The cycle is repeated as water evaporates from bodies of water, from the soil and leaves of plants and grass that have absorbed water from soil or rain. The annual floodplain is likely to be flooded each year.

Intense Rain the Cause of Floods Flood is caused when the river capacity is exceeded, the banks overflow, and floodplain is generated. In the United States, the Mississippi watershed is the largest. Reservoirs and dams can be built to store the excess water to make it available for drinking, agriculture, or the generation of hydroelectric power.

Flood is the inundation of an area adjacent to water in a flowing river or stream that creates a new floodplain or enlarges an existing floodplain. The problem is greater when the floodplain engulfs existing homes or structures. A floodplain is formed by overbank flow or by lateral migration of meandering streams or bends. The causes of floods are

- Intense rainstorms
- Storm surges from a hurricane
- Failure of a flood protection structure such as a reservoir or dam
- Melting of snow
- A tsunami caused by earthquake

Toxic Effects of Floods

Mining waste and debris from deep in the earth are likely to contain toxic substances and chemicals. Surface water flow from recurring precipitation will wash the poisonous substances into streams, contaminating drinking water. Groundwater will be adversely affected by percolation.

In industrial areas located in urban and rural areas, the air is polluted with carbon dioxide, sulfur dioxide, and other toxic gases. Rainfall carries sulfuric acid, and the acid rainfall percolates in the soil.

The groundwater is also contaminated and flows to streams. Fish can perish in great numbers. Floodwater carries impurities by washing the land surface. The water supply, if not treated with chemicals, can cause diseases on a large scale, especially in villages in Africa, India, or South America and any areas where water treatment facilities are not available.

Agriculture production is also affected by the contaminated acidic water. Storm water management is required. A graduate certificate course offered by Temple University (and other institutions of learning) addresses these issues.

Increased Interaction of Groundwater with Soil Layers during Floods

The decomposition of organic matter and weathered rock forms soil. The topmost layer is made up of leaves, crop residue, and humus. Below the top layer is another layer containing a mixture of organic matter and mineral particles. Underneath the top two layers is the subsoil, containing clay particles. The bottom two layers comprise particles of rock and the parent rock, which can be igneous, metamorphic, or sedimentary. The thicknesses of layers are likely to vary at each site. The contaminated groundwater will have a chemical reaction with the layers and will affect the fertility of the soil. Climate change and global warming have caused increased evaporation, rainfall, and floods. The sea level also rises, causing flooding in coastal towns.

2.2 Erosive Effects of Floods and Tides on Transportation

The adverse effects of floods on transportation are discussed in Chapter 1. More details are provided here. The two types of flows are the riverine flow and coastal tides. Floods in rivers are not a one-time event; they seem to recur every few years. They are mainly generated in perennial rivers and in dry rivers from flash floods, intense rainfall, snowmelt on mountains, or melting of glaciers. Near the coastline, bridges are subjected to daily tidal effects, from both incoming tides caused by ocean energy and the soil erosive effects accompanied by the receding tides, which act in reverse direction toward the ocean. Coastal tide effects are maximum at nighttime, especially on the monthly full moon events. Tidal effects are described in FHWA Hydraulic Engineering Circular no. 25 (HEC-25).

Damage Caused by Riverine Floods

There are three types of flood effects:

1. Scour and erosion of soil at riverbanks and under bridges
2. Inundation of highways due to formation of floodplains
3. Flooding of entire towns near rivers

The first two types are discussed in detail, while preventive actions to be taken are briefly discussed for the third type owing to different topography and drainage methods used by each flooded town.

Disaster management aspects (relocation and relief measures) for the postflood scenario are addressed. FHWA Hydraulic Engineering Circulars and Federal Emergency Management Agency (FEMA) methods apply.

Damage Caused by Tidal Flow

Tidal effects similar to those of riverine flow but resulting from daily tides need to be discussed. Tidal flow is of different magnitude and acts at different times. Some of the countermeasures described in HEC-23 may be common to scour generated by both riverine and tidal flows. Once again, FHWA Hydraulic Engineering Circulars (including HEC-25) and various FEMA methods will apply to the erosion effects. Flood velocity analysis and scour depth computations are required.

2.3 Multiple Types of Riverine Flow Processes

Field inspections and observation of the velocity of flow during floods (which may last from a few hours to more than 12 hours) and immediately after floods are likely to reveal the following processes.

1. Initial soil erosion occurs at the bridge substructure during the first few hours of rainfall. If the water level increases to the bridge deck elevation, pressure flow will also result.
2. There is advanced soil erosion at the bridge substructure owing to obstruction from debris deposition.
3. Reduction of bridge opening size occurs, thereby increasing the kinetic energy and velocity of flow. Scour depth will continue to increase. Scour holes may form under the bridge.
4. After the rainfall has stopped, flow from the bridge upstream will continue because of the potential energy from a higher elevation. Aggradation from debris deposit will increase; scour depth will continue to increase. Additional scour holes may form.
5. Obstruction downstream of the bridge or hydraulic jump will adversely affect the free flow of water, thus increasing the water pressure on the piers and abutments.

6. It is normal practice to consider the peak flow hydrology of the standard probability of 2-, 10-, 25-, 50-, 100-, and 500-year events and the resulting Q values and depths of flow from hydraulic analysis, for each event.
7. Preventive actions and scour countermeasures designed to protect the highways and bridge foundations from scour are discussed in later chapters.

Preventing Soil Erosion from Floods

- Planting crops in rows on riverbanks restricts the land surface for flowing water. The roots bind the soil particles and prevent erosion.
- Strips of perennial alfalfa can be planted adjacent to cornfields to prevent erosion.
- The land surface cover can be made impervious by using pavements and sealed surfaces.
- River training measures and countermeasures are used to protect foundations, scour countermeasures as remediation and selecting applicable countermeasures are discussed in detail in Chapters 3 and 8.

2.4 Basic Concepts of Open-Channel Hydraulics

Scour analysis methods are based on open-channel hydraulics. The basic equilibrium equations for the velocity of water, kinetic energy, and height of the water need to be understood. The fundamental hydraulic theorems for water flow by Bernoulli, using Manning's coefficients, are explained here.

Bernoulli's principle can be derived from the principle of *conservation of energy*. This requires that the sum of kinetic energy and potential energy remain constant. Thus an increase in the speed of the fluid occurs proportionately with an increase in both its dynamic pressure and kinetic energy, and a decrease in its static pressure and potential energy. If the fluid is flowing out of a reservoir, the sum of all forms of energy is the same on all streamlines because in a reservoir the energy per unit volume (the sum of pressure and gravitational potential ρgh) is the same everywhere.

Bernoulli's principle can also be derived directly from Newton's second law. If a small volume of fluid is flowing horizontally from a region of high pressure to a region of low pressure, then there is more pressure behind than in front of the fluid.

Resistance to Flow

This gives a net force on the volume, accelerating it along the streamline. The type of soil will determine its resistance to horizontal shear caused by the fast-flowing water and resulting in the gradual buildup of scour. Free flow is guided by contraction and local scour. Open-channel flow can be classified and described based on the change in flow depth with respect to time and space. The fundamental type of flow is dealt with in the well known textbook *Open-Channel Hydraulics* by V. T. Chow.

Unsteady flow The depth of flow does change with time. It is the most common type of flow.

Uniform flow The depth of flow is the same at every section of the channel. Uniform flow can be steady or unsteady, depending on whether the depth changes with time (although unsteady uniform flow is rare).

Varied flow The depth of flow changes along the length of the channel. Varied flow technically may be either steady or unsteady. Varied flow can be further classified as either rapidly or gradually varied.

Continuous flow The discharge is constant throughout the reach of the channel under consideration. This is often the case with steady flow. This flow is considered continuous and therefore can be described using the continuity equation for continuous steady flow. Because the flow is sufficiently close to steady flow, it can be analyzed as gradually varied steady flow.

Types of Channels

Bridges are generally located where the width of the river channel is minimum. In practice, these are

- Natural channels (an example is a network of rivers)
- Scoured channels in noncohesive soils, formed from recurring rainfall
- Artificial or excavated channels, such as irrigation channels or those used for water supply

Scoured channels and rivers having significant longitudinal slopes are a subject of study of this book, and the modified hydraulic equations are described in detail in the User's Manual of the Hydrologic Engineering Center (HEC-RAS). The hydraulic analysis software and scour computations are prepared by the U.S. Army Corps of Engineers, who oversee all bridge construction over the U.S. rivers.

Cross-Sectional Shapes

For hydraulic analysis, the assumed geometric shapes of cross section are generally trapezoidal, V-shaped, or rectangular. Field surveys will determine the actual shapes based on topography.

States of Flow

The behavior of open-channel flow is governed by the effects of viscosity and gravity relative to the inertial forces of the flow. Surface tension makes a minor contribution. Depending on the effect of viscosity relative to inertia, as represented by the Reynolds number, the flow can be laminar, turbulent, and transitional.

Fluid Properties

These include the Froude number, Reynolds number, and viscosity. Other related topics include the classic energy problem in open-channel flow, hydraulic jump, Manning's formula, Chézy's formula, and the standard step method.

Applications of Bernoulli's Flow Equilibrium Equation

A common form of Bernoulli's equation, valid at any arbitrary point along a streamline, is

$$\frac{v^2}{2} + gz + \frac{p}{\rho} = \text{constant}$$

where v = fluid flow speed at a point on a streamline

g = acceleration due to gravity

z = elevation of point above reference plane, with positive z direction pointing upward, so in direction opposite to gravitational acceleration
p = pressure at chosen point
ρ = density of fluid at all points in fluid

The fundamental equations are applications of the laws of conservation of mass, energy, and momentum. The well-known continuity equation, developed by Bernoulli, assumes that between two adjacent cross sections, mass or flow entering = mass or flow leaving. Thus,

Potential head + pressure head + velocity head = constant between two cross sections

Any friction between water and soil is neglected.

$$Z_1 + \frac{p_1}{\rho g} + \frac{u_1^2}{2g} = Z_2 + \frac{p_2}{\rho g} + \frac{u_2^2}{2g} = \text{constant}$$

However, additional equations including Manning's roughness coefficients are used.

A computer software program such as HEC-2, WSPRO was used in the past but HEC-RAS is currently preferred. A minimum of four cross sections along the length of a straight bridge and at upstream and downstream locations is used. For meandering channels, additional cross sections are required in keeping with HEC-18 guidelines.

Reference for Equations of Flow See V. T. Chow, *Open-Channel Hydraulics*, 2008.

Software Application The Army Corps of Engineers software (HEC-RAS, River Analysis System) and alternate hydraulic analysis software are discussed in Chapter 6. A minimum of four cross sections is selected at the openings located at the two ends of the bridge and one upstream and one downstream of the bridge. Additional sections may be required when there is a change in the geometry.

HEC-RAS, for example, performs open-channel analysis for steady or unsteady, one-dimensional, gradually varied flow in both natural and artificial river channels. It also solves the scour analysis equations based on HEC-18. In addition, it is based on sediment transport-mobile bed modeling, and it performs water temperature analysis. The HEC-RAS (Version 4.1) software supersedes the earlier HEC-2 river hydraulics package, which was a one-dimensional, steady flow water surface profiles program.

The hydraulic model and the input variables used are described in detail in Chapter 6. An independent scour analysis using an Excel spreadsheet based on the updated 2012 HEC-18 equations can be developed to verify the earlier version of the HEC-18 equations used by HEC-RAS. The WSPRO software uses Bernoulli's standard equations.

2.5 Hydraulic Issues of Bridges on Waterways

Major Factors Affecting Magnitude and Rate of Scour

The following physical factors need to be considered:

1. River configuration types
   a. Straight
   b. Braided or multichannel
   c. Meandering

2. River bottom types
   a. Aggrading
   b. Stable
   c. Degrading
3. Soil types: To determine the grain size distribution for a scour analysis, samples from the riverbed need to be tested in accordance with ASTM D422. Soil types may be broadly classified as noncohesive and cohesive materials. *Noncohesive* materials include
   a. Gravels
   b. Sands
   c. Silts

   The threshold of movement depends on
   d. Particle size
   e. Density
   f. Shape
   g. Packing
   h. Orientation of bed material

   Cohesive materials include silts and clays. Most fine-grained sediments possess some cohesion. The clay content is of great importance. Cohesive sediments typically require relatively large forces to detach the particles and initiate movement, but relatively small forces to transport the particles away.
   i. Silts
   j. Clays

Types of Soils with Scour Problems Table 2.1 shows the sediment type and comparative percentage of scour, which varies in each soil type according to resistance offered.

1. Soil profiles for typical scour-critical bridges
   The soil profile for a particular bridge site should be based on the boring logs. While detailed geotechnical and borehole testing needs to be carried out to obtain site-specific information, the present study will utilize existing maps and information available with the state and the U.S. Geological Survey (USGS).

   Different soil and rock materials will exhibit different rates of scour. The kind of geologic material, coupled with the intensity and duration of a flood, will determine the scour depth. The magnitudes of scour for cohesive and noncohesive soils are different, but scour takes much longer in cohesive soils, resulting in longer bridge life.
2. Soil types are broadly classified as follows:
   a. Noncohesive (e.g., gravels, sands, and silts)
   b. Cohesive (silt, clay) materials

3. The threshold of movement of particles of both cohesive and noncohesive materials depends on the following:
   a. Particle size
   b. Density
   c. Shape
   d. Packing and orientation of bed material
4. Noncohesive sediments (e.g., sands, gravels, and silts) have a granular structure. Such soils are considered to be the most susceptible to scour. Individual particles are susceptible to erosion when the applied fluid forces (drag and lift) are greater than the stabilizing forces (gravity and friction with adjacent bed particles).
5. Most fine-grained sediments (clay, silty clay, and clay mixtures) possess some cohesion, the clay content being of great importance. Cohesive sediments typically require
   a. Relatively large forces to detach the particles and initiate movement
   b. Relatively small forces to transport the particles away

The bed material comprises sediments (alluvial deposits) or other erodible material. If bed materials are stratified, there is a greater risk of scour breaking through the more resistant layer into the less resistant layer. A survey of U.S. bridges indicates that bridges founded in sand have the greatest number of scour problems, as summarized in NCHRP 24-7. The clay content in the soil increases cohesion, and relatively large forces are required to erode the riverbed. Higher pulsating drag and lift forces increase the dynamic action on aggregates until the bonds between aggregates are gradually destroyed and aggregates are carried away by the flow. One approach estimating scour in cohesive soil is to couple the erosion rate with the cumulative duration of flows that exceed the threshold velocity for particle movement.

Professor Briaud at Texas A&M University has proposed the SRICOS (*s*cour *r*ates *i*n *co*hesive *s*oils) method of scour measurement, described below.

- In cohesive soils such as clay, the local scour magnitude and the contraction scour magnitude may be similar, but scour takes place considerably later than in the noncohesive sand.
- Scour analysis methods are different for cohesive and noncohesive soils.
- Bridge foundations supported by cohesive soils resist erosion for a much longer period than is usually calculated, resulting in a longer life of the bridge. The bed material is comprised of sediments (alluvial deposits) or other erodible material. If bed materials are stratified, a conservative approach needs to be adopted regarding the risks of scour breaking through the more resistant layer into the less resistant layer.

Cohesive sediments require relatively large forces (pulsating drag and lift) to detach particles and initiate movement, but relatively small forces to transport particles away.

Streambeds that either consist of bedrock or contain a high percentage of oversize cobbles and boulders are the most scour-resistant materials. To determine rock quality, careful evaluation is needed to assess factors such as strength, fracture frequency,

Sediment Type	Percentage
Sand	48
Cohesive	19
Mixed	13
Gravel	10
Bedrock	5
Uncertain	5
Silt	0
Total	100

TABLE 2.1 Soil Types with Scour Problems (NCHRP 24-7)

weathering, slaking, and stratification. If bed materials are stratified, the more resistant layers will retard scour while other intervening layers will not.

A breakdown of scour problems by soil type is given in Table 2.1. However, since geologic conditions vary widely across the country, the statistics for each state will vary.

Scour due to Obstruction to Flow at Abutments and Piers

Hydraulic conditions and rates of erosion are vastly different at abutments and piers at any bridge site. The extent of erosion at abutments may be minimized by placing them away from the riverbanks. Piers are located in the middle of peak flood zones, where flood velocity is the highest. The direction of flow is at right angles to the pier, which acts basically as an obstruction, with the water flowing on both its sides. Hence, the foundation all around a pier may be scoured. Or the foundation with only one side exposed to the flow, as in the case of an abutment, may be scoured.

Scour Analysis

From a review of 45 scour-critical bridges in New Jersey, it is seen that the total scour depth is higher at pier locations than at abutments or wing walls. Total scour at bridge footings is primarily the sum of degradations and aggradations, local scour, and contraction scour. *Degradation* is a general and progressive (long-term) lowering of the channel bed due to erosion over a relatively long channel length. Local scour is due to the increase in local flow velocities and turbulence levels because of obstruction to water flow, caused by bridge piers and abutments. Contraction scour results because of the increased water velocity in the bridge opening caused by the decrease in cross-sectional area of the waterway at the bridge crossing. Scour analysis of bridge footings should be carried out based on the guidelines and equations presented in HEC-18. Scour depth should be measured from a reference line 1 ft 0 in above the top of the footing. If the eroded elevation is located at a higher elevation than 1 ft 0 in above the top of the footing, the higher elevation will be considered.

Scour Analysis of Bridge Piers and Abutments in Cohesive Soils This can be carried out based on NCHRP 516, Pier and Contraction Scour in Cohesive Soils, by the method developed by Briaud et al. (2004). The countermeasures for bridge footings can be designed according to guidelines presented in this handbook based on scour analysis data, such as total scour depth and peak flow velocities. A detailed scour analysis at a bridge site may require the following steps:

- Study the bridge hydraulics to screen bridges as scour-critical and non-scour-critical.
- Study past floods and obtain flood data from any gauges installed near the bridge sites.
- Perform a field survey to obtain discover the topography, river boundaries, obstructions to flow, catchment area, vegetation, and terrain.
- Perform a hydrologic analysis.
- Perform a geotechnical investigation for types of soil and foundations.
- Examine aggradation, degradation, contraction, and local scour.
- Set up a hydraulic model to calculate scour depths due to degradation, local scour, and contraction scour.

The design of countermeasures at a site may require the following steps:

- Perform a feasibility study on the use of appropriate countermeasures based on alternatives, economics, site conditions, permit considerations, and constructability.
- Consider the alternatives for river training, structural planning, and deep foundations for new bridge designs.
- Select the preliminary design of countermeasures and prepare cost estimates.
- Review the countermeasure design for compliance with structural and environmental permit requirements.
- Develop the technical specifications, construction methods, and construction schedule.

Bank Protection

The most common method of bank protection involves the use of rock riprap. Factors to consider in the design of rock riprap protection include the stream velocity, the angle of the side slopes, and the size of the rock. Filter blankets of smaller-gradation bedding stone or geotextiles are used under riprap to stabilize the subsoil and prevent piping damage. Riprap bank protection should terminate with a flexible cutoff wall.

The designer should specify a minimum blanket thickness of 18 in (460 mm) for embankment protection and 24 in (610 mm) for slope protection along stream banks and for streambeds. Refer to FHWAHI-90-016, Highways in River Environment, and HEC-11, Design of Riprap Revetment. Also see "Hydrology and Hydraulics," *Delaware DOT Bridge Design Manual,* May 2005.

Topographic Survey

The designer must request a topographic survey to estimate the flow direction from the contours. Any specific information needed for the hydraulic checklist or information in addition to that normally required must be included in the designer's request.

Recording and Coding Guide for Field Inspection

In the absence of scour analysis, the Coding Guide of the National Bridge Inspection Standards (NBIS) shall be used to classify the bridge if it is scour-critical. Items 61, 71, and 113 are of concern during underwater inspections. Item 113 is used for scour rating.

2.6 Role of Movable Bridges to Facilitate Navigation in Rivers

There are restrictions from environmental agencies, such as the placement of piers in riverbeds, resulting in longer spans. It is difficult to design and construct deep foundations in flowing water. Maintaining bridge substructure underwater and regular painting of corroded girders are required.

Wide rivers permit navigation of ships and need to be kept open for commerce. This may not be possible depending on the height of the ships under the bridge. Both highway and railroad bridges are located on wide rivers. Hence movable bridges are required, because they can be raised many times a day. Because of wide rivers and the maximum allowable lifting span, wide piers are required. Maintenance and inspection of the movable bridges become difficult in the flood season. To ensure minimum delays to the movable bridge and to maximize on-time performance, safe and reliable infrastructure is required, especially in the flood season.

Technical support in regard to movable bridge inspection, maintenance, and construction is required. The engineer's duties include the following:

- Provide maintenance plans and specifications as necessary.
- Troubleshoot movable bridge problems both on-site and off-site as necessary.
- Develop and provide inspection training.
- Attend quarterly movable bridge inspections.
- Provide technical expert experience in material requests related to movable bridges.

Consultants engaged in design, inspection, and construction activities on movable bridges should be knowledgeable about the hydrology of wide rivers in order to close down the movable bridges during peak floods. Their responsibilities include the following:

- Knowledge of mechanical, electrical, and electromechanical motors and motor controls from modern equipment to systems that are more than 100 years old
- Review of hydrologic and hydraulic calculations
- Design of countermeasures
- Engineering estimates
- Specifications and special provisions
- Construction procedures
- Construction schedules
- Coordination through oral and written communications

Hydraulic Design of Highway Culverts Use HY-8, Culvert Analysis, software based on HDS-5 for the computations. The average velocity computed on the gross waterway will be the representative velocity for open-span structures, furnished by computer analysis for water surface elevations.

Examples of highly erodible soil can be found in all areas of the state. Areas of loamy deposits, which are highly sensitive to erosion, are prevalent in Delaware.

County SCS soil maps may aid in judging the in situ material.

The designer must consider the downstream erosion potential in evaluating and sizing the structure. Under some conditions, any additional erosion would be intolerable. Thus, risk considerations should be included in the site study. It should be recognized that stream banks erode regardless of the presence of a highway crossing. Any alteration of erosion potential by a structure must be closely evaluated in judging the adequacy of a design.

Streams naturally tend to seek their own gradient through either degradation or aggradation. *Degradation* is the erosion of streambed material, which lowers the streambed. *Aggradation* is the transport and deposition of the eroded material to change the streambed at another location. The effect of the structure on degradation or aggradation of a stream must be evaluated in the bridge crossing design. The designer should evaluate the stability of the bed and banks of the waterway channel, including lateral movement, aggradation, and degradation, using HEC-20, Stream Stability at Highway Structures.

Channel Modifications

A *channel change* is the physical relocation of the streambed channel. A *channel improvement* is the clearing and dressing of stream bank and overbank slopes, or excavation of overbank and stream bank areas but with no change of the streambed profile.

The primary objective in the design of a highway stream crossing is to avoid interruption in the behavior of the stream. Channel modifications should be made only where necessary to accommodate stream flow and where regular maintenance can sustain the improvements or no maintenance is needed.

The preferred procedure for dealing with channel changes is as follows:

- Establish the nature of the existing stream (slope, section, meander pattern, stage discharge relationship).
- Determine the limits for changes in the various stream parameters.
- Duplicate existing conditions where possible, within established change tolerances.
- Evaluate the constructability, considering water table elevations, streambed materials, and site conditions.

Erosion can be minimized by installing countermeasures. Stone lining or shielding of soil is the oldest form of countermeasure. A *countermeasure* is defined by HEC-23 as "a measure incorporated at a stream/bridge crossing system to monitor, control, inhibit, change, delay, or minimize stream and bridge stability problems and scour."

The type of countermeasure is dependent upon the nature of scour, contraction or local, clear water or live bed, aggradation or degradation, and meander or debris accumulation. Scour-critical bridges across the United States are currently being retrofitted using different standards for countermeasures. Their design procedures depend on individual bridge owners, which comprise a large number, representing hundreds of city, county, and state government agencies. Design guidelines are applied differently in different states for the old bridges.

The author carried out research on the subject for a joint publication with Professor Anil Agrawal of CUNY, to develop a handbook for scour countermeasures (Agrawal et al., 2007). The project was sponsored by the Bureau of Research of the New Jersey Department of Transportation. The detailed report is now available on their website for general use:

http://www.state.nj.us/transportation/refdata/research/reports/FHWA-NJ-2005-027.pdf

Some of the countermeasure details provided in this chapter are based on the handbook. The latest ideas, ingenuity, and contributions from individual researchers from publications listed at the end of the chapters are acknowledged.

Scour at Abutments and Piers

Hydraulic conditions and rates of erosion are vastly different at abutments and piers at any bridge site. The extent of erosion at abutments may be minimized by placing them away from the riverbanks. Piers are located in the middle of peak flood zones, where the flood velocity is highest. The direction of flow is at right angles to the pier, which acts basically as an obstruction, with the water flowing on both its sides. Hence, the foundation all around a pier may be scoured. On the other hand, the foundation may be scoured only on the side exposed to the flow in the case of an abutment.

2.7 Debris Accumulation Upstream of Bridges

Debris accumulation changes both the geotechnical and hydraulic characteristics of a bridge. Debris may accumulate at the upstream or downstream end of a bridge or under a bridge.

Debris consists of indigenous material deposited at the bridge obstruction from continued floods and long-term aggradation. Figure 2.2 shows common procedures for cleaning up streams that may become clogged due to bridge obstructions after floods.

The material transported by water flow varies according to the demography and type of terrain. A great majority of bridges are located on narrow stream widths; therefore the effects of debris may be more extensive. Basic procedures for debris removal are shown in Figure 2.1.

It has been observed that sediments, small stones, gravel, leaves of trees, tree branches, bushes, broken pieces of timber from furniture, boats, etc., and fragments of metals may substantially block a bridge opening. Depending upon the size of the debris accumulation, the waterway opening area may be reduced. Hydraulically, the reduction

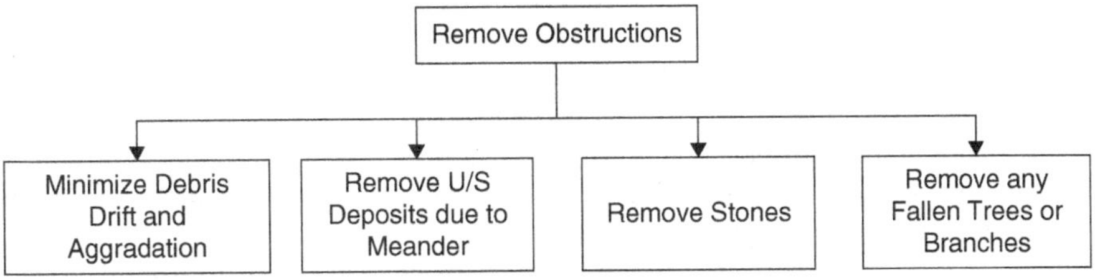

FIGURE 2.2 Procedures for clearing streams of obstructions.

in area will increase contraction scour. However, a significant depth increase from sediment transport may resist and delay the contraction erosion process.

Debris accumulation in bridge openings may result in

- Overtopping
- Failure of roadway embankments because of increased local scour at piers and/or abutments
- The formation of pressure flow scour

The blockage of an opening through the bridge structure causes significant increase in flow, which in turn causes an increase in the contraction scour through the bridge if the entire bridge opening is affected.

Countermeasures for Debris Accumulation

The most effective countermeasure against debris accumulation during a high-flow flood is monitoring and debris removal. A bridge monitoring crew should closely monitor signs of debris buildup, especially near the lower chord. If, during the course of monitoring, a large difference in the water surface elevation is noticed from the upstream to the downstream side of the bridge, it may be a sign of a severe debris blockage developing under water. The structural loading caused by this situation could lead to failure of the bridge. Debris accumulation changes both the geotechnical and hydraulic characteristics of a bridge.

Debris may accumulate

- At the upstream
- At the downstream end of a bridge
- Directly under a bridge

Debris consists of indigenous material deposited at the bridge obstruction from continued floods and long-term aggradation. The material transported by water flow varies according to the demography and type of terrain.

Formation of Scour Holes

Experimental studies have shown that debris accumulations cause larger and deeper scour holes. This happens as a result of the significant increase in the downward velocity below the debris and increase in both the horseshoe vortex size and the contact area of the vortex. The increase in both contraction and local scour near the bridge structure could possibly damage or cause failure of the structure due to undermining of the pier footing or the abutment toe. Hence, removal of debris accumulations and the need for debris control structures are an essential part of any scour countermeasure design of a bridge.

Generally, medium and large floating debris causes a significant problem at bridge structures because of their entrapment near bridge piers. Ice and debris can cause the water flow to plunge downward against the streambed and increase pier scour. Figure 2.3 shows large floating debris accumulation under a bridge along Lumberton-Vincentown Road in New Jersey. Figure 2.4 shows debris accumulation under the Frenchtown-Uhlerstown Bridge in Frenchtown, New Jersey.

The effect of debris is greater at pier footings than at abutments. The effect is similar to freezing of ice around a pier and ice forces due to impact from moving ice.

Figure 2.3 Large floating debris accumulation under Lumberton-Vincentown Road during July 13–14, 2004, flooding.

Figure 2.4 Debris accumulation at a New Jersey bridge in 2006 floods.

However, since the type and magnitude of debris vary for each bridge site, it may not be possible to develop an analytical approach similar to that for floating ice for the analysis of debris loads.

Removing debris from the nose of a pier, for instance, can prevent significant additional scour during that flood event. Debris removal during high flow is usually accomplished from the bridge deck by using a bridge inspection truck, crane, or excavator arm.

Alternative site-specific approaches may be needed at bridges with a large bridge deck overhang because of the difficulty in reaching the pier nose under the deck and in pulling the debris away from the pier against the flow. The following measures are recommended to address debris accumulation problems in New Jersey:

1. A flood watch and inspection should be carried out immediately after a flood to evaluate the extent of debris and litter deposit at a bridge site.
2. The site should be cleaned up and all such debris removed.
3. The riverbed should be dredged to preflood conditions.
4. For new bridges, the bridge superstructure and the general elevation of approach roadways should be above the maximum flood level of a 100-year or critical flood.
5. For streams that carry a large amount of debris, the elevation of the lower chord of the bridge should be at least 2 ft above the freeboard for a 100-year flood.
6. If there is a hazard of ice and debris buildup, multiple pile bents should be avoided. For scour estimation, a bent pier should be evaluated as a solid pier.
7. Specially designed debris deflectors can be installed on the upstream side of the bridge.

Figure 2.5*a* and *b* shows displaced riprap after a flash flood in Burlington County, New Jersey. Use of riprap in New Jersey is restricted, and riprap is not being used as a permanent countermeasure.

2.8 Planning and Design Issues to Restrict Scour at Scour-Critical Bridges

Site-Specific Data for Scour Analysis and Design

The following is the summary of site-specific factors that directly contribute to soil erosion. These factors need to be identified for scour analysis and countermeasure design:

1. Geology and types of soils: geotechnical investigation is to be performed.
2. Hydrology: meteorological data are required.
3. Hydraulics: Manning's coefficient to model waterway for HEC-RAS software should be used.
4. Location and width of waterways: field survey is required.
5. Longitudinal slope: field survey is required.

(a)

(b)

Figure 2.5 (*a*) Example of drift of riprap in a Burlington County, New Jersey, bridge (reduced channel width due to displaced riprap on one side of channel only). (*b*) A culvert inlet blocked by displaced riprap. (Photographs by author.)

6. Environmental conditions: data for marine and plant life are needed as well as NJDEP permit approval.
7. Bridge characteristics, size of opening, and site conditions must be known.
8. Footing type and depth: unknown foundations require boreholes.
9. If considerable erosion has already taken place and the riverbed elevation is below the top of the footing, hydraulic analysis should be based on the new channel profile by considering the new opening size.

Planning Defects and Narrow Openings

In the past there were no scour analysis criteria or software when a bridge was planned. As per AASHTO specifications, scour analysis criteria are now required. Old bridges are more likely to have planning defects than new bridges. Effects of floods include both aggradation and degradation.

Examples of original planning defects are narrow openings (represented by A in the stream gravity flow equation given below) that affect the velocity of flow. At higher velocities, soil erosion is likely to be greater and foundations placed at shallow depths are vulnerable to settlement. There is a considerable increase in velocity due to the increase in discharge or reduction in flow area under the bridge. All flow in open channels is driven by gravity. The simplified discharge equation of flow under a bridge is conceptually $Q = AV$. Expressed as velocity, $V = Q/A$, where velocity V is in feet per second, discharge Q is in cubic feet per second, and A is the area of the bridge or culvert opening in square feet.

1. The original design size of opening A may be inadequate.
2. The river has meandered over a long period, and direction of flow is at a skew, thereby reducing the area of flow A.
3. Debris may accumulate during floods, reducing the size of the opening.
4. Heavy storms, increased snow and subsequent melting, global warming of glaciers, and increased use of water by population increase will increase discharge Q.

When both the discharge Q increases and area A decreases simultaneously, the magnitude of the increase in velocity will make the bridge scour-critical. To calculate discharge Q, velocity V can be calculated by the well-known Manning equation (for flow through open channels). It can be represented as

$$\text{Velocity of flow } V = (1.486/n)R^{2/3}S^{1/2}$$

where n = roughness coefficient
R = hydraulic radius = A/P and is a measure of channel efficiency
A = cross-sectional area of flow, P is the wetted perimeter, and
S = slope of hydraulic grade line and is the average channel bed slope when the water depth is constant.

In the Gauckler–Manning formula,

$$\text{Velocity of flow } V = (k/n)R^{2/3}S^{1/2}$$

where k = conversion factor for SI units. The greater the hydraulic radius R, the greater the efficiency of the channel and the greater the volume it can carry. For channels of a given width, the hydraulic radius is greater for the deeper channels.

Scour at Bends due to River Meander

Special consideration must be given to scour for bridges located on bends. S. T. Maynord's investigations (Maynord, 1989, 1995) are helpful to determine flow depths and the alternative countermeasures for these conditions are applicable at abutments and piers. When scour occurs at the confluence of two rivers, flow depths can be estimated by using Ashmore and Parker's (1983) or Klaassen and Vermeer's (1988) equations, which will be applicable at the abutment and pier. Note that the magnitudes of scour due to thalweg effects and migration of bed forms are typically small and are usually neglected.

AASHTO Section 2.6, Hydrology and Hydraulics Commentary C2.6.1

The AASHTO Task Force on Hydrology and Hydraulics, in cooperation with the FHWA Rural Technical Assistance Program (RTAP), has assisted in writing this AASHTO *Model Drainage Manual*. This comprehensive drainage manual provides guidance on most aspects of highway drainage. Design theories, concepts, manuals, policies, and procedures are condensed.

The *Model Drainage Manual* has been developed to give the designer a basic working knowledge of hydrology and hydraulics with example problems. The manual is organized with Chapters 8 to 13 covering hydrology, channels, culverts, bridges, energy dissipaters, storage facilities, and storm drainage systems. It also incorporates guidance and references from the AASHTO *Drainage Guidelines*, which is a companion document to the AASHTO *Model Drainage Manual*.

Hydrologic, hydraulic, scour, and stream stability studies are concerned with the prediction of flood flows and frequencies and with the complex physical processes involving the actions and interactions of water and soil during the occurrence of predicted flood flows. These studies should be performed by an engineer with the knowledge and experience to make practical judgments regarding the scope of the studies to be performed and the significance of the results obtained. The design of bridge foundations is best accomplished by an interdisciplinary team of structural, hydraulic, and geotechnical engineers.

The AASHTO *Model Drainage Manual* also contains guidance and references on the following:

- Design methods for evaluating the accuracy of hydraulic studies, including elements of a data collection plan
- Guidance on estimating flood flow peaks and volumes, including requirements for the design of interstate highways as per 23 CFR 650, Subpart A, Encroachments
- Procedures or references for analysis of tidal waterways, regulated streams, and urban watersheds
- Evaluation of stream stability
- Use of recommended design procedures and software for sizing bridge waterways
- Location and design of bridges to resist damage from scour and hydraulic loads created by stream current, ice, and debris
- Calculation of magnitude of contraction scour, local scour, and countermeasures thereto
- Design of relief bridges, road overtopping, guide banks, and other river training works
- Procedures for hydraulic design of bridge-size culverts

Information on the National Flood Insurance Program (NFIP) is contained in 42 USC 4001-4128.

AASHTO Superflood Provisions C2.6.2

The assessment of hydraulics necessarily involves many assumptions. Key among these assumptions are the roughness coefficients and projection of long-term flow magnitudes, e.g., the 500-year flood or other superfloods. The runoff from a given storm can be expected to change with the seasons, immediate past weather conditions, and long-term natural and human-made changes in surface conditions.

The above factors make the check flood investigation of scour an important, but highly variable, safety criterion that may be expected to be difficult to reproduce, unless all the designer's original assumptions are used in a postdesign scour investigation. Obviously, those original assumptions must be reasonable given the data, conditions, and projections available at the time of the original design.

2.9 Bridge Performance Based on AASHTO (LRFD) Load Combinations for the Extreme Conditions

The extreme-event limit states relate to flood events with return periods (usually 100 years) in excess of the design life of the bridge (usually 75 years). The aim should be to design the bridge for all times and for all occasions. There are generally three such limit states that may involve consideration of the effects of scour at bridges:

1. A flood event exceeding a 100-year flood (the check flood for scour or superflood is used to evaluate scour for this event, a 500-year flood is recommended for the check flood for scour)
2. An earthquake
3. A vessel collision with the bridge

In addition to the above, there are conditions relating to scour that the designer may determine as significant for a specified watershed, such as ice loads or debris logging operations.

Scour with Vessel Collision Substructures must be designed for an extreme vessel collision load by a ship or barge simultaneously with scour. The substructure should be designed to withstand the following two load-scour (L-s) combinations:

4. Load-scour combination 1:

   L-s (1) = vessel collision assumed at (½ × long-term scour)

   where vessel collision is assumed to occur at normal operating speed.

5. Load-scour combination 2:

   L-s (2) = minimum impact vessel at (½ × 100-year scour)

   where minimum impact vessel is defined in LRFD with related collision speed.

Computer Software for Hydraulic and Scour Analysis Construction drawings for scour countermeasures must be based on detailed designs. For detailed design, approved commercial software may be used. A spreadsheet may be developed and used in lieu of software.

Case Study of Modern Approach in Planning Route 46 Bridge by the Author

For the design of the Route 46 Bridge in New Jersey, the following recommendations are made:

1. Direction of abutment skew is now made parallel to the direction of river flow to minimize scour.
2. Based on hydraulic analysis, the opening size is increased to minimize overtopping flood.
3. Use of integral abutments and integral approaches makes the bridge highly resistant to longitudinal forces.
4. Steel girders should be replaced by prestressed spread box beams to prevent corrosion.
5. Use of deep sheet piling sections minimizes the scour of abutment and wing walls.

Maintenance of Bridges on Waterways Due to increased corrosion of steel bridges as a result of daily evaporation, the life of bridges is adversely affected. During floods, countermeasures should not be displaced. Sediment deposits in some rivers require dredging and clearing stones under the bridges before it is too late. Hence frequent inspections may be required.

Bridge Deck Drainage

Bridge deck drainage systems direct runoff off the bridge and away from the travel lanes. Scuppers or complex piping systems to carry roadway runoff are not always required. When these systems are not properly designed or routed, ancillary problems such as poor visibility, icing, and corrosion of structural members can result. Use the rational method to determine the runoff.

In a storm, the bridge deck receives rainfall directly. The volume of rainwater gets added to the floodwater in the river below by deck drainage and from the approaches to the bridge. The well-known *rational formula* is

$$Q = ciA$$

where Q = peak discharge, ft^3/s or cfs
c = rational method runoff coefficient
i = rainfall intensity, in/h
A = drainage area, acres

The rational method runoff coefficient c is a function of the soil type and drainage basin slope. (See Table 2.2.) The rainfall intensity i is typically found from intensity, duration, frequency curves for rainfall events in the geographical region of the bridge location.

The duration is usually equivalent to the time of concentration of the drainage area. The storm frequency is typically stated by local authorities depending on the impact of the development. A 10-, 25-, 50-, or even 100-year storm frequency may be specified.

Surface Cover	Runoff Coefficient c
Bridge decks	0.75–0.95
Concrete approaches	0.7–0.95

Table 2.2 Rational Method Runoff Coefficients

Bridge deck designs should conform to roadway standards for crown cross slope and superelevation. These standards are as follows:

- Crown cross slope: 2 percent
- Superelevation: 6 percent maximum, but 8 percent maximum for rural roadways

Refer to the FHWA publication HEC-21, Bridge Deck Drainage Systems. This utilizes design nomographs, based on grade, cross slope, design speed, design rainfall intensity, and bridge deck width, to determine an estimate of the bridge length without scuppers. Because bridge deck drainage systems can cause maintenance problems, their use should be avoided or at least minimized. Once it is determined that scuppers are required, additional criteria for scupper type, spacing, and location are provided.

The 10-year storm should be used as a minimum for the design of deck drainage. Refer to HEC-21 (FHWA-SA-92-010), Design of Bridge Deck Drainage. All scuppers in the sag surface of deck should have the same capacity.

Provisions should be made to collect water and to prevent it from running on the bridge. Several significant issues should be considered to provide the best possible design:

- Adequate longitudinal slope to move water off the bridge should be provided.
- All free-falling scupper outlet pipes should extend a minimum of 8 in (200 mm) below the nearest superstructure element within 10 ft of the scupper.
- The drainage from a scupper outfall pipe should not be allowed to free-fall to the ground if in so doing other deleterious effects (i.e., erosion, icy conditions, etc.) can result.
- Scupper downspout pipes, where needed, should always be attached to the exterior surface of the substructure and never placed within the concrete substructure elements. Water may freeze in the pipes and break the concrete.

Report Card on Transportation Needs by ASCE Committee

The American Society of Civil Engineers (ASCE) Report Card identifies the structurally deficient and functionally obsolete bridges and suggests ways and means to resolve the outstanding issues. The author was a member of the Report Card Drafting Committee which reviewed the 2014 status of Pennsylvania bridges for safety, durability, and innovative design.

Figure 2.6 shows the author presenting a certificate of appreciation to the director of ASCE/SEI Jim Rosberg for updating ASCE-7 design loads for applications in International Building Code (IBC). SEI Committee members for Philadelphia, Dr. Vijay Gupta of AEG and Dr. Nazhat of NJDOT, are also present.

FIGURE 2.6 The author as chair of the Structural Engineering Institute of Philadelphia, presenting a certificate of appreciation to ASCE/SEI director Jim Rosberg, for updating ASCE-7 design loads.

2.10 Project Funding Requirements and Regulations

FHWA SHRP2 Program

The second Strategic Highway Research Program (SHRP2) is an FHWA-administered program, authorized by Congress in 2005, to identify solutions that will have a lasting effect on the nation's highways. The Technology Catalog has detailed information on geoconstruction and ground modification techniques. For the R02 project, the GeoTech-Tools product is a Web-based tool. SHRP2 solutions provide new research-based tools and innovative products and processes that will create greater efficiencies and a smarter use of state and federal investments of taxpayer dollars.

GeoTechTools (www.GeoTechTools.org) is a comprehensive design aid for the design and construction of earth retention and structural foundations, transportation earthwork, general ground improvement, or pavements.

Funding Issues Are the Major Constraints

The ASCE urged Congress to take immediate action to avert insolvency in the trust fund. The Highway Trust Fund is normally funded by revenue collected from the federal gas tax. But the gas tax has not been increased since 1993, and soaring road-building costs have dwarfed receipts by $20 billion per year in recent years. Options that have been discussed include these:

- Raise the federal gas tax.
- Implement more tolls.
- Supplement gas tax revenue with money from closing corporate tax loopholes.
- Transfer money from the General Fund.

Partnership by Funding

The FHWA report gives insights for developing and implementing transportation infrastructure projects using PPP (public-private partnership, or P3) approaches. It is aimed at both the early practitioners of PPP projects as well as those agencies just beginning to consider the possibility of instituting some form of PPP arrangement for a particular project currently stalled for lack of available resources.

This PPP User Guidebook describes the many participating groups, stages of development, and institutional factors associated with surface transportation PPPs. It considers the full life cycle of PPPs, from development to execution to performance reporting. It identifies and discusses statutory, regulatory, financial, and institutional issues that should be addressed to successfully implement and manage a PPP project.

The guidebook suggests a general process for developing transportation project PPPs and strategy for overcoming impediments faced by public- and private-sector partners during the contract development and project implementation phases. The guidebook also offers general lessons learned from prior or current transportation PPP projects in the United States and other countries.

The common element of a PPP is that public sponsors of transportation projects engage the private sector to a greater degree in the performance of certain functions previously handled by the public sector. This can range from contract maintenance to life-cycle finance, development, operations, and preservation. The FHWA is encouraging its counterparts at the state and local government levels to consider the use of PPP approaches to accomplish more projects in their work programs.

The PPP User Guidebook is intended to assist sponsors and providers of transportation projects in taking the necessary steps and precautions to promote successful delivery of PPP projects while protecting the public interest, especially the ultimate users of the facilities so developed.

Public-private partnerships (P3s) are an effective tool for state and local governments to accelerate the delivery of much-needed services and infrastructure to the public, while fostering new jobs and future economic development. With state and local capital budgets severely constrained in the current economy, more and more agencies are looking for private-sector financing and participation to deliver critical infrastructure projects. Topics include:

- The pros and cons of long-term P3s' identification of existing market sectors and new regions to watch
- Entry into project team arrangements
- Forecast of revenue and risk

The Interstate 595 widening marks Florida's first P3 effort (ENR dated 06/20/2013 by Aileen Cho). The Florida Department of Transportation used a design-build-finance-operate-maintain contract to reconstruct 10.5 mi of Broward County's I-595 corridor.

The $1.8 billion Interstate 595 widening project also came with plenty of complexity. The design-build contractor Dragados United State completed construction.

Going with a public-private partnership and the design-build approach shaved at least 10 years off the project.

Pennsylvania Project Funding and Future Needs

Due to increasing vehicle fuel efficiency, Pennsylvania now collects less fuel tax revenue per mile traveled than it has at any time in the past. When adjusted for inflation, the total gas tax paid by the average driver has decreased 60 percent since about 1970 while the cost of construction and maintenance continues to increase with inflation.

Pennsylvania's Transportation Funding Plan, Act 89, signed into law in 1913, is an unprecedented, comprehensive package that will bring much-needed additional investment to the Commonwealth's transportation system.

Recommendations for Project Selection

- Encourage and support the creation of P3 projects.
- Advocate for an increase in the motor fuel tax.
- Advocate for high-occupancy toll (HOT) lanes or other user-preferred tolling options such as express toll lanes (ETLs).

TEA-21 and the 2005 IBRC Funding Program

The Transportation Equity Act for the 21st Century (TEA-21) was enacted on June 9, 1999 as public law. TEA-21 authorizes the federal surface transportation program for highways, highway safety, and transit. It was initially for a 6-year period but was extended later. The FHWA TEA-21 website summarizes its content, progress, reports, and so forth. Related publications are available online.

The Innovative Bridge Research and Construction (IBRC) Program was established with the passage of TEA-21 and was initially funded by Congress for 6 years. The program is intended to demonstrate the application of innovative material technology in the construction of bridges and other highway structures, and it has two components.

- The larger component provides funds for repair, rehabilitation, replacement, or new construction of bridges and other highway structures, using innovative materials.
- The smaller component is intended to support research and technology transfer activities related to the program's goals for the construction component.

Statutory References

Section 5103 of TEA-21 authorized the components of funding shown above to be available to the states for projects to demonstrate innovative materials relating to the repair, rehabilitation, and construction of bridges and other highway structures.

Federal Share

The FHWA determines the federal share for each project by considering

1. The size of the project
2. The amount of funds requested for the project
3. Available funding for the program

It is the goal to fund as many projects as possible, at a 100 percent federal share; however, some projects may be funded at a lower federal share.

Eligibility

TEA-21 addresses the federal share of the cost of repair, rehabilitation, replacement, and new construction of bridges and other highway structures that demonstrate the application of innovative materials.

- Funds are available for bridge projects that meet one or more of the seven program goals listed in TEA-21, Transportation Research Provisions.

Surface Transportation Research Strategic Planning

According to TEA-21, strategic planning activity determines national surface transportation research and technology development priorities. TEA-21 requires the secretary to develop an integrated surface transportation research and technology development strategic plan to

- Identify the general goals and objectives for surface transportation R&D.
- Establish the roles of the department and other federal agencies in achieving the goals
- Set forth a strategy for carrying out the plan
- Include funding requirements, interagency coordination procedures, and indicators to evaluate program effectiveness

DOT is given the leadership role in the process; other participants are federal agencies performing surface transportation research and technology development and academic institutions, industry, and other private- and public-sector organizations engaged in such R&D. The project may be on any public roadway, including state and locally funded projects. Funds are available for the costs of preliminary engineering, repair, rehabilitation, or construction of bridges or other highway structures as well as the costs of project performance evaluation including instrumentation and performance monitoring of the structure following construction. These funds may be used for the federal share of the cost of the repairs, rehabilitation, replacement, or new construction on the "innovative materials and accelerated bridge construction" portion of the project.

Project Selection Criteria

The project shall meet the criteria of each program, in addition to the recommendations described above:

- The development of new, cost-effective, innovative material highway bridge applications

- The reduction of maintenance costs and life cycle costs of bridges, including the costs of new construction, replacement, or rehabilitation of deficient bridges
- The development of construction techniques to increase safety and reduce construction time and traffic congestion
- The development of engineering design criteria for innovative products and materials for use in highway bridges and structures
- The development of cost-effective and innovative techniques to separate vehicle and pedestrian traffic from railroad traffic
- The development of highway bridges and structures that will withstand natural disasters, including alternative processes for the seismic retrofit of bridges
- The development of new nondestructive bridge evaluation technologies and techniques

The following criteria are used to evaluate the submitted candidates for selection:

- Projects ready for or near the construction phase, which are given priority consideration
- Projects with designs that are repeatable or have widespread application

Each year, a memorandum is sent from the FHWA headquarters Office of Bridge Technology to the FHWA division offices requesting the submission of candidate projects for a given fiscal year's funding. The state transportation departments coordinate with local and federal agencies and transportation authorities within their respective states to develop viable candidate projects.

Structure description. A description of the structure should be provided including type (e.g., continuous prestressed concrete box girder), number of spans, total length, and total width (out to out).

Innovative material. The innovative material (e.g., high-performance steel) to be used in the project should be specified.

Program goals. The states' submissions should show how the application of the innovative material will meet one or more of the IBRC program goals.

Estimated costs. The estimated cost of the project should be provided as follows: cost of the entire project, cost of the innovative portion of the project, and amount of federal IBRC discretionary funds requested.

Projects that stress innovative technology to monitor, measure, and report on engineering and operational performance of bridges, particularly those with high-performance materials, range from material degradation to cracking and/or loss of section to catastrophic failure. The state may transfer IBRC funds from an eligible, funded project to one or more of its submitted eligible candidates as it may deem appropriate. Funds may be transferred to a different bridge for an innovative project of the same type.

A bridge consists of the foundation, the substructure and the superstructure. The kinetic energy and impact of floods cause erosion of soil and apply dynamic forces mainly on the piers and abutments. The composite action between substructure and superstructure also contributes to flood resistance. Recent development in integral abutments will also provide greater stability against floods than, when the superstructure is separated from the superstructure, by the conventional bearings. Hence the type of girders and their connections with the piers and abutments also play a role in the overall stability of the bridge against wave action.

Box girders and tub girders need to be investigated against I-girders, for stability against lateral forces.

2.11 Conclusions

1. *Use of hydraulic analysis and load combinations with hydrostatic forces to study scour*: The basic concept of open-channel hydraulics and the application of Bernoulli's equilibrium equations was discussed. The issue of erosion from obstruction to flow was addressed. Bridges on waterways, scour-critical bridges, and factors affecting scour have been discussed.

   AASHTO (LRFD) load combinations for extreme conditions have included scour as a requirement for design. The issues of obstruction to free flow of a river flowing at speeds greater than 10 ft/s were discussed. During peak flood events of 50, 100, and 500 years, the water level is likely to be high, spilling over the embankments. In planning new bridges, the bridge deck elevation should be higher than the high flood level (HFL), and pier widths should be kept to a minimum to allow the natural river opening size to remain unchanged. Abutments should be kept a few feet behind the river width.

   Multiple types of riverine flow were considered. It is important that the hydraulic analysis model take into account all parameters of bridge geometry, to compute floodwater elevation and the increased velocity.

2. *Use of auxiliary bridge and movable bridges*: Provision and feasibility of an auxiliary bridge may be considered where overtopping of the deck is frequent and the bridge is being closed. On navigation channels, the use of movable bridges will eliminate any freeboard problems and allow safe passage of ships even when the water surface elevation is high.

3. *Wider bridge opening and use of deep foundation*: Abutments should be kept a few feet behind the river width. Deep foundations on piles or drilled shaft may be preferred. In planning, span lengths in proportion to the precast girder length and construction material can be selected from the ranges of clear spans readily available from the selected precast factory manufacture tables. Debris accumulation upstream, under the bridge, and downstream can cause overtopping and increase the velocity of flow.

4. *Adopting European scour practice*: Recommendations from the European practice to prevent scour were addressed.

5. *Funding provisions to be preferably based on ASCE report card*: The well-known TEA-21 regulations were described in detail. The ASCE Report Card identifies the structurally deficient and functionally obsolete bridges and suggests ways and means to resolve the outstanding issues. The author was a member of the Report Card Drafting Committee.

   Project funding is the bottleneck, as bridge maintenance and replacement require a great deal of money. This sensitive aspect was discussed in detail in Section 2.10, Project Funding Requirements and Regulations.

2.12 Bibliography

AASHTO, *Drainage Guidelines*, 4th ed., Washington, D.C., 2007.

AASHTO, *Model Drainage Manual*, 3d ed., Washington, D.C., 2005.

Agrawal, A., Zhihua, Y., and Khan, M. A., *Handbook of Scour Countermeasures*, Region 2, Transportation Research Center, Final Report No. FHWA-NJ-2005-027, NJDOT Bureau of Research, Trenton, NJ, December 2006.

Aref, A. J., and Alampalli, S., "Dynamic Behavior and Damage Detection of a Fiber Reinforced Plastic Bridge," Advanced Technology in Structural Engineering, Structures Congress, Philadelphia, PA, May 2000.

Ashmore, P., and Parker, G., Confluence Scour in Coarse Braided Streams, *Water Resources Research*, Vol. 19, Issue 2, April 1983.

Balazs, G. L., and Borosnyoi, A., "Prestressing with CFRP Tendons," *Proceedings of the International Conference, High Performance Materials in Bridges*, United Engineering Foundation, Inc., Kona, HI, July 29–August 3, 2001.

Bane, R., "Strengthening Masonry Arch Bridges," *Scanner*, American Society of Highway Engineers, Summer 2007.

"Bridge Rehabilitation Laser Scanning Sample Projects," Meridian 3D Laser Scanning Modeling Analysis (www.meridian-3d.com), Meridian Associates, Boston, 2007.

Chin, D. A., *Water-Resources Engineering*, Prentice-Hall, 2000.

Chow, V. T., Maidment, D. R., and Mays, L. W., *Applied Hydrology*, McGraw-Hill, 1988.

Chow, V. T., *Open Channel Hydraulics*, Blackburn Press, Caldwell, NJ, 2008.

Construction Handbook for Bridge Temporary Works, American Association of State Highway and Transportation Officials, Washington, D.C., 1995.

"Corrosion Protection of Steel Bridges," *Steel Bridge Design Handbook*, National Steel Bridge Alliance, Report no. FHWA-IF-12-052, Vol. 19, November 2012.

Design of Bridge Deck Drainage, HEC-21, FHWA-SA-92-010, May 1993.

Dunker, K., and Dajin, L. "Foundations for Integral Abutments," *Practical Periodical on Structural Design and Construction*, American Society of Civil Engineers, February 2007.

Engineering News-Record, Issue: 06/24/2013, FHWA Programs, Hydraulic Engineering, Washington, D.C., December 2014.

FHWA, Hydraulic Engineering, Culvert Hydraulics, HY-8 Software, Version 7.3, Washington, D.C., August 2014.

Interstate 595 Widening Marks Florida's First P3 Effort, A. Cho, June 2013.

Field Manual for Bridge Painting Inspection, FHWA-RD-98-084, FHWA, Washington, D.C., October 1997.

Gu, X., Chen, Z., and Ansari, F., "Fiber Optic Distributed Crack Sensor for Concrete Structures," *Proceedings of Structures Congress*, American Society of Civil Engineers, Philadelphia, PA, 2000.

Huston, D., *Structural Sensing, Health Monitoring and Performance Evaluation*, CRC Press, Boca Raton, FL, 2008.

Implementing the Results of the 2nd SHRP, Special Report 296, TRB, Washington, D.C., 2009.

Kibriya, T., "Performance of Blended Cement in High Strength Self-Compacting Concrete," *Proceedings of Structures Congress 2006*, American Society of Civil Engineers, St. Louis, MO, May 2006.

Klaassen, G. J., and Vermeer, K., "Confluence Scour in Large Braided Rivers with Fine Bed Material," Proceedings of International Conference on Fluvial Hydraulics, Vituki, Budapest, 1988.

Laman, J. A., Reichard, K. M. and Dollard, A. J. "Long Gage-Length Interferometric Optical Sensors for Condition Assessment in Bridge Structures," *Proceedings of Structures Congress,* American Society of Civil Engineers.

Lindsley, R. K., Franzini, J. B., Freyberg, D. L., and Tchobanoglous, G., *Water-Resources Engineering,* 4th ed., McGraw-Hill, 1992.

LRFD Bridge Construction Specifications, 2d ed., American Association of State Highway and Transportation Officials, Washington, D.C., 2004.

LRFD Bridge Design Specifications, 3d ed., American Association of State Highway and Transportation Officials, Washington, D.C., 2004.

Manual on the Use of Self-Propelled Modular Transporters to Remove and Replace Bridges, FHWA, Washington, D.C., June 2007.

Maynord, S. T., Ruff, J. F. and Abt, S. R. "Riprap Design," *ASCE Journal of Hydraulic Engineering,* vol. 115, no. 7, 1989.

Maynord, S. T., "Gabion Mattress Channel-Protection Design," *Journal of Hydraulic Engineering,* ASCE, Vol. 121, No. 7, 1995.

McCuen, R. H., *Hydrology Analysis and Design,* 2d ed., Prentice-Hall, 1998.

Myers, J. J., "A Worldwide Survey of Environmental Reduction Factors for Fiber Reinforced Polymers (FRP)," Department of Civil Engineering, Penn State University, University Park.

National Transportation Safety Board, "Safety Recommendation, (I-35W) Highway Bridge over the Mississippi River in North Minneapolis, Minnesota," Washington, D.C., January 2008.

National Cooperative Highway Research Program (NCHRP) Report 516, "Pier and Contraction Scour in Cohesive Soils," J.-L. Briaud et al., Texas A&M University, College Station, TRB, Washington, D.C., 2004.

OMICS Group, 2nd International Conference on Hydrology & Meteorology, Mohiuddin Ali Khan, Session Chair, "Recent developments in the regression methods for computing peak flood discharges," Raleigh, NC, August, 2013.

OMICS Group, 3rd International Conference on Hydrology & Meteorology, Mohiuddin Khan, Keynote Lecture: "Protection of Bridge Foundations Against Scour Using Partially Grouted Riprap," HICC, Hyderabad, India, September, 2014.

Pennsylvania Department of Transportation, *Design Manual Part 4* (Pub. 15M), Volume 1. Part A Policies and Procedures, Chapter 7, Drainage Structures, Scour and Culverts, Harrisburg, May, 2012.

PCI Committee on Bridges, "Seismic Design of Precast Concrete Bridges State-of-the-Art Report," Precast/Prestressed Concrete Institute, Chicago, IL, 2013.

Rahrig, P. G., "Galvanizing Plate Girders," *Modern Steel Construction Journal,* AISC, Chicago, IL, August 2002.

Rajabipour, F., and Weiss, J., "Linking Health Monitoring in Concrete Structures with Durability Performance Simulations," *Structures Congress,* American Society of Civil Engineers, 2006.

SHRP2 RO2, Geotechnical Solutions for Soil Improvement, Rapid Embankment Construction and Stabilization of the Pavement Working Platform, TRB, November 2014.

Singh, V. P., *Elementary Hydrology,* Prentice-Hall, 1992.

Smith, J. W., and Bright, S. "Upgrading Orthotropic Bridge Decks with Fiber Reinforced Composites," *Proceedings of the International Conference, High Performance Materials in Bridges,* United Engineering Foundation, Inc., July 29–August 3, 201, Kona, HI.

Straser, E. Sohn, H., Kiremidjian, A., and Law, K. H., *Structural Standards Manual*, Maryland Department of Transportation, State Highway Administration, Office of Bridge Development, Baltimore, MD, March 2006.

Sung, H. P., *Bridge Inspection and Structural Analysis, Handbook of Bridge Inspection*, 2nd ed., Trenton, NJ, 1999.

Sung, H. P., *Bridge Superstructure, Design and Rehabilitation*, Trenton, NJ, 2005.

Teal, R., "Fabrication with High Performance Steel Grade HPS 70W," Metals Consultant, Averill Park, NY.

User Guidebook on Implementing Public-Private Partnerships for Transportation Infrastructure Projects in the United States, Final Report, Prepared for Office of Policy and Governmental Affairs, AASHTO, Washington, D.C., July 2007.

Using Corrosion Inhibitors to Avoid Damage to Reinforced Concrete, Use of Corrosion Inhibitors in Concrete, James M. Gaidis and and Arnie Rosenberg, Journal of Materials Performance, National Association of Corrosion Engineers, Concrete Corrosion Inhibitors Association, Potomac, MD, January 2004.

Voo, J., Foster, S. J., Gilbert, R. I., Gowripalan, N., "Design of Disturbed Regions in Reactive Powder Concrete Bridge Girders," *Proceedings of the International Conference, High Performance Materials in Bridges*, July 29–August 3, 2001, United Engineering Foundation, Inc., Kona, HI.

CHAPTER 3

Scour Countermeasures as Remediation

3.1 Scour Mitigation Using Various Types of Countermeasures

The study of bridge failures reveals the many dangers of not making bridges safe enough against foundation settlement or movement due to high flood velocities. Bridges with narrow openings and erodible soils have suffered collapses.

This chapter presents the background of investigating the need for protective measures and discusses the importance of detailed design of countermeasures. The majority of bridge work in the United States aims to maintain the network of thousands of bridges on rivers by making the foundations safer during occasional floods.

Objectives

Chapters 1 through 9 and solved examples in Appendices A and B in this book can be used for scour analysis according to HEC-18 and selection and design of countermeasures using HEC-23 Guidelines. In addition, coastal erosion and coastal countermeasures guidelines are addressed in HEC-25. Chapter 8 addresses additional details of countermeasures and the many case studies of scour-critical bridges performed by the author.

Scour computation procedures are addressed by HEC-18, Maryland ABSCOUR software, Melville's textbook bridge scour and guidelines developed by NJDOT Section 46 (2002) and Section 39 (2007).

Appendices A and B give solved examples using spreadsheets by the author and based on HEC-18 scour equations and use of ABSCOUR software. Melville's equations in addition to the New Jersey DOT *Bridge Design Manual* section on scour at bridges developed by the author.

Chapters provide additional information on countermeasure design.

In addition, the case studies of three bridges located in New Jersey are presented, in a joint research paper by Mr. Salim Baig, P.E., and the author with the head of the bridge inspection department of the New Jersey DOT. The paper was published in the proceedings of an international conference on scour and erosion (held at Texas A&M) in 2002. The goals of the research paper were to prevent or minimize failures of highway structures located on scour critical rivers:

- To help plan a new bridge site on a river
- To reduce the scour risk by avoiding countermeasures altogether, by having deeper foundations of abutments, wing walls, piers, and retaining walls

- Harnessing the river to reduce flood velocities, thereby minimizing their failures

Practical considerations require increasing the factor of safety against damage to the substructure. Some common types of countermeasures are riprap, gabion baskets, concrete blocks, and sheet piling.

The HEC-23 scour countermeasures matrix lists hydraulic and structural aspects of countermeasures. Selection of the right type of countermeasure is important. Guidelines given in HEC-23 are followed. HEC-23 is an FHWA publication and all the states in the United States are expected to follow the guidelines or those modified by the states to meet their flood and specific physical needs. The author investigated nine scour-critical bridges in New Jersey as project engineer and as chief bridge engineer. The investigations were followed by 2-year research projects each with City University of New York, New Jersey Institute of Technology, and Temple University.

The Need for Effective Countermeasures

Scour countermeasures are defined by HEC-18 as "those features incorporated after the initial construction of a bridge to make it less vulnerable to damage or failure by scour." In general, the countermeasure approach typically used is to provide some form of revetment adjacent to the substructure elements, or often over the full width of the waterway opening. This armoring layer, which can be either rigid or flexible is used to inhibit the scour and erosion of the finer soil materials that comprise the channel bed. If designed in accordance with HEC-23, this revetment can protect the bridge from most flow conditions.

In some cases, however, this approach may be used to provide a temporary measure of protection. Other longer-term (and more costly) solutions may be required to eliminate the bridge's vulnerability to scour. In addition, the specific nature of the scour problem at each individual bridge needs to be addressed. Sometimes the size of the opening may not be adequate, or lateral stability of the waterway may be a controlling issue. Design methods for countermeasures are described in detail in later chapters. An introduction to the types of countermeasure is given here.

Definition According to AASHTO LRFD Specifications (Section C3.7.5), "Scour is the most common reason for the failure of highway bridges in the United States." Scour is an interesting subject related to the impact of water in motion. Scour is the result of the erosive action of running water excavating and carrying away material from the bed and banks of the stream.

A *countermeasure* is defined by HEC-23 as a measure incorporated at a stream–bridge crossing system to monitor, control, inhibit, change, delay, or minimize stream and bridge stability problems and scour. Erosion can be minimized by installing countermeasures. Stone lining or shielding of soil is the oldest form of countermeasure.

Erosion is a very old subject. It is gradually being placed on a scientific basis, using scientific disciplines such as these:

1. Hydrology of peak floods
2. Bridge hydraulics
3. Soil cohesion

4. Scour analysis
5. River training and countermeasures

Existing design guidelines presented in hydraulic engineering circulars are applied differently to old and new bridges. The design procedures depend on a large number of individual bridge owners, representing hundreds of city, county, and state government agencies. Scour-critical bridges across the United States are therefore currently being retrofitted using different standards for countermeasures.

3.2 Modifications to FHWA HEC-23 Countermeasures Matrix

There are constraints in implementing several aspects of the recommended countermeasures for piers and abutments. State codes for New Jersey and Pennsylvania have made modifications to the types of countermeasures to be used or to their construction details based on local experience of any past failures and the frequency and magnitude of floods. It has been observed that not all the listed matrix countermeasures are applicable for use in New Jersey.

The master countermeasures matrix utilizes some of the European practices in this discipline. The matrix is based on assuming the maximum duration of floods. The matrix lists the countermeasure types (in rows) versus countermeasure characteristics (in columns). The HEC-23 countermeasures master matrix was modified and placed in Bibliography in this chapter (under New Techniques). The reference includes NJDOT Handbook of Scour Countermeasures prepared by Agrawal, Khan and Zhihua so that it can be readily applied to New Jersey conditions. Other states may make changes according to their specific conditions.

Countermeasure characteristics are classified into three groups:

1. *Functional applications*: Functional applications are the computed or observed scour conditions, such as local, contraction, and stream instability conditions. All the listed types are applicable to New Jersey conditions as an example.
2. *Suitable river environment*: The suitable river environment group lists a wide range of physical data for hydraulic and geotechnical conditions related to the river.
3. *Maintenance frequency and costs.*

The original HEC-23 master matrix covers structural countermeasures, river training measures and monitoring in addition to the commonly used armoring countermeasures. Additional design guidelines for filters, which are essential to the effectiveness of countermeasures, such as in combination with riprap, gabions, and so forth, are also given in HEC-23.

However, note that the types of rivers, span length of openings, volumes and sources of water, and types of soil vary within each state and sometimes within the same county. Some rivers are narrow while others are wide. Some are meandering and undulating while others are straight, and so forth.

Countermeasures for Mountain Rivers

As discussed earlier, the Forest Service has developed guidelines for mountain rivers in the hilly areas; the steep longitudinal slopes increase the velocities several times, according to Manning's and Chezy's equations. Minor modifications to the master matrix have been made by some states. For example, in New Jersey, riprap is not considered a permanent solution. Similar deviations from HEC-23 guidelines are being made by some states to the general countermeasures matrix. Considerations for deviations are based on:

- Potential high costs in relation to the risk and chances of failure
- Constructability issues in deep rivers
- Environmental concerns regarding pollution and wildlife conservation during construction
- The scour type based on the bridge span and dimension of the openings
- The skew angle of attack with the alignment of the bridge
- Downstream features, such as lack of spillway, that affect flood velocities

A countermeasure matrix facilitates preliminary selection of feasible alternatives prior to a more detailed investigation. The matrix is based on engineering factors, environmental factors, and cost. Countermeasures are organized here into three groups based on their functionality with respect to scour and stream instability.

Group 1: Hydraulic countermeasures

Group 1A: River training structures

- Transverse structures
- Longitudinal structures
- Arial structures

River training measures include guide banks which are placed so as to move scour away from the abutment foundation. Spurs or bendway weirs are installed at a bend that is migrating toward a bridge abutment. Spurs redirect the flow away from the abutment.

Group 1B: Armoring countermeasures

- Revetment and bed armor (rigid, flexible and articulating)
- Local armoring

Group 2: Structural countermeasures

- Foundation strengthening
- Pier geometry modification

Group 3: Monitoring

- Fixed instrumentation
- Portable instrumentation
- Visual monitoring

Each countermeasure must be selected on the basis of a scour analysis for each specific site.

Armoring Progress in Factory Manufacture of Countermeasures The guidelines recommended in this book can be supplemented with the available design data of any manufacturer's development work. Detailed specifications for several countermeasures that are proprietary in nature may be obtained directly from the vendors, such as Maccaferri USA (Williamsport, MD), who develop new products after research and experimentation. In some cases software is being made available for analysis and design.

3.3 Applications of FHWA HEC-18 and HEC-23 Procedures

The procedures developed by Hydraulic Engineering Circulars represent the current state of the art in the United States and serve as a practical basis for any similar scour study.

Investigation of Factors Affecting Stream and Foundation Stability It was found by the author that the bridges studied by the author in New Jersey, Pennsylvania, and Maryland are scour-critical and the bridge footings appear to be unstable. Abutment walls can be skewed to the direction of flow, and the existing spread footing elevation is placed above the computed scour depth, which makes it vulnerable to scour. There are no existing river training measures or foundation armoring.

Streamlining HEC-18 Procedures These guidelines, based on field experience and experimentation, are recommended for uniform applications on all scour-critical bridges and used in reference to similar in-depth studies.

Organizing Data Collection Because of the variety of in-house and field data required for completing scour analysis, a New Jersey Department of Transportation (NJDOT) memorandum was adopted for New Jersey bridges and Design Manual No. 4 for Pennsylvania bridges. For Maryland bridges, the SHA provisions are applicable.

Importance of Investigating "Unknown Foundation" in Scour Analysis Test boreholes were driven to provide important missing data for unknown footings and unknown rock elevations, if any. This led to a reliable and economical design of countermeasures based on soil conditions and response to floods. Ground penetration radar (GPR) does not always seem to distinguish between foundation concrete and surrounding soil.

Use of Structure Inventory and Appraisal (SI&A) Sheet Preliminary or phase I bridge coding for New Jersey bridges was checked and revised based on extensive phase II in-depth studies (a comprehensive program developed by NJDOT). This approach warns of immediate danger of collapse of footing. Accurate determination of inventory coding helps in future planning and funding allocations for substructure repairs or for replacement of the bridge.

Design of Countermeasures For the Guinea Hollow Road Bridge in New Jersey, strengthening the footing by constructing an apron wall and installation of armoring are recommended. For the Water Street Bridge in New Jersey, the cost of both underpinning the east abutment footing and providing countermeasures for both abutments will approach the cost of a replacement structure. Hence replacing the 63-year-old bridge appears to be a better choice in economic terms. Interim repairs and replacement of the bridge are, however, recommended.

Distant Monitoring Continued monitoring of the footings and of the main channel, during routine biennial inspections, will be necessary. Timely monitoring after floods is also required. Structural health monitoring (SHM) methods and remote sensors and Lidar methods can be adopted.

Federal Government Sources Related to Scour Technology On July 6, 2012, President Obama signed into law the Moving Ahead for Progress in the 21st Century (MAP-21) Act. MAP-21 also has a penalty for states that are in noncompliance with NBIS.

The NBIS definition of *underwater inspection* notes that an underwater inspection generally requires diving and cannot be accomplished visually by wading or probing.

The FHWA has discovered several issues regarding noncompliance with the NBIS:

- Scour-critical bridges without plans of action (POA) are identified in 23 CFR 650.313(e)(3). The Code of Federal Regulations (CFR) is the codification of rules published in the Federal Register by the departments and agencies of the federal government. NBIS is codified in 23 CFR 650 Subpart C.
- For some scour-critical bridges the POA was not properly implemented.
- Bridges were not being load-rated for state legal loads and/or routine permits.
- Unqualified team leaders performed inspections.
- Inadequate or nonexistent inspection procedures and overdue inspections were problems.
- Some critical inspection findings were not addressed.

The metrics (or specific measures) required by current NBIS regulations are examined to assess each state's compliance with the NBIS. Each of the 23 metrics is annually assessed by FHWA and assigned one of four compliance levels—compliant, substantially compliant, noncompliant, or conditionally compliant—based upon specific thresholds or measures for each compliance level for each metric. These specific thresholds or measures are contained in the NBIS Oversight Program document entitled "Metrics for the Oversight of the National Bridge Inspection Program (April 1, 2013)." Some of the applicable regulations for river bridges are listed here:

Metric 5: Qualifications of personnel—underwater bridge inspection diver, 23 CFR 650.309(d)

Metric 8: Underwater inspection frequency—lower-risk bridges, 23 CFR 650.311(b)

Metric 9: Underwater inspection frequency—higher-risk bridges, 23 CFR 650.311(b)

Metric 17: Inspection procedures—underwater, 23 CFR 650.313(e) (e)(1)

Metric 18: Inspection procedures—scour-critical bridges, 23 CFR 650.313(e)

Title 23, Highways Chapter I, FHWA, Part 661 Indian Reservation Road Bridge Program

IRR Bridge Program (IRRBP) The IRRBP is a nationwide priority program for improving structurally deficient and functionally obsolete Indian Reservation Road (IRR) bridges. The activities of projects include replacing, rehabilitating, and seismically retrofitting, painting, or using other environmentally acceptable, minimally corrosive anti-icing and deicing compositions. Scour countermeasures for structurally deficient or functionally obsolete IRR bridges, including multiple pipe culverts, should be installed. IRRBP funds can be transferred to the Indian Tribal Agreement under the FHWA/Tribal program agreement. Section 661.23 describes what these IRR bridge funds can be used for, and Section 661.25 describes the criteria for bridge eligibility.

3.4 Case Studies of Scour-Prone and Scour-Critical Rivers

The results of case studies of eleven New Jersey scour-critical bridges and four bridges in Pennsylvania that were investigated by the author are presented here.

A number of bridges on wide rivers required countermeasures, such as the Route 46 Bridge in New Jersey, the Schuylkill River Bridges in Philadelphia, Pennsylvania, and the important Potomac River Bridge in Maryland. These bridges are located in different watersheds and in different counties.

New Jersey has a large network of brooks, creeks, streams, and rivers. Examples of perennial rivers with high flood velocity are the Raritan and Passaic Rivers. On the west side of the state is the Delaware River and on the east side is the Atlantic Ocean with tidal inlets. There are numerous streams, which are seasonal and which become active during storms. Morris and Hunterdon Counties have the largest number of bridges on rivers subjected to scour. The case studies show that the bridges can be "salvaged" by foundation repairs, or would need replacement.

In New Jersey, riprap alone is recommended not as a permanent countermeasure, but as emergency shielding only for a period of 5 years or longer, after regular evaluation from underwater bridge inspection reports. From past experience of flash floods it was seen that riprap can be used as secondary local armoring, in conjunction with primary structural countermeasures or with river training measures.

A field survey of the scour critical bridges in New Jersey studied by author's team, which included licensed structural, hydraulic, and geotechnical engineers, was carried out during March and April 2001. Field survey results showed that no gauge data for peak flood discharges were available. The Stankowski method of hydrologic studies, as applicable to New Jersey rivers, and HEC-RAS hydraulic analysis methods are used. Contraction and local scour depths are computed using Excel spreadsheets, which were developed by the author (and included in Appendix A) based on the latest HEC-18 methods. Conceptual design of remedial measures using the HEC-23 countermeasures matrix is presented. Various techniques to evaluate the unknown foundations were investigated. In his article "Evaluating Unknown Bridge Foundations, Is It Worth It?" William A. Horne discussed the methods of evaluating unknown foundations at the 1998 International Water Resources Engineering Conference, ASCE, Memphis, Tennessee.

The factors that affect stream stability include the location of the bridge, either close to a river bend or at a skew angle to the direction of flow. The factors that affect foundation scour include the use of shallow foundations (placing the footing elevation above the computed scour depth) or any deficiencies in planning, such as the absence of river training measures and lack of foundation armoring. The methodology for scour studies is in accordance with FHWA procedures and represents the current state of the art in the United States.

Stream stability and bridge scour were recognized as design criteria in the AASHTO LRFD Bridge Design Specifications 1998, Section 2.6 (Heibaum and Trentmann, 2010). The scour sufficiency rating needs to be evaluated.

The scour report consists of a detailed field survey, substructure information, scour analysis results, hydraulics-related findings, and countermeasure recommendations.

Primary countermeasures: These will be selected based on the type of scour. Scour countermeasures will be economical solutions for particular types of streams, soil conditions, and environmental permit requirements. They may be classified as follows:

- Monitoring countermeasures using fixed and portable instrumentation
- River training measures to control channel instability
- Hydraulic and armoring countermeasures such as riprap, grouted riprap, and grout-filled bags
- Structural countermeasures such as foundation strengthening, sheet piles, concrete apron walls and curtain walls, and pier geometry modification

A new decision matrix will show a diagram for the mathematical expressions of the hydraulic design, stability analysis, factors of safety, construction practices, and specifications for testing.

Secondary countermeasures: Geotextiles and filters are secondary countermeasures, but when combined with a primary countermeasure such as rock riprap, gabions, or artificial riprap, they act as an additional armoring by making the armoring system more effective. To optimize countermeasure design and reduce costs, the flood velocity and projected scour depth may be reduced by river training measures such as construction of guide banks and spurs and growing vegetation on the bank. The following construction procedure for installing countermeasures should be followed:

1. Make repairs prior to implementing countermeasures (including removing debris from the channel bed and filling scour holes).
2. Build a temporary cofferdam and dewater filling the existing gap below the bottom of the footing at the abutment with concrete, for approximately a 20-ft length or as required.
3. Place grout bags at the east abutment (including geotextile layer).

3.5 Methods for Investigating Conditions of Existing Footings

As discussed in "Evaluating Unknown Bridge Foundations, Is It Worth It?" (Horne, 1998), routine data collection basically consists of the following:

1. The location and number of required channel cross sections
2. Channel characteristics to be surveyed for the HEC-RAS hydraulic analysis software
3. Probe of unknown footing dimensions for abutments, pier, and wing walls
4. Soil samples for testing
5. The type and size of the stream
6. The bed material type and the classification of the roughness coefficient (to be used in the hydraulic analysis)

Bridge No.	Hydrologic	Survey	Hydraulic	Geotechnical	Foundation Details
1000 041	FEMA studies available	Number of channel sections required	Verification of existing HEC-2 analysis	Soil properties required (samples obtained from field)	Test pits required

TABLE 3.1 Currently Available and Recommended Methods of Obtaining Existing Bridge Information

7. Type of scour and condition of the banks
8. Plotted size and location of scour holes formed during floods, and sizes and location of sediment deposits.

Information that is currently available and recommended methods of obtaining additional required information can be found in Table 3.1. For existing bridges, Table 3.1 shows the documentation and the hydrologic, hydraulic, and geotechnical information required prior to countermeasure selection and design. Inspection reports are also required.

Figure 3.1 shows post-flood scour of an abutment foundation, where timber piles are exposed to severe erosion.

FIGURE 3.1 Abutment foundation and timber piles exposed due to erosion.

Geotechnical Investigations

A geotechnical report based on field and laboratory investigation of the type of soil is required. The information is required for the selection of the type of countermeasures to be installed. The following steps are needed:

- Review subsurface information given in the geotechnical report.
- Evaluate historic scour-related conditions and potential scour holes at the bridge site.
- Laboratory tests for grain size samples are required for soil classification.
- From streambed photographs, estimate the distribution of stones in the riverbed.

Investigating Existing Spread Footings Direct methods such as probes, augers, rotary drilling, or digging test pits adjacent to footings will be used. To identify areas of loose sediment, organic debris, or scour holes, the soil probe may be carried out adjacent to existing footings.

Investigating Existing Deep Foundations Geotechnical analysis will be used to determine the foundation elevation and density of subsurface strata by using techniques such as borings, standard penetration sampling, seismic refraction, and resistivity surveys. A geotechnical engineer should be consulted. Nondestructive, ground-penetrating radar or acoustical emission techniques may be used.

Evaluating Rating Hierarchy for Priority of Repairs

See the FHWA Hydraulics Revision of Coding Guide, Item 113, Scour-Critical Bridge, April 2001.

Table 3.2*a* shows the assignment of applicable scour code following preliminary information according to the degree of scour. The scour code is defined in FHWA

Item 113 Scour-Critical Bridges

The listed column of codes in Table 3.2a will identify the physical conditions of the bridge for the purposes of repairs or replacement and for funding allocation.

Minnesota Bridge Scour Code Definitions

Table 3.2b is available in Minnesota Bridge Department publications and explain the categories of deficiencies in terms of assigned code designations. Monitoring generally requires measuring the riverbed elevation at the bridge to determine if

- The bridge should be closed.
- Corrective action should be taken immediately (i.e., remove debris).
- Protection should be installed as soon as practical.

Code	Description
N	Bridge is not over a waterway.
U	Bridge with "unknown" foundation that has not been evaluated for scour. Until risk can be determined, a plan of action should be developed and implemented to reduce the risk to users from a bridge failure during and immediately\after a flood event (see HEC-23).
T	Bridge over "tidal" waters has not been evaluated for scour, but is considered low-risk. Bridge will be monitored with regular inspection cycle and with appropriate underwater inspections until an evaluation is performed ("Unknown" foundations in "tidal" waters should be coded U.)
9	Bridge foundations (including piles) are on dry land well above floodwater elevations.
8	Bridge foundations are determined to be stable for the assessed or calculated scour condition. Scour is determined to be above the top of footing by assessment (i.e., bridge foundations are on rock formations that have been determined to resist scour within the service life of the bridge), by calculation, or by installation of properly designed countermeasures (see HEC-23).
7	Countermeasures have been installed to mitigate an existing problem with scour and to reduce the risk of bridge failure during a flood event. Instructions contained in a plan of action have been implemented to reduce the risk to users from a bridge failure during or immediately after a flood event.
6	Scour calculation and evaluation have not been made. (Use only to describe cases where the bridge has not yet been evaluated for scour potential.)
5	Bridge foundations are determined to be stable for assessed or calculated scour condition. Scour is determined to be within the limits of footing or piles by assessment (i.e., bridge foundations are on rock formations that have been determined to resist scour within the service life of the bridge), by calculations, or by installation of properly designed countermeasures (see HEC-23).
4	Bridge foundations are determined to be stable for assessed or calculated scour conditions; field review indicates action is required to protect exposed foundations (see HEC-23).
3	Bridge is scour-critical; bridge foundations are determined to be unstable for assessed or calculated scour conditions: • Scour within limits of footing or piles • Scour below spread-footing base or pile tips
2	Bridge is scour-critical; field review indicates that extensive scour has occurred at bridge foundations, which are determined to be unstable by • A comparison of calculated scour and observed scour during the bridge inspection or • An engineering evaluation of the observed scour condition reported by the bridge inspector in Item 60 of Recording and Coding Guide
1	Bridge is scour-critical; field review indicates that failure of piers and abutments is imminent. Bridge is closed to traffic. Failure is imminent based on • A comparison of calculated and observed scour during the bridge inspection or • An engineering evaluation of the observed scour condition reported by the bridge inspector in Item 60 of Recording and Coding Guide
0	Bridge is scour-critical. Bridge has failed and is closed to traffic.

Table 3.2a Applicable Scour Code (Assigned from Preliminary Information) Item 113 Scour-Critical Bridges of the Standard Recording and Coding Guide Which Is Widely Used

Minnesota DOT Code	Literal Title	Description
A	Nonwaterway	Bridge is not over a waterway.
B	Closed—scour	Bridge is closed to traffic; field review indicates that failure of piers and/or abutments due to scour is imminent or has occurred.
C	Closed—not scour	Bridge is closed to traffic for reasons other than scour.
D	OBS scour—immediate protection required	Bridge is scour-critical; field review indicates that extensive scour has occurred at bridge foundations. Immediate action is required to provide scour countermeasures.
E	Culvert	Culvert structure. Scour calculation, evaluation, and/or screening has not been made.
F	No evaluation found, known	Bridge structure scour calculation, evaluation, and/or screening has not been made. All substructure foundations are known.
G	No evaluation found, unknown	Scour calculation, evaluation, and/or screening has not been made. Bridge is on unknown foundations.
H	Found above water	Bridge foundations (including piles) are well above floodwater elevations.
I	Screen—low risk	Bridge is screened and determined to be low-risk for failure due to scour.
J	Screen—scour-susceptible	Bridge is screened and determined to be scour-susceptible.
K	Screen—limited risk	
L	Stable evaluation	
M	Stable—scour above footing	
N	Stable—scour in footing and pile	
O	Stable—action required	Bridge foundations determined to be stable for scour conditions.
P	Stable due to protection	
R	Critical—monitor	
U	Critical— protection required	

Table 3.2*b* Minnesota DOT Codes (January 2007)

3.6 Streamlining of Structural Elements for Efficient Planning

Compared to a vertical wall abutment, spill-through and integral or semi-integral abutments have many advantages. Similarly, wing walls, if curved or angled to the direction of flow, perform better for hydraulic flow. If the number of piers is kept to a minimum, there will be less obstruction to flow and less scour. Some planning considerations are as follows:

1. Sloping (spill-through) abutments cause significantly less scour than a vertical wall abutment.
2. Integral abutments with a single row of piles and semi-integral abutments (which are similar to stub abutments on piles but with composite connection) have a lower scour depth as calculated from standard scour analysis equations.
3. Angled wing walls (typically set at 30° to 75° to the longitudinal flow direction) or curved wing walls improve the hydraulic performance of vertical wall abutments. Under normal circumstances, angled wing walls are adequate. Wing walls at 90° to the longitudinal flow direction are also acceptable if turbulence due to separation of flow is unlikely to be a significant problem.
4. For bridges on scour-critical rivers, the number of piers selected should be as small as possible. This can be achieved by selecting longer spans. Minimum obstruction from the pier group during a peak flood event will prevent water overtopping the deck.
5. The best hydraulic performance is given by rectangular piers having a wedged-shaped nose (known as cutwaters).
6. Where the river may change its angle of approach over the life of the structure, the circular pier, or a series of circular piles with a pile cap above the high-water elevation, may be more appropriate.
7. Where debris accumulation is likely to be a problem, debris deflectors should be used.
8. Where the bridge deck could become submerged by an extreme flood, it may be appropriate to streamline the underside of the bridge deck by rounding the upstream and downstream faces to facilitate passage of debris.

3.7 Selection of Flood-Resistant Countermeasures

The author has carried out hydrologic and hydraulic studies for scour in several states, including Massachusetts, New York, New Jersey, Pennsylvania, Delaware, and Maryland. Each state has developed its own unique procedures according to the conditions existing, except that the FHWA recommended procedures are generally being followed. There are many techniques of controlling erosion at bridge sites, such as river training measures to slow down the flood velocity, monitoring and flood watch to take immediate action such as detour and shutdown, and plantation to stabilize the sides of banks, etc.

1. A variety of available countermeasures are listed in HEC-23. The countermeasures matrix is selected and designed to resist each type. As mentioned, several countermeasure tables are presented in HEC-23. For selection, the most common requirements are:

   a. A study of erosion science (HEC-18) shows several components of scour resulting from obstruction:

      - Local scour at abutments and piers
      - Contraction scour

- Aggradation
- Degradation
- Lateral channel instability during floods leading to meander and change in the direction of flow
- Vertical channel instability resulting in breach of levees in the vicinity of the bridge and overtopping
- The demands of each type will require a different type of resistance offered by use of a special material

2. River contamination and encroachment result from the foreign materials adversely affecting the ecology, such as marine life, habitats, and plant life. A separate approach to selection is required.

Construction Costs Related to Countermeasures

To help plan for the long-term installation of scour countermeasures, realistic estimates are required for the individual scour-critical bridges. While the individual in-depth scour reports provided estimated costs for each bridge, there were some discrepancies in these values. This appeared to be a result of differences in the unit-cost values used as well as the inclusion of various items, such as excavation, in some of, but not all, the estimates prepared by the various consultants. In addition, the estimates appeared to be lower than what was found when contracts were put out to bid for countermeasure installation.

The countermeasures recommended in the consultant's reports were reviewed, and it was determined that the variations in the types of scour countermeasures discussed above appeared most often. These variations include the following:

Stone riprap

- 24 in deep with a median stone size (D_{50} or d_{50}) of 1.2 ft
- 36 in deep with a median stone size (D_{50} or d_{50}) of 1.8 ft

Rock-and-wire mattress

- 9-in-deep gabions
- 12-in-deep gabions

Concrete slab

Countermeasures in Ascending Order of Cost

An initial cost consideration for repair of scour-critical bridges can be based on following scour countermeasures in ascending order of cost:

1. Development of bridge inspection and scour monitoring programs, closing bridges when necessary
2. Providing riprap at piers and monitoring
3. Providing riprap at abutments and monitoring
4. Constructing guide banks (spurs and dikes)
5. Constructing river training countermeasures and channel improvements

6. Strengthening the bridge foundations
7. Constructing sills or drop structures (check dams)
8. Constructing relief bridges or lengthening existing bridges

Prerequisites to Selection of Countermeasures

Selection of an appropriate countermeasure for a bridge footing depends on numerous factors, including river characteristics, local soil profile, type of scour, costs, environmental constraints, and constructability limitations at a site.

It should be emphasized that the selection and design of countermeasures should be based on hydraulic and scour analyses, geology of the area, and site-specific situation such as the importance and remaining life of the bridge.

A designer must apply engineering judgment in examining the results obtained from scour, hydrologic, and hydraulic data. Hydrologic and hydraulic data should include

1. Performance of the structure during past floods
2. Effects of regulation and control of flood discharges
3. Hydrologic characteristics and flood history of the stream and similar streams
4. Redundancy of the bridges (continuous versus multispan simply supported)

The information in Table 3.3 will guide the engineer in selection of the countermeasure that will impact the environment the least.

Countermeasures Based on Types of Scour

The effectiveness of a countermeasure depends on the type of scour. When the magnitude of total scour is based on several types of scour acting simultaneously, the countermeasure is selected based on the governing or predominant component of scour. The HEC-23 modified countermeasures matrix shows countermeasure selection based on types of scour.

Countermeasure Characteristics

These are listed in HEC-23 countermeasures matrix as the countermeasures group, namely, group 1: hydraulic countermeasures, group 2: structural countermeasures, and group 3 monitoring; and *functional applications* required to resist local scour, contraction scour, and stream instability.

Suitable River Environment

This category describes the characteristics of the river environment for which a given countermeasure is best suited or under which there would be a reasonable expectation of success. This category could also indicate conditions under which experience has shown the countermeasure may not perform well. The river environment characteristics that can have a significant effect on countermeasure selection or performance are

- River type
- Stream size (width)
- Bend radius
- Flow velocity

Techniques	Water Quality Impact due to Chemicals in Concrete or Rust in Wire	Air Quality Impact during Construction	Impact on Fauna due to Lack of Fish Channel and Habitat Disturbance	Impact on Flora	Impact on Fish Spawning
Armoring Countermeasures					
Riprap, dumped or hand-placed	L	L	L	L	L
Riprap on geotextile layer	L	L	L	M	L
Grouted riprap	M	L	M	L	M
Grout-filled mattress	M	L	M	L	M
Wire-encased gabion mattress	M	L	M	M	M
Articulated concrete mattress	L	L	M	L	M
Cable-tied blocks	M	L	M	M	M
Concrete armor tetrapods	M	L	M	L	M
Structural Countermeasures					
Sheet piling	L	M	L	L	M
Extended footing	M	L	M	L	M
Curtain walls	M	L	M	L	M
Pier geometry modification	L	L	L	L	M
Monitoring					
Fixed instrumentation	L	L	L	L	L
River Training Measures					
Impermeable spurs (buried groins)	L	L	L	L	L
Guide banks	L	L	M	L	H
Stream barbs	L	L	M	L	M
Plantation	L	L	L	L	L
Use of turbidity barrier	M	L	L	L	H
River diversion using cofferdams	L	L	H	H	H
Bendway weirs	L	L	M	M	M

L = low; M = medium; H = high.

Table 3.3 Comparative Study for Selection of Countermeasure Techniques with Least Impact on Environment

- Bed material
- Ice and debris load
- Bank condition
- Floodplain (width)

For each environmental characteristic, a qualitative range is established (e.g., stream size is wide, moderate, or small) to serve as a suitability discriminator. While most characteristics are self-explanatory, both FHWA well-known publications HEC-20 and "Highways in the River Environment" (HIRE) provide guidance on the range and definitions of these characteristics of the river environment.

Maintenance

The maintenance category identifies the estimated level of maintenance that may need to be allocated to service the countermeasure. The ratings in this category range from low to high and are subjective. The ratings represent the relative amount of resources required for maintenance with respect to other countermeasures within the matrix shown in Countermeasures Matrix Table. A low rating indicates that the countermeasure is relatively maintenance-free, a moderate rating indicates that some maintenance is required, and a high rating indicates that the countermeasure requires more maintenance than most of the countermeasures in the matrix.

Countermeasure Installation Requirements

A designer's responsibilities do not end after the countermeasure design has been completed. The following important issues need to be resolved for the successful installation of countermeasures.

1. Permitting identification
2. Right-of-way identification
3. Relocation of utilities
4. Construction coordination

To determine a realistic estimate for installation of the countermeasures, a methodology is required to utilize the limited data available at each bridge (countermeasure type and area) and determine a total cost of the installation. Unit prices for the individual countermeasure elements were identified, where available, from the DOT's Bid Price Report for Standard Items.

Installation/Experience by State

This category identifies DOTs for which information was available on the use of a particular countermeasure. These listings may not include all the states that have used a particular countermeasure. Information on state use was obtained from three sources: a National Cooperative Highway Research Program questionnaire (University of Minnesota survey for NCHRP Project 24-7); Brice and Blodgett (1978); and correspondence with DOT staff. It is expected that additional information on state use will be obtained as this matrix is distributed and revised. Certain countermeasures are used by many states. These countermeasures have a listing of "widely used" in this category. Both successful and unsuccessful experiences are reflected in the listing.

Design Guidelines Reference

Reference manuals that provide guidance in countermeasure design have been developed by government agencies through research programs. The FHWA has produced a wealth of information through the federally coordinated program of highway research and development. The design guideline reference column in the countermeasures matrix (Table 2.1 of HEC-23) identifies reference manuals where guidance on the design of the countermeasures can be obtained. The references are symbolized by numbers in this column. The numbers correspond to the number of the references listed on the second page of the matrix.

3.8 Factors Affecting Detailed Design of Countermeasures

Solved examples of countermeasure design are listed in the Appendices A and B. Factors affecting detailed design of countermeasures are:

1. Natural issues such as existing soil geology
2. Physical factors, such as the width of the bridge opening and traffic volume on the bridge
3. Economic considerations such as the existing condition of the bridge and the life of the proposed countermeasure compared to the remaining life of the bridge
4. Availability of resources for monitoring and field inspection frequency and of the underwater inspection
5. Priority of funding for repairs due to flood damage
6. Constructability

Armoring Countermeasures

Armoring is the most commonly used method in New Jersey and throughout the United States. Hydraulic countermeasures are primarily designed to either modify the flow or resist erosive forces caused by the flow. They are organized into two groups: river training structures and armoring. Their performance is dependent on design considerations such as filter requirements and edge treatment.

From a review of more than 50 NJDOT phase 2 in-depth reports for scour-critical bridges, it is seen that a majority of bridges located on rivers will be subjected to 100-year flood velocity (resulting generally in peak flood velocity of 6 to 8 ft/s or less). If the peak flood duration is not long enough to cause extensive erosion, then armoring countermeasures should be sufficient for velocities between 6 and 8 ft/s or less.

Armoring is distinctive because it resists the erosive forces caused by a hydraulic condition. Armoring countermeasures do not necessarily alter the hydraulics of a waterway, but act as a resistant layer to hydraulic shear stresses, providing protection to more erodible materials underneath. The countermeasures generally do not vary by function, but vary more in material type. Armoring is classified by two functional groups: (1) revetments and bed armoring and (2) local armoring.

1. *Revetments and bed armoring* are used to protect the channel bank and/or bed from erosive and hydraulic forces. They are usually applied in a blanket-type fashion for aerial coverage. Revetments and bed armoring can be classified as

either rigid or flexible and articulating. Rigid revetments and bed armoring are typically impermeable and do not have the ability to conform to changes in the supporting surface. These countermeasures often fail due to undermining. Flexible and articulating revetments and bed armoring can conform to changes in the supporting surface and adjust to settlement. These often fail by removal and displacement of the armor material.

2. *Local scour armoring* is used specifically to protect individual substructure elements of a bridge from erosion by vortices created by obstructions to the flow. Generally, the same material is used for revetments and local armoring. Some important aspects of local armoring are as follows:

   An important aspect is erosion. During a flood, erosion may take place more commonly in the vicinity of the river constrictions, such as at bridges and culverts. For protection against scour at bridge and culvert sites, the most commonly used countermeasure is bed armoring. It is applicable at the following locations:

   a. *Abutment* (full height, stub, integral, and spillthrough) with spread footings, pile caps, piles, and drilled piers
   b. *Piers* with spread footings, pile caps, piles, and drilled piers
      (1) Wall piers
      (2) Column bents
      (3) Pile bents
   c. *Wing walls* for abutments with spread footings, pile caps, and piles
   d. *Headwalls* of culverts
   e. *Wing walls* for culverts with spread footings
3. *Armoring:* The failure of a riverbank or an embankment upstream or downstream of the bridge may lead to change in the direction of flow of the river and increased scour at the bridge due to change in hydraulic conditions. The primary function of armoring is reduction in velocity and energy dissipation, both upstream and downstream of the bridge.
4. *Revetments:* The erosion of banks immediately upstream and downstream is likely to contribute to increased erosion of bridge foundations. The following off-bridge and off-culvert locations are likely to be eroded after a major flood. Revetment may be required at locations such as these:
   a. Banks of rivers, streams, and channels
   b. Embankments at the intersection of the bridge and the roadway
   c. Streambeds adjacent to pier and abutments, where scour holes are formed
   d. Upstream and downstream of culverts and the toe of aprons
5. *Commonly used armoring* at culvert headwalls includes
   a. Rock riprap
   b. Concrete pavement or concrete apron
   c. Concrete armor units
6. *Armoring around wing wall:* The height of a wing wall decreases at locations away from the bridge because of the reduced severity of scour. Wing wall scour

is maximum at the interface with an abutment, where its magnitude is the same as that of the abutment. The thickness of armoring may be reduced linearly if the change in wall height is 2:1. Due to ease of construction, the same type of armoring as for abutments should be used.

7. *Armoring countermeasures combined with river training:* Experience has shown that mandatory armoring countermeasures alone may not be adequate and a combination of optional river training measures and armoring may necessary for the higher-velocity rivers.

By providing river training measures, less pressure will be put on the armoring mechanism. Accordingly, the effectiveness of the system will be increased. Since large investments would be involved, economic considerations become important. Hence, cost reductions may be adopted in the design detailing by optimizing the depth and width of the armoring mechanism which are provided as revetment. Using scaling factors in the next section, riprap or gabion blankets may be used more economically.

Riverbank Protection It is normal practice to protect 100 to 300 ft of riverbanks by revetment upstream and downstream of bridges and culverts. The upstream and downstream countermeasures differ from bed armoring in that they have a smaller thickness and are longer. Their sizing takes into account correction factors for stability, gravity, and the angle of repose of the riprap. In addition to mattresses, the continuous framework of articulated concrete blocks and grout bags has been used for revetment. Filters should be used when concrete blocks and grout bags are employed. HEC-23 Design Guideline 12 provides examples of revetment designs. The common types of revetment in use are:

1. Dumped riprap
2. Wire-enclosed riprap mattress
3. Articulated concrete block system
4. Grout-filled mattresses
5. Concrete pavement

It is recommended that some type of river training measure be considered as complementary to armoring when

- The flood velocity exceeds 10 fps
- The bridge carrying traffic volume exceeds an average daily traffic (ADT) of 500

Minimum Countermeasure Requirements

If the projected (computed) scour is small or negligible, theoretically a design of a formal countermeasure will not be required. Such cases arise

1. When a spread footing is located or placed on bedrock or when a spread footing is located or placed below the total scour depth
2. When an additional pile length equal to the projected scour depth is provided

3. When pile stiffness exceeds the minimum required and the exposed length of pile due to erosion can safely act as a long column

Although a minor surface erosion of soil will not pose a danger to footings, a soil cover or protection to the concrete footing or piles is still required. An adequate soil cover needs to be maintained for

- Frost resistance (minimum frost depth requirement)
- As-built cosmetic appearance
- Unforeseen error in the scour analysis data or computations

Several countermeasures are used for protection of bridge abutments by different state DOTs and federal agencies. Two typical approaches for protecting bridge abutments from scour are (1) to mechanically stabilize the abutment slopes with riprap, gabions, cable-tied blocks, or grout-filled bags or (2) to align the upstream flow by using guide banks, dikes or spurs, or in-channel devices such as vanes and bendway weirs.

Design Guidelines for Riprap at Bridge Piers

The use of riprap blankets alone is recommended only for existing bridges that are due for replacement in a few years. Use of riprap blankets should be restricted to low and medium flood velocities not exceeding 10 fps. The following guidelines should be followed for the design of ripraps:

1. *Sizing:* The HEC-18 equation [Richardson and Davis (1995)] is recommended for sizing riprap at pier footings. However, the maximum rock size should not exceed $2D_{50}$, where D_{50} = median stone diameter in feet.
2. *Thickness of mat:* The recommended thickness of a riprap mat is $2D_{50}$. The top of the riprap should be flush with the bed at low flow. The minimum thickness is 12 in.
3. *Layout of riprap mat:* The width of a riprap layer adjacent to a footing W_R around the pier should be not less than the following:
   a. Width of scour hole at pier.
   b. $2W$ (W = width of pier at base) or $2W/\cos\beta$ when $\beta > 15°$.
   c. 1.5 ft + $y \cot\varphi$ from the face of the footing, where y is the design scour depth at any abutment or pier and φ is the angle of natural repose for the soil, as obtained from the geotechnical report. Place riprap around footings with the slope starting at a distance that is a minimum of 1.5 ft from the vertical face of the footing.
4. *Gradation:* The distribution of riprap sizes is described under "General Guidelines for Riprap" (CIRIA Report No. 61), one of the earliest publications on riprap detailing.
5. *Placement of riprap:* Riprap around bridge piers can be installed either from a bridge deck or from the bed and banks of a stream.
   a. *Installation from bridge deck:* Riprap is installed from a bridge deck by dumping from trucks and spreading by loader. Since stones are placed

irregularly, they are unstable. Also, there are voids between the stones through which fine particles of soil travel under water pressure and cause erosion. The success rate with dumped riprap at many existing sites may be misleading since the bridge may not have been subjected to peak design floods of 100 years. For large bridges, installation can be done either by machine placing from a dragline or by buckets from the bridge deck.

b. *Installation from bed and banks of stream:* In this method, by hand placing and packing, a compact, mortarless masonry type of construction can be achieved. Stones packed into a close interlocking layer will minimize the size of voids. This method of installing riprap results in a stable configuration and in uniformity of stone size distribution. The quality of construction is better than with the dumping method. A fish channel made within the top stone layers can be maintained. Although this method is more expensive than other methods, its use is recommended.

Design Guidelines for Riprap at Bridge Abutments

Design guidelines for riprap at bridge abutments in this section are based on the following:

- "Stability of Rock Riprap for Protection at the Toe of Abutments Located at the Floodplain," 1991.
- "Design of Riprap Revetment," Hydraulic Engineering Circular-11.
- "Design Guidelines 8 and 12" of HEC-23.

Side Slopes

For preventing slump failure, the side slope is a significant factor in the stability of riprap. It is desirable to decrease the steepness of abutments, thus increasing the stability of the riprap on the slopes. The recommended minimum value for side slopes varies from 1:2 to 1:1.5.

Extent of Riprap Protection

Vertical-Wall Abutments For vertical wall abutments without wing walls, wing walls at 90°, or splayed wing walls, the width of a riprap layer WR adjacent to a footing at the river side of an abutment should be the greatest of the following:

- Width of scour hole.
- $2W$ (W = width of abutment at the base) or $2W/\cos\beta$ when $\beta > 15°$.
- $X + 18\text{ in} + y\cot\varphi$, where X is the width of abutment footing, y is the design scour depth at abutment, and φ is the angle of natural repose for the soil, as obtained from the geotechnical report. Place riprap around the footings with the slope starting at a minimum distance of 1.5 ft from the vertical face of the footing.

Spill-Through Abutments For spill-through abutments, extend the riprap around the abutment and down to the expected scour depth. The launching apron at the toe of the abutment slope should extend along the entire width of the abutment toe and around the sides of the abutment to a point of tangency with the plane of the embankment slopes. The apron should extend from the toe of the abutment into the waterway a distance equal to twice the flow depth in the overbank area near the embankment, not exceeding 24.6 ft.

The Maryland State Highway Administration has developed minimum riprap blanket thickness, minimum D_{50} size, and approximate D_{50} weight for class 1, 2, and 3 types.

Empirical Equations for Sizing Riprap at Piers

Riprap is the most commonly used countermeasure, and there are many design approaches to compute depth or thickness of stone layers:

1. Faraday and Charlton (1983)

$$d_{r50} / y = 0.547 \text{ Fr}^3$$

2. Parola and Jones (1989)

$$\frac{d_{r50}}{y} = \frac{C \text{ Fr}^2}{(S_s - 1)}$$

where $C = 1.0$ for rectangular pier and $C = 0.61$ for round-nose piers

3. Richardson and Davis (1995)

$$d_{r50} = 0.692(f_1 f_2 V)^2/(S_s - 1)2g$$

where $f_1 = 1.5$ for round nose, 1.7 for rectangular nose; $f_2 = 0.7$ for 0.9 for a pier near the bank in a straight reach to $f_2 = 1.7$ for a pier at bend in the main current; and S_s is specific gravity of riprap > 1

Lateral extent:

$$\text{Width} > 5b$$

$$\text{Thickness } t_r > 3d_{r50}$$

Top of riprap at bed level

$$\text{Grading } d_{r50} \geq 0.5\, d_{r,\max}$$

4. Lauchlan and Melville of the University of Auckland (1999) proposed

$$d_{r50} / y = 0.3S_f\,(1 - y_r/y)^{2.75} \text{ Fr}^{1.3}$$

where S_f = factor of safety = 1.1 and y_r = placement depth below bed level.
Lateral extent:

Width = b to $1.5b$ in all directions from pier face

Synthetic filter = 75% of lateral extent

Thickness = $2d_{r50}$ to $3d_{r50}$

A factor for level of placement y_r

An earlier work on riprap design specifications is by C.R. Neill of Roads and Transportation Association of Canada in 1973,

Lateral extent: Project around nose of pier by distance = 1.5*b*

$$\text{Thickness } t_r = 2d_{r50}$$

The placed riprap bottom level is below expected general scour level.

5. PennDOT Design Manual DM 4.

Empirical Equations for Sizing Riprap at Abutments *Note:* For assumptions made, notations used, the units and derivation of equations by various authors, the Bibliography should be consulted.

1. Brown and Clyde (FHWA Publication, HEC-11 [1983]) proposed

$$d_{r50} = 0.006V^3(S_f/1.2)^{1.5} \;/\; K_{sl}^{\;1.5}(y)^{0.5}$$

where S_f = stability factor, varying from 1.6 to 2.0 for abutment protection.

2. Pagan-Oritz (FHWA Research Report No. fhwa-rd-91-057 [1991])

Vertical-wall abutment:

$$d_{r50} = \left[\frac{1.06V_2^2 y_2^{0.23}}{(S_s - 1)g}\right]^{0.81}$$

where V_2 = mean velocity in contracted bridge section and y_2 = flow depth in contracted section.

Spill-through abutment:

$$d_{r50} = \frac{0.535V_2^2}{(S_s - 1)g}$$

3. Richardson and Davis (1995), HEC-18

$$\frac{d_{r50}}{y_2} = \frac{K_s \mathrm{Fr}_2^2}{(S_s - 1)}$$

where K_s = shape factor = 0 to 89 for spill-through abutments and K_s = 1.02 for vertical wall abutments.

4. Commonwealth of Pennsylvania, Department of Transportation, *Design Manual*, Part 4

3.9 Use of Rock Riprap as Temporary Armoring

Riprap is defined as a layer of natural rock. Because of its sheer dead weight, it acts as a shield and protects the soil underneath. It prevents direct contact of soil with the erosive forces that are generated at high flood velocities. Riprap prevents the strong vortex motion at the front of a pier from entraining bed sediment and forming a scour hole. Riprap is flexible, and it adapts itself to the new profile after settlement of the soil.

The ability of a riprap layer to settle into a developing scour hole and armor the base of the scour hole is an important factor in the protection offered by riprap. Riprap blends well with the river environment compared to other countermeasures. Its availability, economy, ease of installation, and flexibility are considered highly desirable characteristics.

The performance of riprap has not been verified for different conditions over a long period of time. For high flood velocity, the stones at the boundaries of riprap layer may get dislodged. In such cases partial grouting will prevent movements and is preferred.

It is both cost-effective and environmentally acceptable for the short term only. Properly sized riprap in relation to a peak flood velocity placed over geotextile and constructed according to standard detailing procedures has shown satisfactory results.

With adequate monitoring, bridge foundations with riprap installations may be maintained indefinitely. Eventually, bridges with such foundations may be replaced. Riprap may be used alone or in combination with guide banks, sacrificial piles, or structural countermeasures. In cases of river meander, riprap alone will not be effective, and guide banks may also be required.

General Guidelines for Riprap

NCHRP Report 568, "Riprap Design Criteria, Recommended Specifications, and Quality Control," presents detailed guidelines on all aspects of riprap design and construction. The following general design guidelines apply to both piers and abutments for the design of riprap.

1. *Records of foundations:* Records of construction drawings showing footing sizes and elevations of existing bridges may not be available to plan the extent of riprap and excavation. Probes or boreholes may be drilled adjacent to the abutment and pier walls to locate the footings.
2. *Quality of stones:* Only natural fieldstone or quarry stone should be used for riprap installations. The stone should be hard, angular or of approximately rectangular shape. The quality should be such that it does not disintegrate on exposure to water or weathering.
3. *Excavations:* The top layer of riprap should be at least 6 in below a riverbed to avoid encroachment of the river or dislodging of stones by floating debris, ice, or currents. Prior to placing the riprap, excavation of a ditch of specified dimensions per the design should be carried out in layers. The depth of riprap should be at least below the contraction scour depth. Abutments and piers need to be supported during excavation. Underpinning of footings may be required. Footing widths or raking piles may inhibit excavation. Adequate precautions should be taken to avoid damage to existing foundations. For the safety of construction personnel, the slope of excavation should be properly planned. Depending on the angle of repose of the soil, slopes should conform to OSHA standards.
4. *Equipment:* Excavation machinery, drilling equipment, and small cranes for lifting heavy stones may be required.
5. *Maintenance:* Since riprap is buried, no maintenance is required. However, if riprap gets exposed or is scattered on riverbed except after heavy floods, remedial work should be planned.

The following maintainability issues should be considered with riprap installations:

a. Under-bed installation increases durability.

b. More maintenance is required with dumped riprap.

c. It is difficult to repair ripped geotextile or locate damage to riprap.

d. Cleanup is difficult after failure.

e. Gravel filters are easier to maintain than geotextile.

6. *Riprap for pile caps and exposed piles:* Since pile clusters are generally closely spaced, it may not be feasible to place riprap around them. Instead tremie concrete may be used under a pile cap. Riprap may then be placed around the pile cap.

7. *Design parameters:* Specific design procedures for piers and abutments are based on the size of the riprap, gradation, layer thickness, horizontal extent, placement techniques, filters, and type of equipment. A set of scour calculations based on HEC-18 procedures should be prepared. Scour depth increases with the increase in Froude number, and is directly proportional to velocity. The depth of riprap will increase proportionally to both the approach velocity and the Froude number.

Scour Data at Abutment/Pier

For detailed design of riprap, the following data is required:

- Computed scour depths flood data
- Discharge frequency
- Long-term scour
- Short-term scour
- Contraction scour
- Local scour
- Total scour
- Maximum elevation
- Maximum flood velocity, 50-year, 100-year, 500-year

  If 500-year discharge is not available, use $Q_{500} = 1.7Q_{100}$.

  The following flow parameters are required to calculate the size of rock:

  a. Flow velocity V

  b. Velocity magnification factor K

  c. Specific gravity of riprap S

  d. Water depth d

8. *Rock size and gradation:* The median size of riprap D_{50} shall be determined as:

   Minimum D_{50} size of riprap shall be R-6 and maximum D_{50} size shall be R-8 per National Crushed Stone Association (NCSA) rock size and gradation.

   Nominal placement thickness (in): Maximum 48 in., Minimum 12 in.

As described in HEC 11 (Brown and Clyde, 1989). For protection of pier footings using riprap,

$$D_{50} \frac{0.692}{S-1}\left(\frac{V^2}{2g}\right)$$

where

D_{50} = median rock size, m (ft)
V = velocity at the exit of the dissipator, m/s (fps)
S = riprap specific gravity

9. *Exceptions to depth of riprap:* Riprap may not have to be installed to the full scour depth in the following circumstances:
   a. If computed scour depth is high, it may not be feasible to provide riprap depth equal to full scour depth. Deep excavations adjacent to spread footings are not permissible. In such situations, the thickness of riprap may be reduced to $d = 0.5y$, where y = maximum scour depth, provided d is not less than 3 ft, a filter layer is used, and, if effective, monitoring is done.
   b. If the excavation depth is not excessive and does not cause instability of footing, a higher value of d between $0.5y$ and y may be used. However, the size of the stone will not be reduced and will be based on the calculated velocity. If the design depth d is greater than the available depth between the riverbed elevation and the bottom of the footing or bedrock is not available within the depth d, then an alternative countermeasure should be considered.
10. *Durability:* The following durability issues should be considered with riprap installations:
    a. Broad band of failure threshold potential
    b. Catastrophic failure if riprap is exposed
    c. Abrupt failure of geotextiles
11. *Effectiveness of riprap installations:* Effectiveness of riprap depends on
    a. The seal around piers
    b. Reduced tendency of rock dispersal
    c. Degradation of granular filters

    It is more effective if tied into the abutment countermeasure when the pier is located within three pier diameters of abutment footings.
12. *Constructability of riprap installations:* The following constructability issues should be considered during riprap design:
    a. Excavation required
    b. Difficulty of sealing geotextile to pile bents
    c. Limited ability to preexcavate due to pier footing and/or pile geometry

d. Specialized construction techniques for geotextile placement

e. Gravel cushion on geotextile to avoid rupturing

f. Performance dependent on construction sequence

13. *Costs:* Costs may vary according to local conditions. The current estimated cost is $150 to $200 per square foot. Long-distance freight charges for riprap may increase the unit cost by 10 percent. The following cost issues should be considered with riprap installations:

    a. Greater cost of geotextile compared to granular filter

    b. Preexcavation costs

    c. Disposal costs

    d. Reduced stone costs

    e. Traffic disruptions

14. *Riprap detailing*

    a. Construction drawings should be prepared. Conceptual sketches for the layout of riprap with details for riprap placement at abutments and piers, based on this handbook, should be used.

    b. In addition to hydraulic data, construction drawings should show tables summarizing flood elevations, flood velocities, and scour depths.

    c. Maximum side slope is 1 V: 2 H. Where excavation is difficult, 1 V:1 H may be used with fractured rock, where V = Vertical and H = Horizontal sides of the slope.

15. *Construction permits:* Stream encroachment and other applicable permits must be accounted for. Refer to guidelines given for permit applications in the New Jersey Department of Environmental Protection's *Stream Encroachment Technical Manual.*

16. *Computer software:* The following software may be used:

    a. Riprap Design System (WEST Consultants, Inc., San Diego)

    b. Riprap Design Software (U.S. Army Corps of Engineers)

Modes of Failure of Riprap

To prevent the failure of riprap during floods, the following common causes of failure must be considered and controlled during the riprap design.

1. Inadequate rock size and gradation
2. Absence of filter layer
3. Internal slope failure
4. Placement of riprap at steep slopes
5. Poor toe design
6. Degradation of channel

Riprap failure mechanisms are affected by riprap size. From extensive studies under peak flood conditions, the following modes of failure of riprap have been identified:

1. Riprap blanket
   a. *Hydraulic erosion failure:* Causes include inadequate stone size, steep side slopes, inadequate gradation, and removal of stones by impact. Hydrodynamic forces of flowing water may dislodge individual stones.
   b. *Winnowing failure:* Underlying finer material is removed through the voids of the riprap. This type of failure can be avoided by increasing the number of layers.
   c. *Edge failure:* Due to channel scour, the toe of the riprap blanket is undermined, and scour holes are formed. Making the layer of riprap sufficiently thick can prevent edge failure.
2. Sloping riprap
   a. *Slump failure:* The causes include steep side slopes and the presence of impermeable material that acts as a fault plane when subjected to excessive pore pressure. Due to shear failure of the underlying material, a rotational-gravitational movement of material may occur along a rupture surface with a concave-upward curved shape. Riprap stones cannot withstand the horseshoe vortex associated with the scour mechanism. This type of failure can be avoided by increasing the median size of the stone.
   b. *Modified slump failure:* The causes include steep side slopes and disturbance of material in the lower riprap layers. Mass movement of material may occur along an internal slip surface within the riprap layer.
   c. *Transnational slide failure:* The causes include steep side slopes, excessive pore pressure, and toe failure due to undermining. A downslope movement of a mass of riprap with the fault line on a horizontal plane occurs.
   d. For a spill-through abutment, the initial failure zone begins at the armored floodplain downstream of the contraction near the toe. For a vertical wall abutment, the initial failure zone occurs at the upstream corner of the abutment (Pagan-Ortiz, 1991).

Limitations on the Use of Riprap

1. *Monitoring:* Riprap should be used as a countermeasure only if accompanied by field inspection that occurs immediately after floods and by the use of monitoring equipment during floods.
2. *Critical velocities:* If a 100-year flood velocity exceeds 10 ft/s, riprap should not be used.
3. *Scour depth:* If the calculated scour depth is high and excavation to place riprap under the riverbed would endanger the stability of soil adjacent to the footing, then riprap shall not be used.

4. *Economic considerations:* If riprap is not available locally or at a reasonable distance, it may not be economically feasible. In such situations, other alternatives may be considered. Also, if the cost of hand placement of riprap is high, other less expensive countermeasures may be considered.
5. *Dumped riprap:* Truck-dumped riprap can easily get dislodged during floods and washed away due to high velocities. It is less stable than hand-placed riprap, and its use is therefore not recommended.

Traffic and Utilities Issues on the Use of Riprap

1. *Site access:* Adequate access to the site should be provided for trucks to deliver riprap.
2. *Right-of-way:* Construction easements and right-of-way access may be required for the duration of construction.
3. *Detours:* Detour, lane closure, or nighttime work may be necessary. Coordination with traffic control would be required. Emergency vehicles and school bus services should not be affected by lane closures.
4. *Utilities:* Relocation of any utilities at the sides of an abutment or a pier may be necessary for the duration of construction. Coordination with utility companies would be required.

The two alternate methods for scour analysis other than HEC-18 that are applicable to New Jersey hydrology, hydraulics, and soils are the Melville equations for abutments and piers and the Maryland State Highway Authority (MD SHA) ABSCOUR Program for Piers.

A temporary countermeasure is used for emergency situations or as an interim countermeasure, until permanent countermeasures are installed. New Jersey's experience with rock riprap as a primary permanent armoring has not been very successful. For bridges and culverts, riprap is recommended as temporary or interim countermeasure until a permanent countermeasure is constructed. The use of riprap for riverbanks, retaining walls, and wing walls as a primary countermeasure is still acceptable.

Countermeasures Alternative to Temporary Riprap

Riprap, though a widely used scour countermeasure, has limited application in conditions marked by very high flow velocities and high intensities of bed sediment movement. In certain situations, required riprap sizes may not be readily available. Hence, the following countermeasures may be used instead of the conventional use of riprap.

1. Armoring countermeasures
   a. Gabions and Reno mattresses
   b. Grout-filled bags and mats
   c. Cable-tied blocks
   d. Tetrapods and related units
   e. Grade control structures
   f. Grouted concrete, pavement, and flexible bed armor

2. Flow-altering countermeasures (with relatively few field applications)
   a. Sacrificial piles
   b. Upstream sheet piles
   c. Collars and horizontal plates
   d. Flow-deflecting vanes or plates
   e. Modified pier shape or texture
   f. Slots in piers
   g. Suction applied to bridge pier

Flow-altering countermeasures should be used only in combination with primary countermeasures to improve their effectiveness.

3.10 Applying Modern Techniques to Contain Erosion

A *scour countermeasure* is defined as a measure incorporated at a stream–bridge crossing to monitor, control, inhibit, change, delay, or minimize stream and bridge stability problems due to scour.

Commonly Used Scour Countermeasures

Countermeasure instructions:

- Conduct monitoring by using underwater inspection through divers.
- Use remote sensors for continuous monitoring.
- Install rock riprap either dumped or broken rock to protect footings.
- Extend the footing structure to support the slope or protect it from erosion.
- Place gabions and Reno mattress, wire mesh baskets, and mattress filled with loose stones as shielding.
- Construct guide bank (spurs and dyke) structure to support the slope or protect it from erosion.
- Use channel lining for pavement protection, with reinforced concrete and bituminous concrete, and for protection of channel bed and banks.
- Provide temporary detour of traffic during construction for bridge closure.
- Use fabric bags filled with concrete and stacked to produce a protective layer.
- Use check dams and install sills or drop structures.

Artificial Riprap Alternatives Such as Tetrapods and Toskanes

The tetrapod shape is designed to dissipate the forces of incoming waves in coastal areas and in rivers with high flood velocities (See Fotherby, 1992). Tetrapod shape is designed to dissipate the force of incoming waves by allowing water to flow around the units rather than against them. The units interlock mutually to dissipate the dynamic forces. Other concrete shapes such as A-jacks and dolos are also used. Lagasse (2007) and Fotherby and Jones (1993) have addressed the problem of scour at piers.

Toskanes at Piers The report (Ruff and Fotherby [1995]) found that toskanes have no significant advantage over riprap stones. The following is the empirical design formula. The manufacturing company of toskanes may also have developed their own formulas:

$$\frac{d_e}{y} = \frac{0.255\ \mathrm{Fr}}{(S_s - 1)\left(\frac{b_p}{y}\right)^{0.5}}$$

where d_e = equivalent spherical diameter of the toskane units
y = scour depth
S_s = specific gravity of stone
b_p = projected width of pier
Froude number $\mathrm{Fr} = V_{v / (g\,y)}^{\ 0.5}$
V_v = adjusted approach flow velocity

For shallow foundations such as spread footings and riverbanks, the alternative countermeasures used in some states are as follows:

Concrete-filled mat: Porous fabric bags filled with high-strength mortar are also effective.

Jetties: Walls to support the slope or protect the bank from erosion are popular in coastal areas.

Flexible revetment: Cable-tied or articulated blocks serve as flexible revetments.

Artificial armoring:

- Precast concrete blocks: concrete blocks of a cellular shape placed as revetment
- Retard (timber and sheet piles, trees)

Sloping retaining walls: Cast in reinforced concrete, masonry or timber walls support the slope or protect it from erosion.

Concrete grouted riprap: Standard riprap with full or partial concrete grout is more effective than loose riprap.

Item No.	Countermeasure	Description
1.	Hard point	In-place soil stabilized with cement
2.	Bulkhead wall	Supports the slope or protect it from erosion
3.	Channel improvements (channelization)	Dredging to increase channel width
4.	Debris basin	Depressions formed to collect debris
5.	Underpinning	Foundation strengthening by columns
6.	Collar plates	Attached to pier to deflect flow
7.	Vegetable planting	Trees planted to prevent bank erosion

Table 3.4 Description of Miscellaneous Types of Countermeasures

Sacrificial piles: Steel, timber, or concrete piles are driven upstream to reduce velocity. Piles deflect the high-velocity flow. Their effectiveness depends upon

- The geometric arrangement of piles
- Their number
- Their protrusion

Soil cement: In-place soil can be stabilized with cement, to serve as grout.

Flow-deflecting plates: Metal plates are connected to concrete piers to deflect the flow of water.

Cable-tied and articulated concrete blocks: Concrete blocks and slabs interconnected with steel cables or articulated are available from precast factories.

Braced piles: When braced in the transverse direction, the pile group provides stable foundations.

Relief bridge: It increases the opening by reconstructing the abutment. Provide a new opening by adding span at approaches. Water flow is facilitated due to additional opening, and scour is reduced.

Vanes: Obstructions placed upstream to redirect or reduce flow also help in reducing kinetic energy.

Tetrapods: Factory-manufactured artificial concrete blocks similar to manufacture of A-Jacks or articulated concrete blocks replace large size riprap stone blocks.

The estimated flow velocity, durability, and cost of the countermeasures are factors in selection of the type of units and method used.

- New Jersey DOT Soil Erosion and Sediment Control Standards: Soil Erosion Sediment Control Manual 2009. It is the purpose of these standards to supply the engineer with information helpful in designing soil erosion and sediment control measures for both temporary and permanent conditions. The manual describes both vegetative standards and engineering standards for soil erosion control.
- Construction Industry Research and Information Association (CIRIA), and CUR/RWS Report 169, Center for Civil Engineering Research & Codes, 1995.

3.11 Applications of Geotextile and Granular Filters

General Use of Filters as Armoring

The use of a filter, or alternatively a geotextile (filter cloth), is particularly important in ensuring that finer material does not leach through or winnow around the riprap. No filter is required for gravel beds. For sand beds, a geotextile filter cover equal to the width of a pier W from the face of the pier in each direction should be used.

Specialized construction techniques are required for geotextile placement. A 3-in-thick gravel cushion placed on geotextile is recommended to avoid rupture.

Filters are a secondary countermeasure, but when combined with a primary countermeasure such as rock riprap, gabions, or artificial riprap, they act as additional armoring, thereby making the armoring system more effective. The filter is used to ensure that underlying fine sediment particles do not leach through the voids of the individual stones in riprap and other countermeasures. Filter prevents the migration of subsoil particles through the countermeasure. It allows water flow across the soil boundary and avoids buildup of an unacceptable head across it. At deep placement depths within the bed, the dominant failure mode for the riprap layer becomes leaching. To prevent this loss of material, the presence of some form of filter beneath the stone layer would be beneficial. In the case of interconnected gabions on a sand-bed river, a geotextile filter should be placed underneath the baskets to prevent sand leaching. The fabric should provide drainage and filtration, keep fine particles underneath in place, allow for release of pore pressure, and help reinforce the armoring.

Commonly used filters are granular, geotextile, and composite types. The two main references for filters are as follows:

- Center for Civil Engineering and Codes (CUR) 1995
- Holtz et al. (1995)

In addition, the manufacturer's technical specifications for geotextiles need to be consulted. The design parameters are empirical and based on test results for specific conditions. New products for geotextiles, for example, may recommend alternative solutions.

Types of Filters

Since the hydraulic responses of both synthetic and granular filters can be similar, the choice of materials is usually based on practical considerations. In New Jersey, the following types of filters are recommended for use:

1. *Geotextile filter:* This covers a wide range of synthetic grids, meshes, and textiles. A synthetic filter fabric is manufactured for specific applications.
    a. Advantages
        (1) Relatively low cost
        (2) Speed of laying large areas
        (3) Small construction thickness (Volume of riprap is therefore reduced considerably compared to having no filter armoring.)
        (4) Generally provision of a broad band of failure threshold
        (5) Nonwoven types okay with soil variations and allow for settlement
    b. Disadvantages
        (1) Long-term behavior less certain than that of granular filters
        (2) Difficulty of laying in deep water
        (3) Difficulty of placement in high currents
        (4) Requirement to be pressed evenly against the subgrade by the armor layer
        (5) Difficulty in repairing damage
        (6) Difficulty in identifying exact location of geotextile failure

(7) Care needed in laying of geotextile and subsequent placing of armor layer to avoid damage (A bedding layer between geotextile and armor layer may be needed.)

(8) Possibility of openings becoming blocked

(9) Not easily sealed to pile bents (If piles are already exposed, a slurry of riverbed material may be placed underneath the pile cap, or tremie concrete may be used.)

c. Woven type of geotextile

(1) Woven geotextiles are formed using regularly placed fibers oriented at right angles to give uniform hole sizes.

(2) They are generally stronger than nonwoven and can be used as filters for soils of a particular size.

(3) They can also be appropriate where very high porosities are required.

d. Nonwoven type of geotextile

(1) Nonwoven geotextiles are formed using randomly placed fibers, giving a range of hole sizes.

(2) They are generally considered more useful as filters in scour protection situations, because the available hole opening sizes cover a wider range of soil types.

(3) They can also stretch more before failure. By maintaining contact with the subsoil and the armor layer when stretched, its filter function is not compromised.

2. *Granular filter:* Granular filter is placed in layers and requires excavation.

a. Advantages

(1) Deformation, allowing good contact between subsoil and armor layer

(2) Relatively easy repairs and sometimes damage of self-healing type

(3) Durability

b. Disadvantages

(1) Excavation possibly required to lay it

(2) Accurate placement difficult in deep water

(3) Difficulty of placement in high currents

(4) Careful placement needed to achieve required thickness

(5) Careful control of grading needs

(6) Multiple layers possibly needed to meet filter requirements

(7) Difficulty of obtaining required grading locally

3. *Composite filter:* Composite filter is composed of sacks or mattresses and is useful for deep water or high-current conditions.

(1) Can be useful in deep water and high current conditions where mattresses or sacks can be more readily placed than loose granular material or light geotextiles

(2) Can be useful in protecting a geotextile from damage by large riprap

4. *Sand-filled mats:* These are geotextile mats filled with sand or fine gravel. The weight of sand ensures that mats can be laid without movement by currents, while the mat provides the required filter properties. The sand can act as a secondary filter.

Cost

The cost of material and installation for geotextile is two to three times that for providing a granular filter. The average cost for providing a geotextile layer needs to be assessed. The cost may vary for each county and water depth.

Vendor Details

Details from manufacturers must be obtained. An approved equivalent product may also be used.

Monitoring and Inspection

Geotextile requires more monitoring than the granular filter does. After each flood and for each 2-year cycle, any rupture of textile layer should be investigated. This may become obvious at the edges if riprap has moved.

3.12 Design Guidelines for Armoring Countermeasures Other Than Riprap

Gabions

Gabions are containers or baskets made up of wire mesh that are filled with cobbles or coarse gravel. They are filled on-site or in a plant. A thinner version of gabions is known as a *Reno mattress*. Whereas gabions typically take the shape of a brick or sausage, Reno mattresses have a short vertical dimension and large lateral extension. In the United States, more than 500 instances of their use were reported by the University of Mississippi in 1996.

The first application of the Reno mattress was to repair a breach of the Reno River in Italy by Officine Maccaferri in 1884. The wire mesh has flexibility, which allows the containers to deform to the bed profile. Compared to riprap, lesser volumes of excavation of soil from the riverbed are required. Gabions have several advantages when compared with other means of bank protection, such as the following:

- Their loose and porous structure reduces their susceptibility to uplift forces.
- They can be stacked easily in stable configurations.
- The flexibility of the wire mesh allows gabions to mold themselves so as to restore their stability and provide adaptability to site conditions. Relatively small rock sizes can be used to provide the protection effectiveness of much larger rock units. Due to its smaller thickness, a smaller quantity of stones is required.

Figure 3.2 shows the use of gabion baskets designed by the author at the beginning bridge location of Lumberton Bridge over Rancocas Creek in southern New Jersey.

Figure 3.2 Gabion baskets designed by the author for Lumberton Bridge over Rancocas Creek, New Jersey.

Gabion baskets have wider applications than riprap, such as at abutment and pier footing locations, on steep slopes of riverbanks (where riprap may be unstable), and upstream or downstream of bridges and culverts, as river training energy dissipaters.

Types of Gabions

The following types of gabions are commonly used as armoring countermeasures: *Gabion sacks* are used when construction in "the dry" is not possible. In the absence of cofferdams, gabion sacks are placed directly in water. The minimum dimension of a gabion sack ranges from 18 in to 3 ft. *Gabion boxes* are larger than the gabion sacks. The minimum dimension of a gabion box ranges between 2 and 4 ft. They are more suitable for higher velocities.

Possible Failures of Gabions

Damage to the wire mesh of gabions, which can result in failure of the entire installation, is a major reliability problem when one is considering the performance of gabions. The damage to the wire mesh may be due to several factors, such as the following:

1. Movement of filling rock in highly turbulent flows leads to wire damage, deformation of the basket, and ultimately unsatisfactory performance of the installation.
2. Long-term corrosion and reduction in strength of the wire mesh lead to installation failure, possibly in combination with the above factor.

3. *Cost:* An estimated current cost of $85 to $95 per square foot of pier area may be used. Actual costs may vary, depending on local conditions.
4. *Construction permits:* Stream encroachment and other applicable permits will be required. Refer to the guidelines given for permit application in the New Jersey DEP's *Stream Encroachment Technical Manual.*
5. *Durability and maintenance:* The following types of failures may occur and may be avoided by good construction practice:
   a. Failure of meshes and stone fallout due to corrosion, abrasion, and damage during construction may result.
   b. Winnowing failure caused by erosion of underlying bed material through the gabions is due to failure of filter layers and inadequate gabion thickness during floods.
   c. Excessive movement of stone within the baskets may occur at high currents due to poor packing.

Articulated Concrete Blocks

Articulated concrete blocks (ACBs) consist of preformed units that interlock or are held together by steel rods or cables or abut together to form a continuous blanket or mat. ACBs comprise a single layer of cellular blocks interlocking with adjacent blocks, with about 25 percent of the plan area open. Interlocking provides strength and can be increased by filling with gravel. ACBs are laid on a geotextile or granular filter.

They are also known as "artificial riprap" and are used as an alternative for rock riprap, especially when there is a short supply of rock riprap or large rock sizes are required to resist extreme hydraulic forces. These units have greater stability compared to riprap due to their interlocking characteristics. ACBs are factory-manufactured precast concrete units and are commercially produced by proprietary firms. They have been used for erosion control in recent years mainly as revetments but less frequently as countermeasures for bridge substructures.

Cable-tied blocks or cable-tied mattresses are the most popular forms of ACBs. They are concrete blocks tied together by steel cables to form a mat. The cables are made of either galvanized steel or polyester. The network of manageable blocks tied together has the capability of resisting mobilizing forces of a severe flood to provide protection against erosion.

Cable-tied blocks have been used successfully by the U.S. Army Corps of Engineers when streams are coarsely bedded and are capable of moving large stones. The block size is designed based on the flood velocity, slope of the riverbed, and side slopes. Spacing between cable-tied block units should be enough to allow sufficient flexibility to the mattress. They are placed by machinery and can be placed under shallow depths of water. However, they are less adaptable for placing in small or confined areas.

Failure Modes of Cable-Tied Blocks Two failure modes of cable-tied blocks that are similar to those for grout-filled mats are the following (McCorquodale, 1993):

1. Overturning and rolling up of the leading edge can occur in the absence of sufficient anchoring and toeing in.

2. The inner mat can lift up at much higher flow velocities when the leading edge is anchored. The seal between the face of the pier, the countermeasure, and geotextile plays a significant role in this mode of failure.

Typically laid-in-place wet placement cost is approximately \$15 to \$16 per square foot. The cost to place cable-tied blocks around a 4-ft × 20-ft rectangular pier should be estimated based on local conditions.

Design Guidelines for Articulated Concrete Blocks

Design guidelines for ACBs for abutments are based on Design Guidelines 4 in HEC-23. Design guidelines for ACBs for piers are based on NCHRP 24-7. Other recommended sources of information on the design of ACBs are HEC-11 and McCorquodale (1993).

Sizing at Bridge Abutments Design Guideline 4 of HEC-23 presents a factor-of-safety approach for the design of ACBs for revetment or bed armor. This approach is developed by considering the stability of a single concrete block on a sloping surface and is directly applicable to the design of ACBs for bridge abutments.

The *failure* of ACBs is defined as finite movement of a block and hence does not include the effects of resistive forces due to cables, rods, or interlocking for conservatism.

For the sizing of ACBs for bridge abutments, design charts developed by the ACB manufacturer should be used.

Layout at Bridge Abutments Layout dimensions of ACBs around bridge abutments should be the same as those for riprap. It is recommended that the spacing between cable-tied block units be enough to allow the mattress a sufficient degree of flexibility.

Extent and Layout at Piers Prior excavation is not needed unless the block height H_{cb} exceeds $0.25y_0$, where y_0 denotes the flow depth under design conditions. The mattress cover should be $4D$, where D is the pier width. Hence, the mattress should extend outward at least a distance $1.5D$ from every face of the pier.

In the event that the angle of attack β exceeds 15°, the cover is taken to be at least $4D/\cos\beta$.

Geotextile Filter at Piers and Abutments Geotextile filter should not be used for gravel bed rivers. For sand bed rivers, the geotextile filter should be fastened firmly under the base of block mattresses. The geotextile material should extend outward 1D from every face of the piers and should never extend as far out as the mattress.

Pier Sealing Requirement The performance of cable-tied blocks depends on the seal between the pier and the block layer. The structural stability of piers should be investigated when the mat is attached to the piers. Cable-tied blocks should be sealed to the pier as per one of the approaches below, depending on site conditions:

Grouted and Partially Grouted Riprap Partially grouted riprap is also an alternative to dumped or hand-placed riprap, in the selection of the method of riprap construction. Full grouting may not be required in following conditions: when the longitudinal slope is small; when the computed velocity of flow for a 100-year flood is not severe; and when the river environment has marine life and needs to be kept clean of chemical additives.

Method of Construction A reconnaissance field visit to the bridge site is required at the beginning of the project to ensure a good understanding of access to the site of trucks, the river conditions, and field data at the bridges, such as the need to cut vegetation.

The practice is to grout only the top layer of the minimum three layers of riprap. The top layer acts as a cap layer. It is subjected to higher shear forces because it is in direct contact with the flowing water. By gravity, the top grouted layer will help to stabilize the lower ungrouted layers. The percentage of partial grouting depends upon the computed peak flow expected and can be less than one-half of the full grouting in volume. The grout mix needs to be designed for maximum shear resistance.

To guarantee a successful application of partially grouted armoring, a number of tests before, during and after installation are required for quality assurance. Segregation and erosion of the grout when poured or dumped are avoided by using special chemical additives or by high-speed centrifugal mixing (colloidal mortar).

Grouting can be done in freshwater and with time to allow the grout to harden and gain strength. Partially grouted riprap involves using a Portland cement–based grout to hold the riprap rocks together.

The grout generally fills about one-third to one-half of the voids in the riprap to be considered partially grouted. The grout increases the rigidity and stability of the riprap and decreases the likelihood for stone displacement which may occur with loose riprap.

Advantages of Partially Grouted Riprap

1. Partially grouted riprap has been proved to be more effective than fully grouted riprap because it allows for flexibility and permeability through the voids. The flexibility of stone layers is not affected since grouted riprap can conform to the soil profile more easily. Like loose riprap, it is most effective when placed on top of a filter layer.

2. A smaller number of partially grouted conglomerates will offer almost the same shear and frictional resistance as fully grouted riprap or loose armoring stones. It will prevent displacement of bed material, but permits movement of pore water. The partial grouting also allows for use of smaller particle sizes since the particles are joined together to form larger "conglomerates." This is advantageous because smaller rock sizes are more common, and thus riprap can be obtained from a wider variety of sources.

   Partial grouting helps in using economical use of rock riprap. Figure 3.3 shows an increased interest in applications of hydrology for the design of effective countermeasures.

3. Another advantage in using partially grouted riprap is the reduction in the plan coverage area of riprap blanket in comparison to area required for loose riprap. For piers, 2.0 times the width of the pier is generally used as the distance from the pier that loose riprap should cover. This number may be reduced to 1.5 when grout is used because of its increased stability properties.

FIGURE 3.3 The author presenting the keynote address on partially grouted riprap at the September 2014 International Conference on Hydrology in Hyderabad, India.

The success of partially grouted riprap has been studied by numerous organizations, including Colorado State University and Braunschweig University in Germany. The latter determined that partially grouted riprap can remain secure and unharmed in flows of 26 cfs (8 m/s). Physical modeling was performed by Colorado State University which produced similar results. Loose riprap was damaged and displaced at flow velocities of 11 fps, while the grouted riprap remains intact.

The risk of mass failure of formation of loose stones into rigid conglomerates can be avoided by regular inspection and maintenance. Environmental issues, however, may be of greater concern. Turbidity and pH changes are the main problems with the use of grout. Monitoring of turbidity is necessary, and use of Sicotan is helpful as an antiwash-out additive. This product is recommended by the Federal Waterway Engineering and Research Institute (BAW) in Germany for riprap placed underwater. The pH changes caused by grout were observed in a study done by the Virginia DOT, where pH levels spiked to 3.9 at some locations around the pier. This value is higher than the allowed maximum of 3.0. The pH level returned to baseline conditions, however, after about 30 min in most locations around the pier.

Design Forces for Grouted Joints Partially grouted riprap needs to be designed for dynamic forces arising from scouring forces due to high velocity flow, uplift forces in turbulent flow, and the impact of floating ice and debris. The grouted joint should be strong enough to resist the kinetic energy due to sharp longitudinal slopes. The slopes should be flatter than $1V{:}1.5H$. Strength of grout should be stronger at the riprap boundaries and at the toe of the blanket for stability.

Soil Bearing Capacity The foundation must have soil bearing capacity sufficient to support either the dry weight of the grouted riprap alone or the submerged weight of the grouted riprap plus the weight of water beneath the maximum water surface elevation, whichever is greater.

3.13 Conclusions

This chapter surveyed the progress made in the newer technology of providing shielding and protection to recurring or unexpected flood events. Foundations which are not located on rock are vulnerable to settlement from soil erosion. Planning of new bridges requires relatively large opening sizes and deep foundations.

Considerable progress has been made in the United States by FHWA and NCHRP in developing and implementing hydraulic engineering circulars such as HEC-18 and HEC-23 for riverine hydraulics and HEC-25 for coastal hydraulics. For scour-critical bridges located in the Northeast which were studied by the author, it was seen that the widely used HEC-18 master countermeasures matrix needs modifications for the feasibility of selected countermeasures according to their physical conditions.

The use of expensive solutions such as river diversion or adding auxiliary bridges can be avoided by using armoring. Peak hydrology and the probability of recurrence of floods and rainfall intensity needs to be forecast more accurately with the help of the weather bureau.

Many states faced with scour critical rivers have also contributed to progress in minimizing bridge failures by trying out experimental techniques.

Proprietary Products

The survey of the types of countermeasures being used by various highway agencies has shown that HEC-23 countermeasures can be successfully used. The latest revision to HEC-23 illustrates the selection and design of a vast variety of countermeasures to counteract the rapidly increasing erosion of soil at bridges. However, the following items need to be addressed:

Consider the relative merits of proprietary products (interlocking block, cable-tied block, articulating block, and mattresses) in relation to the use of riprap for channel protection and as local scour countermeasures.

Suggest techniques to prevent scour at the "joint" between articulating mattresses and a bridge pier when these products are used as a pier scour countermeasure.

Case Studies

Case studies on selected bridges in the northeast United States involved applying hydrology, hydraulics, scour analysis, and countermeasure selection and design. The bridge is scour-critical when the total computed scour depth is below the base of spread footing. In other cases, the nominal depth of countermeasures may be used as shielding. Many states such as California, Minnesota, Florida, Pennsylvania and New Jersey provide standard details for riprap on soil, rock, and so forth. Use of geotextile and granular filters helps prevent the countermeasures from being eroded.

Several states have introduced improved administrative procedures for flood watch and post flood field inspections by training field inspectors in recording and coding guidelines and in emergency fund allocations.

European Practice

The European practice of using types of countermeasures has shown the successful use of alternative countermeasures such as partially grouted riprap. In the United States, vendors are now designing the sizes of countermeasures for given velocities and types of soil. Cable-tied blocks and sheet piles have proved to be more stable.

3.14 Bibliography

Brice, J. C. and Blodgett, J.C., *Countermeasures for Hydraulic Problems at Bridges*, Vols. 1 and 2, FHWA Office of Research and Development, Report No. FHWA-RD-78-162 and 163, Washington, D.C., 1978.

Lagasse, P. F., Clopper, P. E. and Arneson, L. A. "Partially Grouted Riprap as a Pier Scour Countermeasure," ASCE World Environmental and Water Resources Congress, Lakewood, CO, 2008.

Brown, S. A., and Clyde, E. S. "Design of Riprap Revetment," Report No. FHWA-IP-89-016, Hydraulic Engineering Circular-11, 1983.

Clopper, P. E., and Thornton, C. I. "Water Quality Aspects of Partially Grouted Riprap," ASCE World Environmental and Water Resources Congress, Lakewood, CO 2008.

Commonwealth of Pennsylvania, Department of Transportation, *Design Manual*, Part 4 Vol. 1 of 2, STRUCTURES, Procedures—Design—Plans Presentation, Pub No. 15m, April 2000 (Dual Unit).

Faraday, R. V., and Charlton, F. G. "Hydraulics Factors in Bridge Design," Hydraulic Research Station, Wallingford, England, 1983.

Fotherby, L. M., "Footings, Mats, Grout Bags and Tetrapods: Protection Methods against Local Scour at Bridge Piers," M.S. thesis, Colorado State University, 1992.

Fotherby, L. M., and J. S. Jones, "The Influence of Exposed Footings on Pier Scour Depths," Proceedings of National Conference on Hydraulic Engineering, Vol. 1, San Francisco, 1993.

Fotherby L. M. and Ruff, J. F. "Bridge scour Protection System Using Toskanes—Phase 1, PennDOT Report 91-02, 1995.

Heibaum, M., and Trentmann, J., Federal Waterways Engineering and Research Institute, Karlsruhe, Germany, International Conference on Scour and Erosion (ICSE-5) 2010. Conference on Scour and Erosion, held in San Francisco, California, November 2010.

National Cooperative Highway Research Program, NCHRP Report 593, "Countermeasures to Protect Bridge Piers from Scour," Project 24-7, Transportation Research Board, Washington, D.C., July, 1999.

Lagasse, P.F., "Countermeasures to protect piers from scour, ISBN 0-309-09909-9, TRB, Washington, D.C., 2007.

USDA. *Alabama Handbook for Erosion Control, Sediment Control, and Storm water Management on Construction Sites and Urban Areas*, prepared by the Alabama Soil and Water Conservation Committee, Montgomery, AL., 1993.

New Techniques

Agrawal, A., Khan, M.A. and Zhihua, Y. *Handbook of Scour Countermeasures*, New Jersey Department of Transportation, Trenton, 2007.

"Evaluation of Reinforcing Bars in Old Reinforced Concrete Structures," Concrete Reinforcing Steel Institute, Engineering Data Report No. 48, Schaumburg, Illinois, 2001.

Federal Highway Administration, HEC-11, "Design of Riprap Revetment," FHWA Publication No. IP-89-016, 1989.

FHWA, "Highways in the River Environment," FHWA-HI-90-016, February 1990.

Federal Highway Administration, HEC-20, Stream Stability at Highway Structures 4th ed., Publication No. FHWA-HIF-12-004, Washington D.C., April 2012.

Federal Highway Administration, HEC-23, Vol.1, "Bridge Scour and Stream Instability Countermeasures: Experience, Selection, and Design Guidance" Publication No. FHWA-NHI-09-111, 3d ed., Washington, D.C., September 2009.

Federal Highway Administration, "Counter Measures for Hydraulic Problems at Bridges, Analysis and Assessment" Brice, J.C., Blodgett, J.C., Report No. FHWA-RD-78-162, 1978.

Fielding, M., "High-Tech Instrumentation Helps Optimize Project Schedules and Manage Risks," *Journal of Bridge Engineering*, American Society of Civil Engineers, March/April 2007.

Horne, W. A., "Evaluating Unknown Bridge Foundations, Is It Worth It?" Proceedings of the 1998 International Water Resources Engineering Conference, ASCE, Memphis, TN, Part 1, 1998.

Khan, M. A., Baig, S., and Zarriello, J. "Factors Affecting Stream Stability at Scour Critical Bridges in New Jersey," *First International Conference on Bridge Scour*, Texas A&M University, College Station, 2002.

NJDOT, *Soil Erosion & Sediment Control Manual*, Trenton, NJ, April, 2009.

Pagan-Oritz, J.E., "Stability of rock riprap protection at the toe abutment located at the flood plain," FHWA Research Report No. FHWA-RD-91-057, FHWA, Washington, DC, 1991.

Sung H. Park. Bridge Inspection and Structural Analysis, Trenton, NJ 1999.

New Materials

American Iron and Steel Institute, Federal Highway Administration, Navy, "Guide Specifications for Highway Bridge Fabrication with HPS 70W Steel," AISI, Washington, D.C., 2003.

Berke, N., "What's New in Corrosion Inhibitors," *Structure*, July/August, 2003.

Capers, H., "High Performers! New Jersey's First High Performance Steel Bridges," *Second New York City Bridge Conference*, Bridge Engineering Association, New York, 2003.

"Carbon Fiber Prepreg Sheets, Reinforcement and Repair of Concrete Structures," Mitsubishi Chemical Corporation, Tokyo, Japan, 2009.

Construction Industry Research Information Association, (CIRIA) Report No. 61, "General Guidelines for Riprap Design," Thompson, D.M. and Shuttler, R.M., Hydraulic Design Station, Wallingford, UK, December 1976.

DYWIDAG Threadbar Reinforcing Systems, DYWIDAG-Systems International, Bolingbrook, IL, 2008.

Fiber-Reinforced Polymer—Priority, Market-Ready Technologies and Innovations, Federal Highway Administration, Washington, D.C., 2004.

High Performance Concrete, Federal Highway Administration, Resource Center, Washington, D.C., February 2006.

Lauchlan, C. S. and Melville, B. W., 1999, Pier Scour Countermeasures, Report No. 540, Department of Civil and Resource Engineering, The University of Auckland, 1999.
McCorquodale, J.A., "Guide for the Design and Placement of Cable Concrete Mats," Report Prepared for the Manufacturers of Cable Concrete, 1991.
McCorquodale, J.A., 1993, "Cable-Tied Concrete Block Erosion Protection," Hydraulic Engineering '93, San Francisco, CA, Proceedings (1993).
"NDT Tried on George Washington Bridge," *Journal of Bridge Engineering*, American Society of Civil Engineers, September 2000.
Neill, C.R., *Guide to bridge hydraulics*, Project Committee on Bridge Hydraulics, Roads and Transportation Association of Canada, University of Toronto Press, 1973.
Neill, C.R., 1976. Scour Holes in a Wandering Gravel River, Proceedings of Symposium on Inland Waters, ASCE, *Journal of the Hydraulics Division*, 1976.
University of Mississippi, NCHRP Project No. 24-7, "Countermeasures to Protect Bridge Piers from Scour," St. Anthony Falls Laboratory, 1997.

Filters

Center for Civil Engineering Research & Codes (CUR) / Rijks.water.staat (RWS), "Manual on the Use of Rock in Hydraulic Engineering," AA Balkema, NL Report 169, 1995.
Holtz, D. H., Christopher, B. R. and Berg, R. R. "Geosynthetic Design and Construction Guidelines," National Highway Institute, Publication No. FHWA HI-95-038, FHWA, Washington, D.C., May 1995.

CHAPTER 4

Innovative Methods of Flood Disaster Management

4.1 Overview of Scour Repairs

In light of recent progress made in repair methods for steel and concrete materials in the United States and abroad, new repair techniques have been on the rise. In recent years, construction technology has revolutionized modern repairs significantly, and innovative methods will apply to river bridges employing preventive techniques and post-flood innovative repairs.

In this chapter, following market-ready technologies, vendors' products and innovative techniques are addressed:

- The use of new construction materials, remote structural health monitoring, and recent developments in repairs and rehabilitation methods are addressed.
- The aspects discussed are precautionary methods to prevent failures, the use of efficient methods for design and construction of countermeasures, streamlining rehabilitation design procedures, utilizing advancements in concrete technology and ABC methods, integral abutments, developing design codes for river bridges, continuing education in use of modern technology, and training of engineers.
- Use of modern construction materials including advancements in concrete technology, fiber-reinforced polymer (FRP) concrete, HPC, SCC, UHPC, shrinkage compensating concrete, exodermic bridge decks, reactive powder concrete bridge girders, HPS 70W and 100W, anticorrosion paints, use of fly ash, blast furnace slag and silica fume, recyclable materials, and seismic isolation bearings for seismic zones.
- Countermeasure designs such as concrete armor units and grout bags.
- Use of drones for inspection of bridges to help in efficient maintenance. Ground penetration radar techniques for evaluating unknown foundations and deficient decks under the overlay. Modern design and retrofit methods to help in accelerated construction.
- Use of VMS sign structures for warning of floods.

These measures will cut down life cycle costs and the duration of maintenance. Some of the newer methods have not been fully tested, and precautions may be required in their applications. The latest ideas, ingenuity, and contributions from individual researchers and publications are listed in the Bibliography at the end of this chapter. Recent developments in new materials and prefabricated concrete elements for rapid construction are also discussed.

The motto recommended by the Federal Highway Administration (FHWA) is "Highways for LIFE," where L is for long-lasting, I is for innovative, F is for fast construction, and E is for effective and safe.

Recent innovations may be summarized in these categories:

1. Prevention of failure by minimizing identified deficiencies
2. Use of advanced methods of analysis and design
3. Asset management and remote health monitoring (RHM) methods
4. Use of modern construction equipment and techniques
5. Use of new construction materials and systems
6. Closer connection between design documents and construction specifications (which consist of description of work, selected construction materials, construction methods and method of measurements and payment)
7. Continued research efforts in resolving technical issues

4.2 Adopting Precautions to Prevent Failures

A study on bridge failures carried out by the author (reported in Chapter 5) concluded that most failures occur during construction or erection. An accelerated bridge construction (ABC) system must be put in place to avoid future failures.

Use of Drones to Monitor Deterioration of Bridges

Monitoring and inspection of bridges located on rivers are particularly difficult when manual techniques are used. Unmanned aerial vehicles (UAVs) offer a less expensive and more efficient way to study deficiencies such as deck cracks, potholes, and insufficient crown, preventing drainage. Michigan Technological University researchers have successfully obtained high-resolution photographs. Multirotor helicopters equipped with digital cameras can deliver images up to 1 cm in resolution using this technique at low cost. Three-dimensional models of deck surfaces showing cracks in concrete can be obtained.

The study is being funded by the Research and Innovation Technology Administration of the U.S. Department of Transportation (DOT). It will address the deterioration of infrastructure and recurring repair issues at a much reduced cost compared to manual methods and will ensure the safety of bridges.

Development of Diverse Repair Technologies

The author has successfully carried out many rehabilitation projects for concrete and steel highway bridges. General procedures for the repair of old bridges are site-specific and include the following tasks:

1. Field-verify applicable items laid down in inspection reports.
2. Conduct an in-depth evaluation inspection.

3. Investigate any mild defects for possible changes in physical condition.
4. Prepare a standard checklist of deficiencies.
5. Prepare cost–benefit analysis for rehabilitation versus replacement.
6. Perform alternative analysis for selecting the most appropriate solution.
7. Perform emergency repairs.

For each type of repair technique, the following are needed:

- The basic concept behind the technique
- Successful application of the concept to specific engineering problems
- Procedure for field data acquisition and processing

Other recent innovative developments include methods to accelerate project completion, the availability of accurate information, increased productivity, team organization and collaboration with other disciplines, and automated generation of reports in multiple formats for communications.

The following detailed recommendations are based on current practices and the literature review of a large number of publications, including the FHWA, American Association of State Highway and Transportation Officials (AASHTO), the American Society for Testing and Materials (ASTM), and the National Cooperative Highway Research Program (NCHRP). The latest ideas, ingenuity, and contributions from individual researchers, independent publications, and those listed in Bibliography at the end of the chapter are acknowledged.

4.3 Introducing a Fresh Approach in Countermeasure Design and Construction

Comparative Study of Old and New Methods

The following physical causes leading to wear and tear are omnipresent in bridge components:

- Deterioration
- Applied direct stress
- Thermal action and fatigue
- Creep and shrinkage due to changes in the atomic bonds between constituents

Table 4.1*a* shows alternate uses of new methods and construction materials (for rehabilitation as well as replacement). Table 4.1*b* shows basic bridge components and their most common applications.

Construction Methods The most popular types are conventional methods using formwork, modular construction, and the ABC method. The production systems are the conventional consultant-contractor team or design-build system. In recent years the design-build system has become more popular since it leads to speedy construction.

Type of Construction Material	Economical Span Length
Reinforced concrete deck slab	Small < 12 ft
Exodermic deck slab	Small < 12 ft
Fiber-reinforced polymer (FRP) concrete	Small < 12 ft
Precast and prestressed deck panels placed in longitudinal direction	Small < 20 ft
Reinforced concrete beams	Small < 40 ft
Prestressed concrete beams (post-tensioned)	Medium < 120 ft
Grade 36 steel beams	Small < 60 ft
Grade 50 or 50W steel beams	Medium < 100 ft
Hybrid grades 50W and 70W steel	Medium or long span < 250 ft
Grade 70W steel beams	Medium or long span < 250 ft
Grade 100W steel beams	Long span > 250 ft

Table 4.1*a* Alternate Uses of New Methods and Construction Materials

Fixed and Expansion Bearing Types Supporting Single or Multiple Spans				
Bearing Types	Rocker and Roller	Elastomeric Pads	Multirotational	Seismic Isolation
Applications	Maintenance issues	Economical for small spans and easy to replace	Suitable for medium and long spans	Suitable for bridges in high seismic zones
Pier types	Wall types	Frame bents	Pile bents	Tee shapes
Applications	Small and medium heights	Economic design	For river bridges deep foundations eliminate construction of multiple joints	Aesthetic with single column and small cantilevers
Abutments	Full height and cantilever	Stub and cantilever	Spill-through	Integral and semi-integral
Applications	Wall height > 16 ft 6 in	Wall height < 16 ft 6 in	Frame type	Deep foundations with piles required; suitable for bridges over rivers
Wing walls	Splayed	Uniform height	90° return	Mechanically stabilized earthwork (MSE), walls
Applications	Suitable for retaining sloping soil height	Suitable for uniform soil height	Approach slab has a stable foundation	Precast concrete retaining walls tied together with cables
Sign structures	Bridge parapet mounted	Single-column deck-mounted	Multiple columns	Variable message sign (VMS) structures
Applications	Suitable for lower-level traffic	Displaying small sign area	Displaying long sign areas	Electronic messages displayed from long distance

Table 4.1*b* Types of Bridge Components

Rehabilitation Design Procedures

The following factors affect the selection and design of rehabilitation methods and influence the need for innovation:

- Site layout
- Bridge usage
- Deterrence and access control
- Specific bridge features (fixed, moving, suspension)
- Resiliency of bridge components
- Redundancy of the bridge system
- Other bridge-specific items

In addition to the above physical factors affecting rehabilitation of river bridges, the cost of rehabilitation will be based on the importance of the existing costs of assets based on assessment. Replacement may be a better alternative than repairing an existing bridge.

1. *Criticality factor for assets*: The vulnerability factor for assets can be expressed as a characterization of assets into regions of different levels of criticality and vulnerability:
   - Preassessment: resource identification
   - Assessment: vulnerability identification
   - Post-assessment: decision making

   The reconstruction will involve the following activities and specifications:
2. *Plan preparation and presentation*: In plan preparation, actual field measurements should be considered more reliable than the drawings.
3. *Shop drawings*: The shop drawings should be considered more constructible and reliable than bridge plans.
4. *Contract plans*: The contract plans shall be sufficiently detailed to provide an overall view of the bridge, indicating
   - The existing and proposed geometric dimensions
   - Limitations and restrictions
   - Extent and type of work to be performed
   - Construction stages show the use of high-capacity cranes in a marine environment for ABC on the high-traffic Tappan Zee Bridge over the Hudson River, New York.
   - Material information
   - All related information needed to rehabilitate the bridge
5. *Pay Items*: All work shall be accounted for by specific pay items. Pay limits, pay quantities, and pay items should be adequately defined to eliminate ambiguity or confusion. Where applicable, reasons for critical limitations and restrictions should be explained to assist the contractor and the field inspector in adjusting to the field conditions.

6. *Scope and extent of Work*: For rehabilitation projects, the *type, size, and location* (TS&L) plans shall have the normal TS&L plan details, plus a complete scope and extent of work.
7. *Anticipated bridge rating*: The anticipated bridge rating after the proposed work is incorporated shall be shown.

Examples of Innovative Concepts

1. Advanced infrastructure design will utilize ground-penetrating radar (GPR) technology for evaluation of the bridge deck under the overlay. This high-speed nondestructive evaluation doesn't require maintenance and protection of traffic. The cost savings per bridge from using this method is approximately $0.1 million; time saved is 2 months.
2. The effective use of existing shoulders during the stage construction will prove to be a critical factor in the success of the project, as measured by the savings in time of 3 months (schedule) and in cost (design and construction).
3. Advanced relocation of overhead utility lines to their permanent position is accomplished without the use of temporary pole lines. Cost savings is approximately $0.3 million; time saved is 3 months.
4. Early coordination with regulatory agencies such as the EPA and the U.S. Army Corps of Engineers should be accomplished to identify potential project permitting requirements; the design schedule time saving is 3 months.
5. Early identification of issues that may be seasonally sensitive, such as the presence of endangered and threatened species, will help to reduce project delays; the design schedule time saving is 1 month.
6. Road closure versus staging is an issue to be addressed. As an alternative to staging, investigate keeping the bridge open weekdays to accommodate the heavy weekday traffic, and partially close the bridge over multiple weekends.
7. Utilization of precast deck and prefabricated superstructure construction will also accelerate the schedule; the time savings is 2 months.
8. To achieve a reduction in construction staging, investigate the possibility of eliminating the substages required for constructing temporary sidewalks carried by the structure. This will be possible if there is access from below the superstructure to construct the structure-carried temporary sidewalks. As an alternative, consider using a temporary bridge such as an Acrow or Mabey type; time savings: 1 month; cost savings: $50,000.
9. Use of precast concrete modular deck sections in lieu of poured-in-place deck construction, and Inverset-type prefabricated superstructure sections for bridge structures may provide considerable cost savings and a reduction in the construction time of approximately two months.
10. Investigate increasing the strength and service life of the bridge by increasing the beam section properties; this can be done by making the new deck composite and making the existing simply supported beams continuous. The potential increase in service life is 25 years.

In light of the advances in numerical methods and computer techniques, a more accurate analysis and design approach can be used for the planning stage:

- AASHTO load and resistance factor rating (LRFR) procedures are used in place of the load factor design (LFD) method.
- State codes of practice use ultimate loads and the probability approach.
- Advanced methods of analysis and nonlinear finite element methods (FEMs) are used.
- New software applications are employed.
- Design methods for ABC are chosen.
- A plan review checklist and QC/QA document, e.g. that used by the New Jersey DOT and Wisconsin DOT are used.
- A context-sensitive design (CSD) is employed.

Application of Innovative Repair and Retrofit Methods

Commercial products and services focus on mitigation technology for retrofit, restoration, and rehabilitation. The innovative techniques listed below are in the early stages of development. They cover the use of sample project-specific guidelines for final plans and protective coating systems.

- *Smart Japanese paint*: This paint sends out electric signals that are picked up by electrodes (placed on either side of the paint's resin layer) if the structure begins to vibrate. The electrical signal becomes greater as the vibration increases. This paint can evaluate fatigue in old bridges more accurately than strain gauges.
- *Smart paint with special dye*: The National Science Foundation's ATLSS Engineering Research Center has developed "smart paint" with a special dye that outlines a fatigue crack in a bridge as it propagates.
- *Magnified remote cameras*: The computer-based imaging system can provide spatial measurements and surface analysis. It can detect a surface flaw and determine its size, shape, location, and defect details.
- *New vendor products*: These include seismic isolation bearings, Hilti chemical bolts, and so on.
- *Improved drainage systems*: Examples are Ohio Drainage Guidelines (Ohio DOT hydraulic engineering department guide for the hydraulic design of highway drainage facilities) and Guidelines of Tennessee Drainage Manual (TDOT Design Division Drainage Manual, Chapter 5, May, 2011).
- *Innovative techniques*: These include the availability of new repair materials, FRP composites to repair overhead sign structures, Sika CarboDur for general repairs, and SikaWrap for shear strengthening. Sika Corporation's concrete admixture, Sikament 686, can be used in cast-in-place or precast applications as a normal water reducer (ASTM C494 Type A) or as a high-range water reducer (ASTM C494 Type F). Increased workability is provided with no delay in the normal set time.

- Certain aluminum (Al) water-based concrete admixtures, based on amino-carboxylate technology, tolerate cold and hot temperatures. The admixture forms a protective layer on embedded reinforcement that prevents corrosion caused by carbonation, chlorides, and atmospheric attacks. It provides corrosion protection for steel reinforcement, carbon steel, galvanized steel, and other metals embedded in concrete structures.
- Corrosion prevention and maintenance painting of steel use protective coating systems.
- *Corrosion protection for concrete*: Bayer MaterialScience, one of the manufacturers of corrosion protection, has designed thick-film coatings for industrial waterproofing applications, such as concrete and metal bridge decks. Its physical and chemical properties include resistance to wear and abrasion, excellent tensile strength, corrosion resistance for metal substrates, good weatherability, resistance to microbes, and good chemical resistance. Baytec SPR spray systems can be pigmented to any color.
- Fluoroethylene vinyl ether (FEVE) resins were developed in the early 1980s in Japan. Fluoropolymers offer a number of desirable properties, among them excellent stability against uv light and the elements, corrosion resistance, and weather and chemical resistance.
- Fluoropolymer topcoats now are required for use on all bridges in Japan, both for new construction and for repainting. On one of the longest Japanese bridges (a single span of the bridge is 6527 ft), the coating system was a four-coat system consisting of a number of coatings that were both field- and shop-applied.
- *Deicing overlays*: The SafeLane surface overlay is expected to provide a robust surface for more than 15 years of service, plus the much-needed pavement seal to limit chemical and moisture penetration into the concrete bridge deck.

  AASHTO Task Force SafeLane surface overlay acts as a rigid sponge, storing the chemicals and automatically releasing them as conditions develop for the formation of ice or snow. This results in safer roads with better mobility and less maintenance because the overlay helps prevent frost or ice from forming on road or bridge surfaces. The final profile is about 3⁄8 in thick. The recommended method is outlined in AASHTO Task Force 34.
- *Developments in data recording*: Examples are data loggers, data acquisition systems, and measurement and control products. Data-recording computer hardware devices include pen-based tablets, notebook computers, and PCs. For field use they are made more rugged than office computers. Techniques to extract data from laser-scan point clouds into three-dimensional MicroStation drawings have been quite successful.
- Laser technology is used to compute axial forces in cable-stayed bridges.
- *Asset management using robotic devices*: Every transportation agency is faced with management of its assets, which include hundreds of bridges. No two bridges are alike, and the vast varieties range from historic masonry arch bridges to the most modern segmental and cable-stayed suspension bridges.
- Routine inspection and nondestructive testing (NDT) of bridges can be performed by potential robotic devices using inertial navigation, odometer, and

laser techniques. A "manipulator" device will fix the sensors on critical bridge locations.

- *Isolation bearings and retrofit of bearings*: Seismic performance of a movable bridge requires some unique considerations, including the behavior of machinery elements and their tolerance for misalignment during earthquakes, and the issue of the requirements for seismic design when the bridge is in the open position. Surface conditions at the site are highly variable, making site-specific response analysis important. Seismic retrofit design is based on the seismic deficiencies established through the performance evaluation.
- The isolation system consists of lead-core rubber seismic isolation bearings combined with semiactive viscous fluid dampers positioned between the bridge deck and end abutments. Semiactive dampers can offer an effective approach to response modification of seismically isolated highway bridge structures.
- *Fiberglass drains for bridges*: New fiberglass drains for bridges and elevated highways are light, strong, and easy to install. Fiberglass requires fewer supports than other nonmetallic pipe, it resists corrosion, and it does not require painting. This system allows for a high degree of design flexibility.

4.4 Utilizing Advancements in Construction Technology

Modern Construction Equipment

Modern construction equipment includes

1. Megamovers
2. Record length casting bed in Oregon for girders
3. Bridge roll-in construction method to avoid lane closures
4. Innovative construction equipment such as high capacity cranes.

Design-Build Contractual Procedure

Design-build is a method of delivery in which one entity, the design-build team, works under a single contract to provide design and construction services from the initial concept through completion.

It is based on the integrated "master builder" system first used in 1795 B.C. In 1988, the FHWA set up an experimental project which led to the Transportation Equity Act (TEA) enacted for the twenty-first century, dubbed TEA-21, the transportation authorization bill on projects over $50 million as well as Public Private Partnership P12. It led to 6 percent lower costs, 12 percent faster construction time, 33 percent faster project completion, and higher quality in all measured quantities. Today nearly all states authorize design-build in some fashion. For extreme events such as Hurricane Katrina, traditional delivery methods were slow.

The I-35 bridge collapse was catastrophic as well. The new bridge was completed in 14 months, while design alone took 12 months. Project delays can have disastrous effects on public safety and comfort during construction. In design-build, more innovative design and construction techniques and creative solutions are used. Omissions and errors are minimized, and field-generated change orders are eliminated.

Advantages of Design-Build Construction Management

According to an RSMeans study, the method now has over 40 percent of market share. The most important shift from design-bid-build is to work as one team. It is aimed at (1) improving teamwork, (2) reducing bureaucratic procedures, and (3) resolving constructability issues quickly. The traditional approach of client-engineer-contractor is now gradually being replaced by the client opting for design-build contracts and a "turnkey" completion. The prequalified contractor hires the consultant or uses her or his own designers to prepare the contract documents. This method seems to eliminate bureaucracy and is tailored to give a quick turnout based on the contractor's resources.

Rapid Bridge Replacement Using Geosynthetic Reinforced Soil (GRS) Integrated Bridge System (GRS-IBS)

GRS-IBS utilizes closely spaced layers of geosynthetic reinforcement and compacted granular fill material, which provide direct bearing support for beams. About 150 GRS-IBS bridges have been built in 35 states. The Delaware DOT replaced a two-lane bridge using an innovative technology with GRS abutments and prefabricated superstructure elements. A custom-designed instrumentation system was developed to monitor long-term performance.

External and internal stability analyses were carried out using the FHWA GRS-IBS interim implementation guide (2011). The University of Delaware worked with the Delaware DOT to design an innovative system of sensors to monitor performance.

Soil Reinforcement

GRS abutment is constructed by compacting high-quality granular soil and closely spaced geosynthetic reinforcement in a series of thin alternating layers. Concrete masonry unit blocks are used as facing element types. Stress arching between soil reinforcement layers plays a significant role. Unlike MSE structures, pinned connections are not required.

Protection of Foundation

The foundation base of the reinforced soil zone is supported using a reinforced soil foundation, which is created by encapsulating a series of compacted soil layers with geosynthetic material. Over a waterway, riprap scour protection is placed in front of wall-facing blocks at the base of the abutment to prevent soil erosion. A site exploration involving two boreholes, soil sampling, and soil laboratory testing was performed to determine the backfill and foundation soil conditions.

4.5 Accelerated Bridge Construction (ABC) over River and Replacement Methods

Understanding Rapid Construction and Associated Needs

ABC through precasting of bridge components helps to reduce greenhouse gas emissions caused by traffic delays and long-term operation of construction equipment.

ABC Helps in Construction of River Bridges

A textbook published by the author on ABC addresses many innovative concepts which are summarized here. Each concept is a subject in itself. Modern prefabricated construction materials and methods are vastly different from traditional methods and require innovative ideas for making the system safe and efficient. The ABC method can be promoted by understanding and analysis.

The owner's requirements are clear: reduction in schedule, wider decks, reduced seismic effects, increased bridge ratings, longer service life, cost savings, and lower maintenance. Many of these objectives can be achieved by precasting in factories and modular construction. Bridges located on rivers have limited duration available for construction due to the rainy season. Hence promoting rapid construction is desirable.

Promoting Rapid Construction and Accelerated Planning and Design

1. *Alternative design-build construction procedure*: This approach has led to faster turnout. For large projects a design-build-finance-operate-maintain process is most likely to be a complete situation. The simpler design-build method is more common, placing the builder and designer on one team.
2. *Modern construction equipment*: The success of the ABC method is due to powerful equipment. Different erection equipment is required for girders, box beams, trusses, arches, cable-stayed, and suspension cable bridges. Timely availability, a leasing facility, and an experienced erection team will be necessary.
3. *Importance of planning*: An evaluation matrix for construction cost, life cycle cost, environmental and social impact, constructability, future maintenance, inspection, and aesthetics needs to be prepared at the planning stage. Planning must address differences in the ABC method of small, medium, and long spans.
4. *Use of rapid erection method*: The erection contractor shall utilize robotics, cranes such as tower cranes (for maximum lightweight pick of 20 tons and heights greater than 400 ft), lattice boom crawler cranes, mobile lattice boom cranes, mobile hydraulic cranes, and lattice ringer cranes for varying heavyweight pickups and accessories. Also, specialized technology for a superstructure roll-in, roll-out method using self-propelled modular transport (SPMT) is provided in an FHWA publication on Bridge Construction (Manual on Use of Self-Propelled Modular Transporters to Remove and Replace Bridges, June 2007).

Factors Influencing the ABC Method

Efficient structural planning is required to minimize constraints for accelerated construction goals such as erection during extreme weather. The following factors may be considered:

1. To improve the aesthetics, provide an exterior arched beam to hide the fascia girder.
2. Install interactive touch screens featuring bridge information for motorists.
3. Increase rider comfort by using a durable deck overlay protective system to prevent deck cracking, use post-tensioning to prevent cracks in concrete and noncorrosive reinforcement, and improve the crashworthiness of barriers and parapets.

4. Improve bearing performance through the use of multirotational and isolation bearings in flood and seismic zones. Water elevation may reach to the free board elevation.
5. Improve bridge lighting and drainage methods by installing solar roof panels at approaches for bridge lighting and signage.
6. Use grouting methods for improving foundation soil, including liquefaction mitigation to protect foundations in saltwater against corrosion; resistance to earthquakes, liquefaction, and erosion of soil under footings; prevention of fender damage from vessel collision.
7. Construct bioretention ponds that collect and filter runoff from the bridge deck.
8. Introduce ground improvement techniques and modify the soil.

Accelerated-Cure Cast-in-Place Concrete Accelerated-cure cast-in-place concrete uses low-slump accelerated concrete in conjunction with the maturity method to obtain design compressive strengths in 12 to 15 h. The maturity method uses maturity loggers in the deck to analyze the time and temperature profile of in-place concrete to calculate the in-place strength of concrete in real time. The advantages of this approach include lower traffic impacts due to reduced cure times compared to standard HPC mix designs. The use of accelerated-cure cast-in-place concrete allows for deck replacements to be performed over weekend work cycles or 3-day cycles. The strengths and design life of this system are similar to those of conventional cast-in-place concrete.

The disadvantage of this system is that high quality control, particularly in regard to the water content and concrete slump, is essential for concrete performance. Although the use of maturity loggers has been successful in New York State, it is not familiar to most contractors; it has a short history, and for cast-in-place concrete, it is more susceptible to cracking, especially during staged construction.

Enhancing the Environment Using Innovative Methods

The U.S. Congress passed the National Environmental Policy Act in 1969. Its objectives were as follows:

1. Formulate a national policy to encourage productive and enjoyable harmony between people and the environment.
2. Prevent damage to the environment and thereby maintain the health and welfare of people.
3. Enrich the understanding of the ecological systems.
4. Establish a council on environmental quality. This resulted in preserving important historic, cultural, and natural aspects of our national heritage. Also the quality of renewable resources was enhanced, and recycling of resources was made possible. Some of the measures included the following:
   - Use of precast concrete elements with fewer environmental constraints
   - Limiting construction activities to certain months of the year (May to August, for example) for the least environmental impact
   - No disturbance of the wetlands by adopting an innovative top-down method of construction

- Preserving the natural habitat around the bridge such as providing small animal crossings
- Minimizing damage to flora and fauna
- Minimizing side slope erosion of stream banks

4.6 Developing Design Codes for River Bridges for ABC and for Flood Conditions

Besides the bridge designer, it is a requirement that the fabricator and erector shall follow AASHTO construction specifications and "fit-up shall be assumed to be performed under the no-load condition" for which evaluation of erection stress is required. However, the construction equipment load is higher for ABC than for conventional construction, and load combinations need to be identified as given below.

Construction load combinations not covered by AASHTO LRFD bridge design code are summarized: The standard AASHTO LRFD load combination for strength I relates to normal use with live load only; strength III relates to live load and high wind forces and strength V relates to live load with normal wind forces. For LRFD load combinations for live load and wind loads, refer to AASHTO LRFD Specifications.

- For ABC construction loads (strength I condition) and no flood conditions, use all dead loads of bridge components and nonstructural attachments (DC), dead load of utilities and wearing surface (DW), construction equipment load (CEL) such as screed, and construction live load (CLL). Thus

$$\gamma p = \text{LRFD load factor} = 0.9 \text{ to } 1.25$$

$$WS = \text{wind load on structure}$$

  Using AASHTO notations, the construction live load combination without water is

$$\gamma p(DC + DW) + 1.5(CEL + CLL)$$

- For strength III condition, use wind on the superstructure including on forming. The construction wind load combination is

$$\gamma p(DC + DW) + 1.5(CEL) + 1.4(WS)$$

- (For strength V condition) use the wind construction equipment load (WCEL), and the load combination is

$$\gamma p(DC + DW) + 1.5(CEL) + 1.35(CLL) + 0.4(WS) + 1.0(WCEL)$$

  and γp = 0.9 to 1.25, which is less than the 1.5 factor for dead loads used by AASHTO.

Load Combination with Floodwater Load Acting on Girders

In addition to construction loads described above, the impact of static or dynamic flood water forces has not been outlined in AASHTO. The fascia girder takes the brunt of the floodwater in spate. It reduces the momentum of flood flow, and the interior girders have a reduced impact. The long span of the fascia girder is exposed to full width of the flowing body of water. The bearings at the ends of girder may also provide lateral supports.

Long-span bridge girders are significantly deep, depending on the span length. As the girder depth increases, a vertical cantilever action is generated by the composite action of deck slab, provided by shear connectors. The high-magnitude water forces acting on the vertical surface of fascia girder cause local torsion of the girder web. This action is more prominent in webs of steel and prestressed I-girders than in concrete box girders.

An examination of failures of bridges in Mississippi following Hurricane Katrina and in northern New Jersey following Hurricane Floyd shows greater damages to thinner web bridges, from an increase in the shear stresses in the web. Hence adequate shear and bending reinforcement need to be provided. The author was responsible for the replacement design of Peckman's River Bridge on Route 46 in New Jersey.

A composite three-dimensional structural finite-element model of bridge (with flood forces acting up to the level of the deck) will give a better picture of peak stress distribution. A simpler model is presented here. Assuming the water elevation is rising to the top of girder, if depth of girder = d, then

$$\text{Density of water } w = 62.4 \text{ lb/ft}^3$$

Lateral force on girder $P = wd^2/2 = 31.2d^2$ lb/ft acting as a uniformly distributed load (on girder span L). Lateral force P is acting as a line load at de[th2 $d/3$ height from the top of the girder. Force P causes a peak torsional moment of

$$P \times 2d/3 = 20.8d^3 \text{ lb-ft}$$

For girder depth = 6 ft,

$$\text{Moment} = (20.8)(6^3) = 4492.8 \text{ lb-ft} = 53.9 \text{ kip-in}$$

plus a lateral shear force = 1.12 kips.

The high shear stress on the web from the above force and movement has caused girder failures in the past, since the stress from floods get added to those stresses due to permanent vertical dead load shear force and dead load torsion due to the asymmetry of the fascia girder.

Currently, fascia girders are designed not for water loads but for lateral wind loads only, which create less stresses than the stresses from water forces shown above (and generated by high velocities exceeding, say, 10 ft/s).

Specification developments will ensure quality assurance with ABC.

AASHTO Load Combinations for Extreme Loads Although more bridges are damaged in floods in the United States, only earthquakes are included in the AASHTO load combinations. During an earthquake of intensity of magnitude 7.2 on the Richter scale in northern Pakistan which occurred in October 2005, bridges built on rivers were damaged severely (see Khan 2010). Figure 4.1 shows the U.S. team visiting a town by helicopter in the foothills of K-2 Mountains in the north of Pakistan. The town was destroyed by the earthquake and was cut off from the rest of the country as a result of damages to bridges.

FIGURE 4.1 "World-renowned earthquake reconstruction experts from U.S. National Academy of Sciences arrive in Pakistan," Department of State, U.S. Embassy Islamabad, 12/12/2005: William T. Holmes, expert in seismic vulnerability of buildings and loss estimation (extreme left); Marjorie Green, Director Earthquake Engineering Research Institute of California (second from left); Melvyn Green, expert on residential construction in seismic areas (middle); a representative of the U.S. National Academy of Sciences (second from right); Mohiuddin Ali Khan, consulting structural engineer representing ASCE (extreme right).

Use of Integral Abutment Bridges with Flexible Steel Piles Connection detail at the top of piles can be designed for full fixity, partial fixity, and pinned. Use of spread footing, caisson, drilled shaft, or two rows of piles will offer minimum flexibility and will lead to semi-integral abutment conditions.

Orientation Flexible piles, with the weak axis placed normal to traffic flow, will permit longitudinal movement and rotation, thereby releasing beam stresses.

Corrosion Corrosion of the ends of steel girders is also minimal when girder ends are encased in concrete pile cap. Comparative studies of commonly used piles, such as steel H piles, pipe piles, prestressed concrete pile, timber pile, and sheet pile, have shown that steel H pile (with weak axis placed normal to traffic flow) is the most flexible type.

The world's longest fiberglass-reinforced plastic bridge spans about 175 ft and is at a golf course in Aberfeldy, Scotland. The second-longest, about 140 ft, is in Barcelona, Spain, over a series of railroad tracks. The nation's longest and world's third-longest fiberglass-reinforced plastic bridge is located at a northside park in Indiana, built in 2008. The 100-ft-long bridge replaces a concrete one that floodwaters washed out in 2001 at Juan Solomon Park. The bridge provides access to 22 acres of parkland.

4.7 Applications of Modern Construction Materials

Environmental Benefits of Concrete

1. Much of the creation of towns and bridges is now made possible by the use of longer-lasting concrete in the foundations and in the substructure and superstructure.
2. In many structures, exposure to deicing chemicals and marine-sourced chloride is a significant cause of corrosion. The most common procedure for repairing deteriorated concrete involves the removal of the damaged material and replacement with new concrete or mortar. Differences in pH, porosity, and chloride content are a few of the factors that may result in corrosion activity. As a result, "chip-and-patch style" of repairs may fail prematurely in structures exposed to chloride.
3. The concrete industry is continually improving its material quality. Use of concrete in the harshest of environments is an achievement. Limestone, the primary raw material in concrete manufacture, is abundantly available in nature. In addition, silica is available, in the form of aggregates. Alumina and iron oxide are the other basic ingredients in the manufacture of cement. The concrete industry has reduced its CO_2 emissions recently as a result of these improvements in material quality.

Use of New Concrete Materials

Available concrete materials include the following:

- Smart concrete
- Accelerators
- Air-entraining admixtures
- Water reducers
- Super plasticizers
- Pozzolans
- Emulsions
- Antiwashout admixture for underwater concrete
- Use of high-performance concrete (HPC)

Use of Concrete Repair Materials

Available concrete repair materials include the following:

- Composites
- Polymers
- Polymer modified cement
- Epoxy resins
- Polyurethane injection resins

- Shotcrete
- Carbon and glass fiber reinforcement

General Repair Procedures

The following procedures may be applicable and if so, should be adopted:

- Damage and deterioration assessment.
- Load testing.
- Identification of causes.
- Report development.
- Use of NDT techniques.
- Use of a cementitious materials selection process.
- Preparation of surfaces.
- Following the placement methods.
- Performance of crack and joint repairs.
- Use of protective systems for concrete (membranes and waterproofing, sealers, and coatings).
- Repair of corrosion-related deterioration in concrete structures, which offers unique challenges.

4.8 Use of Recyclable Materials Technology

Recyclable Steel and Waste Products

1. Reinforcing steel and prestressing strands have always been recyclable. Steel used in old ships can be salvaged and is sold as scrap. The steel is removed and recycled.
2. Many different materials are being considered for aggregate substitutes, including:
    - Granulated coal ash
    - Blast furnace slag
    - Various solid wastes including fiberglass waste materials, granulated plastics, paper and wood products wastes, sintered sludge pellets and others. The only two materials that have been significantly applied for recycling are glass cullet and crushed recycled concrete.

Use of Recycled Concrete Aggregate (RCA)

RCA is not a cement substitute. Conventional concrete aggregate consists of sand (fine aggregate) and various sizes and shapes of gravel or stones. The coarse aggregate portion of RCA has no significant adverse effects on desirable mixture proportions or workability.

Participation by States In the United States at least 140 million tons of concrete is recycled every year. The Oregon DOT has already taken the lead in this respect on a large scale. Another example is the reconstruction of the Arthur Ravenet Jr. Bridge in Charleston, South Carolina. Over 248,000 tons of concrete was salvaged from demolition of old structures and reused to create 82 acres of reef habitat. California, Texas, Virginia, Michigan, Minnesota, and Utah were identified as being among the highest consumers as well as large suppliers of RCA.

The author was associated with a highway embankment project in the North of New Jersey utilizing RCA for structural fill, which is made abundantly available from the demolition of debris. This resulted in cost savings from the transportation of tons of wasted, but good-quality, aggregates.

4.9 Use of Fiber-Reinforced Polymer Concrete and Products

Introduction

1. The use of FRP bars with HPC is another development in the concrete industry. The first FRP bridge was built in Kansas in 1996. Construction followed in New York, West Virginia, Delaware, and Ohio.
2. FRP bridge decks are used in bridge rehabilitation projects, often because of their relatively low weight and high durability. Related benefits of FRP include rapid construction and advantages in terms of life cycle costs (e.g., corrosion resistance). FRP is a strong material, lightweight, durable, impervious to deicing salts, and weather-resistant, and it has a low life cycle cost. It is a composite material consisting of high-strength threadlike fibers embedded in epoxy or polyester resin. A variety of fiber types and resins can be used as composites. The fibers are glass, carbon, or aramid, which are brittle in solid form but have high strength in fiber form.
3. FRP is available in a variety of forms such as FRP bars, FRP grids, and tendons for prestressing. The combination of plastic and fibers produces great mechanical properties, resulting in a more lightweight material that is durable. It is increasingly being used in strengthening applications and for deck replacements. Longer spans are now possible with lightweight aggregates and FRP, which replace more expensive steel reinforcing bars.
4. Corrosion is the most common type of distress or deterioration in concrete bridges. Replacing a traditional steel bar with an FRP bar eliminates the expansion that comes with corrosion associated with steel embedded in concrete.

Modern Rehabilitation Methods with FRP

1. The applications are as follows:
   - Column wraps to enhance seismic performance
   - Beam wrapping to increase shear capacity
   - Bonded flange plates to increase bending capacity
   - Epoxying FRP rods in grooves cut into the substrate to increase member strength

- Truss strengthening
- Wrapping aluminum columns of sign structures

2. Advantages of FRP:
   - Noncorroding, unlike steel rebars
   - Low maintenance
   - Low fatigue
   - Less dead load, high strength-to-weight ratio
   - Flexural strengthening
   - Benefits of FRP composites recently expanded to include the bridge construction industry
3. Disadvantages of FRP:
   - The cost of an FRP bridge is higher than that of a conventional bridge. The frequency of inspection is greater also. Recommendations for inspection and maintenance are given in ACI 440.2R-02, "The Design and Construction of Externally Bonded FRP Systems for Strengthening Concrete Structures," 2002.
   - Resistance to fire is much lower than for steel rebars.
4. Visual inspection should record observations for
   - Changes in color
   - Debonding
   - Cracking
   - Deflections
   - Peeling
   - Blistering
   - Crazing
5. For sign structure repairs
   - It costs less than full structural support replacement.
   - It allows repairs to be done quickly.
   - It causes less traffic disruption because only lanes beneath repair need to be blocked off. A cost comparison should be made considering all available options. Past price ranges are as follows: decks, $65 to $80 per square foot of deck area; superstructure, $140 to $150 per square foot of deck area.
   - Because of the material's typical low modulus of elasticity, most designs will be driven by deflection limitations and not strength requirements. Although the criterion for deflection is somewhat arbitrary, it is typically kept at 1/800 of the supporting span length.

Use of Carbon Fiber Preimpregnated Polymer (CFRP) and Steel Reinforced Polymer (SRP) Sheets

Carbon fiber–reinforced polymer (CFRP) is an extremely strong and light fiber-reinforced polymer which contains carbon fibers. For strengthening of decks, reinforcement of columns, and repair of retaining walls, the use of carbon fiber sheets is gaining popularity (e.g., supplied by Mitsubishi Chemical Corporation).

1. Recently, a comparison of four techniques for strengthening concrete beams was performed by Harrison et al in 2006. In addition to externally bonded FRP as a successful technique for strengthening concrete members, other techniques such as near surface-mounted (NSM) FRP bars have emerged as viable alternatives. Four composite-based strengthening systems were designed to provide equivalent flexural performance:
   - Externally bonded CFRP sheets
   - NSM prefabricated CFRP strips
   - Externally bonded steel reinforced polymer sheets, known as hardwire
   - Near surface-mounted stainless steel bars
2. Laminated fiber reinforced rubber has also been used for reducing collision impact. The energy absorption capacity of the laminated fiber reinforced rubber installed at girder ends was studied by Japanese researchers, and it compared well to ordinary rubbers. A collision test was carried out between two steel solid bars to assess the energy absorption and the impact force during the collision. Then the numerical simulations of the test results were run by using the mass-spring systems.

   It was concluded that both rubbers are effective in energy absorption during the on; however, the laminated fiber-reinforced rubber can reduce the impact force more than the ordinary rubber can. It was also concluded that the energy absorption can be estimated by using the mass-spring systems for the impact force during the collision.
3. New York State installed its first FRP composite bridge in October 1998. Since FRPs are relatively new, a comprehensive test program, comprised of load tests and visual inspections every 6 months and detailed finite element analysis, is being carried out to monitor the structure and to evaluate its long-term durability.
4. Amjad J. Aref and Sreenivas Alampalli studied the dynamic response of an FRP bridge recently constructed in New York State. The dynamic behavior of the bridge was studied using detailed finite element analysis. These models were then compared with field tests performed on the bridge to validate finite element models.

   According to the authors, in a complex structural system such as a bridge made of laminated FRP materials, several sources of structural degradation may occur during the life of a bridge. For example, damage due to delamination, cracks, and loss of bond between components, and a change in material properties, may often occur in various parts of the structures and cannot be detected by visual inspection.
5. Smith J.W. and Bright S. in 2013 investigated the use of fiber-reinforced composites for upgrading orthotropic bridge decks. The competitiveness of orthotropic bridge decks depends on their high strength-to-weight ratio. However, poor durability of paving materials and fatigue failure of welds have contributed to the high costs of repair, including road user delay costs due to traffic disruption. It is proposed to use a layered surfacing system, combining lightweight asphalt, conventional asphalt, and a layer of glass fiber mesh

embedded just beneath the chip-sealed surface. Fatigue tests indicate that glass fiber reinforcement increases the durability by a factor of at least 10.

6. Balazs and Borosnyoi carried out a comparative experimental study on prestressed concrete beams pretensioned with either CFRP or steel wires. Considerable corrosion of reinforced concrete members accelerated the research on nonmetallic materials such as FRPs for structural applications as noncorrosive reinforcement.
7. Use of FRP decks for bridge rehabilitation has been studied. A bridge in Tippecanoe County is the first in Indiana to be rehabilitated with an FRP deck. County Road 900E over Wildcat Creek is a three-span continuous steel stringer bridge with two concrete approach spans. The FRP deck replacement would take place only on the three main spans.

Monitoring of Polymer Composite Bridges Studies of continuous, long-term monitoring of two polymer composite bridges were carried out in Delaware.

1. Shenton et al. and their team designed a continuous, long-term monitoring system for two polymer composite bridges recently built in Delaware. The Magazine Ditch Bridge has been continuously monitored for more than a year. The monitoring system provides data to investigate the effects of sustained load, environmental factors, and live load on the bridge. Early results show that daily and seasonal temperature changes can induce strains in the bridge that are equal in magnitude to the maximum live load strains.
2. A similar system has been designed for the first state-owned composite bridge in Delaware. A brief overview of the systems and sample results from the data collected for the Magazine Ditch Bridge is presented in Shenton et al., 2000. The monitoring system provides data to investigate the effects of sustained load, environmental factors, and live load on the bridge. Early results show that daily and seasonal temperature changes can induce strains in the bridge that are equal in magnitude to the maximum live load strains.
3. Osman Hag-Elsafi, Sreenivas Alampalli, and their team studied the application of FRP composite laminates to strengthen a 70-year-old reinforced concrete T-beam bridge in New York. Leakage at the joints of this single-span bridge led to substantial moisture and salt infiltration in the superstructure, as manifested in efflorescence, freeze-thaw cracking, and delamination at some locations. The integrity of the steel reinforcing and the overall safety of the bridge were suspected. This concern was heightened by the absence of any documents pertaining to the bridge design, such as bar size, steel type, concrete strength, and design loads. The bridge was strengthened using bonded FRP laminates. Load tests were conducted before and after the FRP laminates were installed to assess the effectiveness of the strengthening system. Test results are also compared with those obtained analytically by using classical approaches.
4. Fiber-reinforced polymer matrix composites (PMCs) are very effective for concrete rehabilitation and deck slab construction. The Virginia DOT used it successfully for superstructure replacement of Tom's Creek Bridge. To inject hairline and wide cracks in concrete, epoxy injection resins are being effectively used. For waterproofing joints and cracks, sealing leaks, and underwater repairs to concrete, polyurethane injection resins are being effectively used.

Lightweight Truss Bridge Rehabilitation Using Composites and FRP Deck

According to the FHWA, there are over 93,000 weight-restricted bridges in the United States. More than 19,000 steel trusses are well suited to lightweight rehabilitation. Existing truss bridges can be made lighter by replacing a heavy concrete deck and numerous asphalt overlays with a lightweight FRP deck. In addition, painting the steel members would increase the life of bridges.

Bridge Details

The 140-ft-span Warren truss bridge carrying New York State's Route 367 over Bentley Creek in Chemung County, built in 1940, found new life in 1999 after the installation of a 13-in-thick, E-glass reinforced vinyl ester resin deck weighing only 32 lb/ft^2. This reduced the dead load by 80 percent and almost doubled the live load carrying capacity of the bridge, from AASHTO live load rating from original HS12 to HS23, a level higher than the original design. A 14-ton weight restriction was able to be removed, and the bridge was reopened to all legal loads. It meets the L/800 deflection requirement (where L is the effective span length of the girder) with an inventory load rating of HS85 (154 tons).

Several innovations were developed in this rehabilitation process:

1. A haunch for each floor beam was uniquely precast to support the deck. This placed the deck at the proper elevation, provided a new load path directly from the deck to the floor beams, raised it above the obsolete steel stringers, and delivered a 2 percent cross slope.
2. The deck was replaced using six discrete deck panels. The modular construction simplified installation and reduced field time.
3. A field splice system was designed to transfer shear and moment between panels. The joints devised rely on FRP cover plates and structural adhesives. The watertight deck protects the steel floor system from the weather.

Research shows FRP repairs are as strong as welded joints. The specification covers restoration of the tensile capacity of secondary sign structural members, such as internal truss diagonals, and not main members, such as longitudinal truss chords. FRP benefits include the following:

- It costs less than full structural support replacement.
- It allows repairs to be done quickly.
- It causes less traffic disruption because only lanes beneath repair need to be blocked off.

4.10 Advancements in Concrete Technology

Salient Features

Concrete bridges are more commonly used for smaller spans. Small steel spans have higher maintenance costs.

1. *Ultra-HP FRC*: Compressive strength reaching 30 ksi is possible as well as a flexural strength of 7 ksi.

2. *HP lightweight aggregates*: Lightweight HPC reduces deadweight, enables longer span lengths, reduces the number of piers in a river, and allows fish to travel with the least obstruction.
3. *Spans*: Spliced girders of varying depth enable lightweight concrete to achieve spans of more than 200 ft.
4. *Overlays*: Use of silica fume and high early strength latex modified concrete (LMC) will open the deck to traffic within 3 h of curing. Silica fume, pozzolan, fly ash, and slag may be used to reduce concrete permeability and heat of hydration. Fly ash and cenospheres are preferred to HPC in bridge decks, piers, and footings. By-products of coal fuel such as fly ash, flue gas desulfurization materials, and boiler slag provide extraordinary technical, commercial, and sustainable advantages.
5. *Self-consolidating concrete (SCC)*: Since vibration time is saved, SCC helps the ABC method; it achieves more workable concrete with lower permeability than conventional concretes.
6. *Rapid-setting concrete*: Nonshrink, multipurpose, high-strength repair mortar is used for concrete repair, plaster repair, mortar beds, formed work, and vertical and overhead applications.
7. *Blended cement concrete*: A blend of portland cement and a combination of silica fume or fly ash are used for enhanced strength and durability. This is used in high-performance (HP) applications with materials such as slag cement.
8. *Fiber-mesh concrete*: Microsynthetic fibers prevent early age cracks during concrete's plastic state.
9. *CFRP concrete*: This concrete is used for repair and retrofit of concrete structures with glass or FRP.

Prefabrication of Components and Connection Details

The contractor must be knowledgeable about the latest technology and availability of new bridge components. Typical connection details must be developed for precast deck panels, piers, and abutments. The quality of the superstructure must be improved by fabricating in a more controlled factory environment. Greater technical service is provided on bridge products such as precast deck units, girders with welded shear connectors, diaphragms, bearings, parapets, and precast pier units. Quick assembly is possible by using systems such as Conspan, Inverset, and Effideck; drop-in deck panels; and post-tensioning in both directions.

Use of High-Performance Concrete (HPC)

1. Advantages of HPC are wider beam spacing and fewer beams, longer span lengths and fewer piers, increased vertical clearance, lower permeability, stronger and more durable concrete, lower initial and life cycle costs, and fewer maintenance requirements. These properties are achieved by special mix design and improved curing.

   With the high-performance concrete composite overlay method, the top of the deck is removed to the underside of the top mat of reinforcement and replaced with high-performance concrete. Placing the concrete below the top mat forms a composite action between two layers of concrete.

2. The benefits of a HPC composite overlay include these:
   - Any spalling or cracking in the top of the existing deck is addressed.
   - Composite action helps to strengthen the deck.

   The disadvantages to the system include the following:
   - Concrete curing time is longer than that of asphalt or polymer overlays, leading to possible maintenance and protection of traffic (MPT) issues.
   - The assumption is that the bottom half of the concrete is in acceptable condition to be salvaged.
   - There is a potential for cracking of the high-performance concrete.

Self-Consolidating Compacting Concrete (SCCC)

1. SCCC is an improved version of HPC with compressive and tensile strengths exceeding those of normal concrete. The ease of placement and reduced demand for skilled labor are the main advantages of using SCCC. It can be used with high range water reducers to achieve high early-strength concrete. Proper curing is critical.
2. SCCC has very high slump and flowability properties without segregation. It uses specialty admixtures which radically modify slump and flowability. Examples of such admixtures are viscous modifiers and high-performance polycarboxylate polymers. Properties of conventional concrete such as compressive strength, cohesiveness, and durability are not affected. Advantages of SCCC include the following:
   - It provides superior appearance and long-term durability.
   - It eliminates the need for vibration.
   - The resulting concrete is totally homogeneous with a uniform surface finish.
   - To further enhance durability, a calcium nitrite corrosion inhibitor may be added to the mix.
   - It attains higher quality control resulting when using higher strength concrete with fc′ values in excess of 9 ksi.
   - Voids and honeycombing in many traditional concrete mixes are avoided.
   - It has been successfully used in new construction projects with difficult placement or finishing.
   - It is effective for concrete repairs of honeycombing.
   - It is effective for strengthening projects when the material must be placed under pressure into confined forms with highly congested reinforcement.
   - Concrete is so flowable that instead of measuring the height of the slump cone, the diameter of the circular blob that pours out of a slump test cone is measured.
   - Concrete pumping is easier. Forms do not have to be vibrated to consolidate concrete.
   - Labor savings are achieved because crews can pour large repair areas using a single pump location.
   - Architectural finishes are easier to achieve.

Concerning the performance of blended cement in high-strength self-compacting concrete, rice husk is an agricultural waste generated in massive quantities from rice processing units worldwide. With no worthwhile use, it creates disposal problems. Its high silica content makes it suitable for use with cement. Tahir Kibriya, of the National University of Sciences and Technology, Pakistan, has investigated rice husk ash and found improved strength and durability.

This experimental study aimed at evaluating the properties of high-strength SCCC made from blended cements using rice husk ash, portland cement, natural aggregates, and sand. Wide-ranging investigations covering most aspects of mechanical behavior and permeability were carried out for various mixes for compressive strengths of 60, 80, and 100 N/mm^2.

Compressive strengths of high-strength SCCC specimen with blended cements were observed to be higher by about 4 to 9 percent than the control specimen, for concrete with 50 percent portland cement blended with 50 percent rice husk ash. Higher elastic moduli and reduced permeabilities were observed along with better sulfate and acid resistance. Better strengths and improved durability of such high-strength SCCC make it a more acceptable material for major construction projects.

Concrete Using Shrinkage-Compensating Cement

Shrinkage-compensating cement is influenced by the amount of binder, the percentage of silica fume, the water-to-binder ratio, and the types of super plasticizer, cement, and aggregate, respectively, on both autogenous and total shrinkage of HPC. Shrinkage can be measured under isothermal conditions and constant humidity of the environment. Service life can be extended by using modern construction materials. Life cycle costs are also minimized for thousands of bridges. Shrinkage development in HPC can be reduced by using shrinkage-compensating cement in lieu of type 1 cement, to help eliminate drying shrinkage cracking of bridge decks.

Ultra-High-Performance Concrete (UHPC)

Ultra-high-performance concrete is proposed as an innovative material for the construction of highway bridge superstructures. The path of innovation can be traced back to the 1980s, when the first big gains were made in increasing the resistance or strength of concrete by a factor of 6 to 10. Developed in France during the 1990s, UHPC has seen limited use in North American bridge projects. Consisting of fine sand, cement, and silica fume in a dense, low water-to-cement ratio mix, this highly moldable material offers a combination of superior properties including compressive strengths up to 30,000 lb/in^2 and flexural strengths up to 6000 lb/in^2, ductility, durability, and a vast range of aesthetic design possibilities.

Advantages

The high-tech versions have different properties that make them more comparable to materials such as stainless steel or aluminum, which are often more expensive. New types of concrete offer the following advantages:

1. They are intrinsically energy-efficient.
2. They provide excellent insulation against wind and water.
3. Their high density means that they store heat during the day and release it at night, preventing bridge decks from freezing in winter.

4. UHPC is denser than conventional concrete, which attributes to its remarkable imperviousness and durability.
5. In addition, UHPC is extremely low in permeability and performs better in terms of abrasion and chemical resistance, freeze-thaw, carbonation, and chloride-ion penetration.
6. To improve ductility, steel or polyvinyl alcohol (PVA) fibers are added, replacing the need for passive mild reinforcing steel.
7. They set much faster.
8. Stronger concrete translates to significant gains for the environment. It can be used more thinly, consuming considerably fewer raw materials than regular concrete. The environmental advantage is clear: zero maintenance, zero painting, and a very long life.

Engineers and builders have far greater flexibility to use the material's long-lasting, thermal, and acoustic properties in pedestrian bridges and at bus stations; this, in turn, contributes to big energy and other environmental savings. White concrete contains titanium dioxide, which acts to keep the concrete clean and simultaneously to destroy ambient pollutants such as car exhaust.

Sustainability

High-tech concrete is just one of the products that have emerged from the research and development laboratories of cement, steel, and chemical firms this decade, and it signals a growing commitment by heavy industry to the notion of sustainability. UHPC mix designs typically include no aggregates larger than sand, and they include steel fibers 0.2 mm in diameter and 13 mm in length. These steel fibers and the special mix design increase the strength and toughness of the UHPC significantly relative to more traditional concretes.

Disadvantages

Low water-to-cement ratios typically used result in difficult casting and curing conditions. There is a need to evaluate the strength and stiffness parameters of as-cast members. The potential for the development of practical quality control techniques for the future implementation of UHPC needs to be considered.

Ultrasonic velocity measurements are used to estimate the bulk elastic modulus, shear modulus, and Poisson's ratio of UHPC, and results are compared with those of traditional destructive methods. The application of ultrasonic testing for the evaluation of early-age material properties and for nondestructive, in situ materials characterization is extremely useful.

Case Study of New Construction Materials

The new Jakway Park Bridge in Buchanan County, Iowa, is one of the first highway bridges in North America to be built with a new generation of UHPC pi-girders. The bridge is 24 ft 3 in wide by 112 ft 4 in long. The UHPC center span is 51 ft 2 in. It is one of the first North American highway bridge projects to incorporate batching of UHPC in a ready-mix truck. The Iowa DOT and the Bridge Engineering Center at Iowa State University designed the bridge; it is a combination of cast-in-place, simple span slabs with a center span consisting of a series of precast UHPC pi-girders. The Jakway Park

Bridge has a clean, balanced, and symmetrical appearance and can certainly be considered an important technological advancement in the bridge building industry. Testing of the section by the Turner-Fairbank Highway Research Center validated the finite element method analysis for flexural and shear capacity in the longitudinal direction. The testing also confirmed that the stress in the transverse direction of the deck was unacceptable for service loading and that a low transverse, live load distribution between adjacent pi-girders would require stiffening.

Exodermic Bridge Deck

An exodermic bridge deck is one that is comprised of a reinforced concrete slab on top of, and composite with, an unfilled steel grid. Exodermic decks are made of composite with the steel superstructure by welding headed studs to stringers, floor beams, and main girders, as applicable, and embedding these headed studs in full-depth concrete.

This system maximizes the use of the compressive strength of concrete and the tensile strength of steel. The system has overall thicknesses ranging from 6 to 9 in while weighing 35 to 50 percent less than a standard reinforced concrete deck of the same span. Reducing the dead load of the deck with this system can increase the live load rating. The concrete component of an exodermic deck can be precast or cast-in-place.

Precast exodermic decks have the benefit of quick installation; they can be erected during short, nighttime work periods, allowing a bridge to be kept fully open to traffic during hours of higher traffic volume. This area is poured at the same time as the reinforced concrete deck when the deck is cast-in-place, or separately when the deck is precast. Exodermic decks require no field welding other than that required for the placement (with an automatic tool) of the headed shear studs. While the design of exodermic decks is patented, the availability of the grid from multiple, independent, licensed suppliers allows it to be considered generic, and it does not need to be specified as a proprietary product. The major advantage of the system, in addition to reduced dead load, is the installation time.

Approximately 750 to 1000 ft^2 of the cast-in-place system can be installed in a standard 8-h period. For example, as reported in the literature, using a precast exodermic system, 1900 ft^2 was installed on the Gowanus Expressway per overnight closing in 2001, and 2000 ft^2 per 7-h closing was installed on the Tappan Zee Bridge in New York in 1997.

The manufacturer suggests using an overlay system and prefers an asphalt overlay over latex modified concrete (LMC) or other concrete overlays to increase the speed of construction. Overlays also allow for accommodating vertical geometric variations as opposed to a bare concrete deck. The major disadvantage of this system is the initial cost. Not accounting for savings due to accelerated construction (reduced maintenance and protection of traffic requirements and shorter construction duration), the cost of an exodermic deck is approximately twice the initial construction cost of a cast-in-place deck (materials only).

Reactive Powder Concrete Bridge Girders

The benefits of using reactive powder concrete (RPC) to carry bursting forces in prestressed bridge girders are significant. RPC has superior structural properties. A compressive strength up to 160 N/mm^2 and a flexural strength up to 46 N/mm^2 can be achieved but later than standard 28 days age. RPC has high shear capacity. The long term behavior needs to be investigated. Tests are reported in the literature on three RPC, 150-MPa, deep beams. They have shown significant potential advantages for using RPC in bridge engineering.

Use of Fly Ash, Blast Furnace Slag, and Silica Fume

Concrete can be made more sustainable by use of cementitious materials such fly ash, which is obtained from coal-fired power plants. Use of fly ash in concrete bridges is on the increase. About 70 million tons of fly ash is produced each year in the United States, but only about 15 million tons of it ends up in concrete products. It's a way to make roads, sidewalks, and bridges stronger and longer-lasting.

Fly ash from power plants and ground-up slag left over from the steelmaking process can replace some of that cement, and it can make the final product stronger. During the past 5 years, the number of concrete suppliers in Pennsylvania using fly ash and slag in mixes has increased from 2 plants to 10. Other states such as Minnesota have switched to using more fly ash and less cement. The typical dosage is about 15 to 25 percent in place of cementing materials, but some places such as Minneapolis use 40 percent or more. Up to 15 percent of portland cement in a typical mix can be replaced with fly ash, or up to 50 percent can be replaced with slag. Silica fume is a key ingredient in high-performance concrete. It significantly increases the life of structures. It is a by-product of silicon and ferrosilicon metal production. *The Silica Fume User's Manual* and a standard reference manual are available from the National Institute of Science and Technology, Gaithersburg, Maryland.

4.11 Developments in Steel Materials and Metallurgy

Revolution in Steel Strength by Using HPS 70W and HPS 100W

Modern bridges carry very high moving loads and span over long distances. ASTM A709 high-performance steel (HPS) 70W (yield strength of 70 ksi) is being used by many states for flanges in hybrid girders exceeding 100-ft spans. The overall costs of construction and life cycle costs are reduced. The author has used hybrid HPS 70W steel girders on a number of bridge projects. Other advantages are enhanced resistance to fracture, elimination of maintenance painting, and improved vertical under-clearance due to resulting shallow girders. Hybrid systems are made possible by keeping web material as 50W steel. Currently, rolled sections are not available in 70W steel, and the availability of a range of plate thicknesses is limited. Greater delivery time needs to be allowed in the construction schedule.

Types of High Strength Girders for Selection

High strength steel is by far, the more durable and stronger material than reinforced and prestressed concrete or timber. Steel bridge girders will last longer and replacements can be delayed eventually costing less whne the owners have hundreds of bridges to look after. Weathering steel requires less painting and is suitable for river bridges also. Modern metallurgy has made possible a wider availability of types and grades of steel.

Use of Hybrid Steel

Steel for girder webs and flanges shall be a combination of ASTM A709 grade HPS 70W manufactured by thermomechanical controlled processing (TMCP) or quenched and tempered heat treatment processing along with ASTM A588/709 grade 50W. All other steel shall be ASTM A709 grade 50W.

This work consists of erecting structural steel members, designed as a hybrid/mix of steel materials consisting of ASTM A709, high-performance grade HSP 70W in combination with grade 50W steel.

Material and Fabrication Specifications: Bridge engineers must be familiar with the specifications of all materials used in construction. Typical specifications are shown below and are required to be shown on construction drawings:

Weld design for HPS 70W Grade Steel: Refer to publication "Guide for Highway Bridge Fabrication with HPS 70W Steel," which is a supplement to ANSI/AASHTO AWS D1.5.

Fatigue Analysis and Live Load Intensity: If the projected average daily truck traffic (ADTT) exceeds 500, the design live load will be HL-912. Fatigue analysis will be based on the projected ADTT. Fracture toughness for steels in civil structures is typically measured using the Charpy V-notch (CVN) test. This test is specified where fracture is a concern. Steel designated CVN shall be impact-tested to exceed the test values of ASTM A709-00.

Compacyness and Bracing Requirements for HPS girders: The following are the compactness and bracing requirements for high-performance steel girders. Earls, Shah, and Thomas experimentally verified nonlinear finite-element modeling techniques for the study of HPS I-shaped bridge girders. Based on research, it appears that the current AASHTO flange compactness and bracing provisions are inadequate when applied to A709 grade HPS 483W bridge girders. Intense interactions between local and global buckling, as observed during the failure of the bridge girders, may be the cause. New flange and web compactness criteria, as well as new bracing criteria, are proposed by Earls et al. for use in the design of A709 grade HPS 483W bridge girders.

In addition to the requirements of American Welding Society (AWS) Code D1.5, all procedure qualifications must be ultrasonically tested in conformance with the requirements of AWS D1.5-95. Evaluation must be in accordance with AWS D1.5-95, Table 9.1, on ultrasonic acceptance, rejection criteria, and tensile stress.

The plates, hardware, and accessories shall be galvanized after fabrication in accordance with ASTM A 123 or ASTM A 1512.

Use of HPS 100W Steel in Bridges The AISI, FHWA, and U.S. Navy HPS Steering Committee have developed HPS 100W steel grade for 100 ksi minimum yield strength (ultimate tensile strength of up to 130 ksi) in bridges. Currently, plate thickness is manufactured from 3⁄16 to 2.5 in.

Nebraska and West Virginia have started using HPS 100W. The HPS 100W has higher toughness, better weldability with a low carbon content, and improved corrosion behavior. It results in a more lightweight bridge and reduces substructure costs. It has great potential for special girders when maximum live load deflection criteria are met. A longer span, fewer girders, and wider spacing may result. Also weathering steel requires minimal painting or maintenance.

HPS 100W leads to rapid construction by reducing the number of girders compared to conventional construction with 50W steel. ABC helps reconstruction of bridges due to difficult field conditions of constructing in water.

Initiatives taken by the FHWA have led to considerable progress in implementing ABC concepts. ABC-related design needs to be made part of AASHTO and state bridge design codes.

Application of the latest techniques in concrete manufacture, composites, HPS, and hybrid materials is feasible. Integrated software covering all aspects of ABC design and drawings may be developed. Deterrents and bottlenecks such as maintenance and protection of traffic (MPT), construction easement, right-of-way, permit approvals, and utilities relocation need to be resolved and administrative procedures simplified to facilitate the ABC method. Certification and training of construction personnel, continuing education of engineers in rapid construction techniques, and construction management ABC courses at universities are recommended.

4.12 Performance of Steel Bridges over Rivers

Existing Steel Girders

For each structural component, the following issues will be addressed:

1. *Structural steel replacement and/or strengthening*
   - Existing steel beams with deck replacement shall be made composite in positive moment regions.
   - To determine the remaining service life, a fatigue analysis of existing steel members to be reused or rehabilitated will be carried out in accordance with AASHTO "Guide Specifications for Fatigue Evaluation of Existing Steel Bridge," "2007 AASHTO LRFD Specifications for Bridge Highways," and the current state LRFD bridge design manual.
2. *Use of high-performance steel.* Use of HP 70W for durability, weight, and cost savings, and strengthening of fractured floor beams will be considered.

   Steel can be bent from overload, collision, earthquake, or fire.

   A repair procedure to straighten plastically deformed regions of damaged steel by applying repetitive heating and cooling cycles is generally used. Each cycle leads to a gradual straightening trend.
3. *Microcomposite reinforcing steel bars, corrosion inhibitors, and latex modified concrete.* Conventional 60 ksi rebars are known to corrode. Microcomposite steel is not coated and is highly corrosion-resistant. Corrosion inhibitors and latex modified concrete can extend the life of bridge decks by more than 20 years.

   The use of corrosion inhibitors with conventional rebars can mitigate the problem by chloride extraction. Corrosion of reinforcing bars is caused by salt penetrating concrete. The chemical corrosion inhibitor additive is designed to protect structural steel reinforcement by preventing oxidation after full- or partial-depth repairs of bridges. The corrosion inhibitor additive allows its products to chemically counteract the corrosion process that takes place at the interface between the iron and the concrete. ASTM has developed ASTM G109 corrosion inhibitor specifications.
4. *Joints, bearings, and devices (JBDs).* A DOT study investigating joint failures listed "lack of designer's awareness" as a key concern. Topics of critical importance are the theory and design of common types of JBD, reliability-based design, installation and maintenance, fatigue and corrosion, modeling, and finite element analysis. Increased awareness of JBDs can enhance the probability that critical components will perform their functions with intended structural control.

4.13 Continuing Education in New Technology Applications for Engineers

Training Engineers in New Technology

It is necessary that construction personnel be given on-the-job training in the use of new technology. Bridge construction costs millions of dollars, and so any mistakes or wastage will be expensive. A learning curve needs to be provided at both senior and junior levels.

- Training aspects should cover tasks such as understanding erection procedures, erection drawings, modern concrete technology, and the use of new steel, processed timber, etc., in bridge construction.
- Continuing education in design and construction and rigorous certification requirements need to be introduced for certain levels of construction personnel. There should be a greater scope of postdesign activities. On each project a preconstruction workshop should be conducted to evaluate possible hiccups and constructability issues. Topics would include equipment locations, construction duration, access, right-of-way, and material availability. To resolve constructability issues and avoid any misinterpretation of contract drawings, designers should be consulted before project construction.
- There should be improved traffic control and maintenance and protection of traffic. The objective of using modern rehabilitation technology and innovations is to minimize life cycle maintenance and recurring costs.

The FHWA has developed a 3-week training program based on the Bridge Inspector's Reference Manual (BIRM). The program consists of a 1-week course on engineering concepts for bridge inspectors and a 2-week course on safety inspection of in-service bridges. The following courses extending from 1 to 3 weeks are usually required for registration as a bridge inspector:

- National Highway Institute (NHI) Bridge Inspector's Training Courses, Part I and Part II
- NHI Stream Stability and Scour at Highway Bridges
- NHI Bridge Coatings Inspection
- NHI Fracture-Critical Inspection Techniques for Steel Bridges
- NHI Bridge Inspector Refresher Training

Measures to Control Floods

Floods are not a new problem but have occurred for many centuries. Their velocities need to be controlled (especially when they exceed 8 ft/s). This is possible in a number of ways, such as by river diversion through canals and tributaries, storing the floodwater for future use for irrigation. Other options are river training measures and making any new bridge openings larger. Areas located close to the river require adequate longitudinal and cross drainage, by using underground sewers and transverse pipes at the foot of embankments, to disperse water to unused dry areas at lower levels. This is not being done. Cost savings can lead to increased risks, but one major flood can wipe out the savings.

Highway and bridge deck elevations are sometimes not made high enough to save on embankment costs. A detailed survey will show the contour elevations and the direction of flow of spilled water. Small pumping stations may be required. Vegetation on sloping ground may be considered to minimize erosion in addition to armoring countermeasures such as riprap.

FEMA directives may be followed for safer foundation construction. A cost–benefit study is required to justify the additional investments in preventing inundation of towns and for preserving highways and bridges against soil erosion.

Following a flood disaster, a one-time investment is preferred to keep life cycle costs to a minimum. A long-term strategy will be used for fixing fallen or damaged houses and bridges since it takes a long time to implement multiple large-scale construction. Relocation options for affected people need to be considered as a last resort.

For bridges, hydraulic, structural, and other types of countermeasures listed in HEC-23 are recommended.

Concrete Armor Units

Although concrete armor units and grout filled bags are listed in HEC-23 recommended countermeasures, these are presented here in greater detail due to the recent surge in concrete products and their low costs: Concrete armor units are also known as *artificial riprap*. They are used as an alternative for rock riprap, especially when there is a short supply of rock riprap. The primary advantage of armor units is their usually greater stability. The increased stability allows their placement on steeper slopes or the use of more lightweight units for equivalent flow conditions compared to riprap. This is significant when riprap of a required size is not available.

Concrete armor units are factory-manufactured precast units and are commercially produced by proprietary firms. Since these blocks are proprietary, standard details of these blocks can be obtained from vendors supplying them. The prominent suppliers of these blocks are

1. Contech Construction Products Inc. (www.contech-cpi.com)
2. Pavestone Company (hydropave.com)
3. Armortec Inc., Bowling Green, Kentucky (prepared by Ayres Associates)

Grout-Filled Bags and Mats

Grout bags are fabric shells that are filled with concrete. Grout mats are single, continuous layers of fabric with pockets, or cells, that are filled with concrete. Grout bags are smaller units that can be stacked in a manner similar to gabions. Hence, grout-filled mats or bags are essentially artificial riprap.

Properly installed grout mats and grout bags reduce scour depth to a degree generally comparable with riprap. In cases of small bridges, bags can be installed where it is difficult to bring in equipment for the placement of riprap. The main body of literature pertaining to grout-filled mats and bags is contained in Fotherby (1992).

Grout bags are more expensive than riprap. Costs in 2007 were estimated to be $25 to $30 per square foot for each layer of grout bags.

Design Guidelines for Grout Bags The design size of a bag or depth of a layer depends upon the following:

1. A design flood velocity of 5 to 10 ft/s is needed.
2. A computed scour depth for contraction and local scour should be 3 to 6 ft.
3. When hydrostatic pressure builds up, the deadweight of bags should exceed the uplift pressure. The mattresses should be provided with filter drains or drain holes for pressure relief.
4. Depending upon the application, bags may vary in capacity, from standard cement bag size to about 5 ft^3, while mattresses are larger, up to 15 ft^3 in volume.
5. Mats must be bound firmly to the pier for good performance. Mats should be installed with their top surfaces flush to the bed.
6. Grout bags should be sized and placed in a manner similar to riprap and underlaid by a geotextile filter with a partial cover or filter layer. Any means to render the surface of bags rough and angular will aid in performance.
7. Properly sized bags are more effective when they extend a single layer of protection laterally than when they are stacked. Efforts should be made to avoid stacking of grout bags.
8. Flexible bags of sand may be preferable to grout-filled bags.

4.14 FHWA TIGER Program Funding

In addition to conventional funding programs available to the owners of bridges for fixing deficient bridges, an FHWA funding program can benefit. For example, ABC comes at a price, since new construction equipment is needed. Construction funding sources are provided by the federal and state governments. The public or the users indirectly contribute through various taxes and tolls charged on highways and bridges. More funds are required when innovative methods need to be introduced since the contractors charge more in order to train their engineers and technicians. The TIGER funding program, detailed below, is one source for the needed funds.

U.S. DOT Transportation Investment Generating Economic Recovery (TIGER) Program

The TIGER program enables the U.S. DOT to use a rigorous process to select rail, transit, port, and road projects. Bridges on rivers may qualify for the grant. The U.S. DOT seeks transportation projects that promise exceptional benefits, explore ways to be completed faster and to reduce construction costs, and improve infrastructure, thus helping to make communities more livable and sustainable.

Since the TIGER program was launched in 2009, Congress has allocated more than $4.1 billion for the grants. On January 17, 2014, President Barack Obama signed the Consolidated Appropriations Act of 2014, which included $600 million for TIGER program funds for fiscal year 2012.

4.15 Conclusions

Floods are not always just local issues, but can create a national crisis. Human welfare and disaster management issues take top priority by the state, and bridge-related issues (which also need millions of dollars to fix) are likely to be placed on the back burner. Restoring highways and bridges as quickly as possible will help the affected community and the authorities in dealing with the disaster and perhaps in saving hundreds of lives. The U.S. Army may be called upon to help in using the river for transportation purposes but may not be involved with fixing the bridge, which is a longer-term process.

Floods affect river bridges adversely. However, currently more guidelines are available for the design and construction of bridges on normal intersections than for river bridges. Hydrologic, hydraulic, and scour analysis subjects are more challenging than those concerning the less complex road bridges. Added issues are emergency monitoring and getting familiar with the vast variety of armoring countermeasures (their selection, design, and construction issues) and the more complex river training measures.

In addition there are environmental issues of overtopping, preserving the marine environment, drainage of water, debris removal, constructability of foundations underwater, and corrosion. Tidal scour for the coastal area bridges is a subject in itself.

In this chapter, an overview of scour causing damages to bridges and other highway structures and the approaches to bridges has been presented. Preventive methods are a must. This is only possible by taking a fresh approach in design and construction and by utilizing advances in construction technology. Repairs and rehabilitation are needed on an emergency basis; therefore ABC methods are required. Prefabricating bridges off-site on dry land and in factories can save construction time and provide better quality control.

Case studies of a number of such bridges that have been were repaired successfully in recent years are presented as reconstruction examples. An important development is seen in the new construction materials such as FRP concrete and CFRP concrete. Ultra-high-performance concrete (UHPC), self-consolidating concrete (SCC), modern paints, and weathering 70 and 100 ksi steels are new materials, all of which have performed well. Recycled materials (such as recycled concrete aggregates and rerolling of steel) are a great asset in availability and can save millions of dollars. Use of robots for underwater construction is being made possible. New techniques for remote monitoring, installing accurate sensors, Lidar, and use of aerial drones are being successfully introduced.

European countermeasure technology is much older than that of the United States. The FHWA seems to have benefited from research and publications of their scientific organizations. However, European funding opportunities seem to be limited, their highway network and traffic requirements are not as intense as in the United States, and their rivers do not extend as far. AASHTO and FHWA specifications published in simpler English and in greater detail than in the past are being adopted on a global basis, for all types of bridges. Each state has a mandate for constructing safe and economical highway bridges. Some states have developed useful guidelines applicable to their needs. U.S. transportation demands with vast networks are extensive and transportation agencies should focus on training engineers in the related technologies and developing scientific and mathematical methods. The introduction of science, technology, engineering, and mathematics (STEM) both at the school and college level when applied will promote civil engineering knowledge and will eventually be helpful in solving engineering issues. Heed the proverb "What you sow, so shall you reap."

Study of Recent Floods in Pakistan: Bridge engineers can help in postflood disaster management by devising new techniques and introducing new procedures. Case studies of recurring floods in Pakistan are again presented. The cumulative Q values at gauged bridges in Pakistan comprise a number of variables:

1. Heavy rainfall (monsoons); peak flows within a few hours of heaviest rainfall
2. Sediment transport (silting) carried down the Himalayan steep slopes and deposited at the bridges
3. Snowmelt from the snowcapped world's largest range of mountains
4. The surge in river water from the eastern Indian rivers, which are also flooded at the same time and merge with the Indus River

The bridge openings are reduced by projections of abutments, piers, and wing walls. The available cross-sectional area A increases the flood velocity V ($V = Q/A$) where Q is the flood water discharge. The concepts and the theoretical approach used in developing regression equations are universal. Verification of measured reliable flood flow results from a large number of gauges will be required at or near the bridges; gauges need to be installed as a first step.

If databases with correct basin characteristics (and other parameters applicable) are furnished, a software package similar to StreamStats can be developed for computing peak floods. Accurate databases are necessary for the physical factors such as the drainage area, longitudinal slope of river, and population density. Accurate databases will help save lives and assets and will prevent destruction of agricultural land in developing countries around the globe. Modern U.S. technology can offer leadership in making bridges safe enough and averting flood disasters.

4.16 Bibliography

Bates, P. D., and De Roo, A. P. J. "A Simple Raster-Based Model for Flood Inundation Simulation," *Journal of Hydrology*, 236: 54–77, 2000.

Chow, V. T., *Open-Channel Hydraulics*, Blackburn Press, Caldwell, NJ, 2009.

Federal Coding Guide, Federal Highway Administration, Washington, D.C., 1994.

"FEMA Firm Item ID 42017C0129F." Map. FEMA Map Service Center, Washington, D.C., 1999.

Hag-Elsafi, O., Kunin, J., Alampalli, S., and Conway, T., "Strengthening of Route 378 Bridge Over Wynantskill Creek In New York Using FRP Laminates," Transportation Research and Development Bureau, NYS DOT, REPORT FHWA/NY/SR-01/135, 2001.

Hamill, L., *Bridge Hydraulics*, E & FN Spon, New York, 1999.

Harrison R., Rasheed H., Peterman, R., and ALkhrdaji, T., "Comparison of four techniques for strengthenig Concrete Beams Using Performance and Practicality Criteria," Proceedings of ASCE Structures Congress, St. Louis, Missouri, 2006.

HEC-1: *Flood Hydrograph Package: User's Manual*. Micromedia, Toronto, Ontario, 1990. Print.

Horne, W. A., "Evaluating Unknown Bridge Foundations, Is It Worth It?" *Proceedings of the International Water Resources Engineering Conference*, Memphis, TN, August 1998.

Jones, J. S., Bertoldi, D. A. and Stein, S. "Alternative scour countermeasures," Proc. Hydraulic Engineering '95, ASCE, New York, 1995.

Khan, M. A., *Bridge and Highway Structure Rehabilitation and Repair*, McGraw-Hill, New York, 2010.

Lagasse, P. F., Zevenbergen, L. W., Schall, J. D., and Clopper, P. E. *Bridge Scour and Stream Instability Countermeasures Experience*, Federal Highway Administration, 2001.

Load Resistance Factor Design (LRFD) Bridge Design Specifications, American Association of State Highway and Transportation Officials, Washington, D.C., 1998.

Manual for Condition Evaluation of Bridges, including all Interims, American Association of State Highway and Transportation Officials, Washington, D.C., 1994.

Melville, B. W., and Coleman, S. E. *Bridge Scour*, Water Resources Publication, Highland Ranch, CO, 2000.

Model Drainage Manual, American Association of State Highway and Transportation Officials, Washington, D.C., 2005. Office of Hydraulic Engineering, Ohio DOT, Location and Design Manual, Volume 2 (guidelines for the hydraulic design of highway drainage facilities), January 2015.

Parker, G., Toro-Escobar, C. and Voigt, R. L. "Countermeasures to Protect Bridge Pier from Scour," University of Minnesota, National Cooperative Highway Research Program, Project 24-7, 1998.

Park, S. H., *Bridge Inspection and Structural Analysis*, Trenton, NJ, 1999.

Recording and Coding Guide for the Structure Inventory and Appraisal of the Nation's Bridges, Federal Highway Administration Report FHWA-ED-89-044, Washington, D.C., 1989.

Richardson, E.V. and Davis, S.R., Evaluating Scour at Bridges, 3rd ed., ASCE, Journal of Hydraulic Division, VA, 1999.

Richardson, E. V., and Lagasse, P. F. *Instrumentation for Measuring Scour at Bridge Piers and Abutments*, Final Report, Phase III, NCHRP Project no.21-3, Transportation Research Board, Washington, D.C., 1994.

Ruff, J. F., and Fotherby, L. M. *Bridge Scour Protection Systems Using Toskanes*, Engineering Center, Colorado State University, Fort Collins, 1995.

Russell, J. S., "Mississippi Flood Failures Demand New Vision," Army Corps of Engineers, Bloomberg, NY, Aug. 23, 2011.

Sayama, T., Ozawa, G., Kawakami, T., Nabesaka, S., and Fukami, K. "Rainfall–Runoff Inundation Analysis of the 2010 Pakistan Flood in the Kabul River Basin," *Hydrological Sciences Journal*, 57:2, 2012.

Scour at Bridges, Technical Advisory 5140.20, Federal Highway Administration, Providence, RI, 1988.

Sheppard, D. M., *Bridge Scour in Tidal Waterways*, Transportation Research Board, Washington, D.C., 1995.

Smith, J. W., and Bright, S., "Upgrading Orthotropic Bridge Decks with Fiber Reinforced Composites," Proceedings of the International Conference, High Performance Materials in Bridges, United Engineering Foundation, Inc., Kona, Hawaii, July 29–August 3, 2001.

Soil Erosion and Sediment Control Manual, New Jersey Department of Transportation, Trenton, 2008.

Sorrell, R. C., *Computing Flood Discharges for Small Ungaged Watersheds*, Michigan Department of Natural Resources and Environment Land and Water Management Division, June 22, 2010.

"Stream Stability and Scour at Highway Bridges," National Highway Institute Course no. 13046, Publication no. FHWA HI-91-011, U.S. Department of Transportation, Federal Highway Administration, October 1999.

Stream Stability and Scour at Highway Bridges, Water Resources Eng., Compendium of Papers, (1991–1998), American Society of Civil Engineers, 1999.

"Stream Stability at Highway Structures," *Journal of Hydraulic Engineering*, Circular No. 20, Federal Highway Administration, 1997.

Technical Manual for Land Use Regulation Program, Bureau of Inland and Coastal Regulations Stream Encroachment Permits, New Jersey Department of Environmental Protection, Trenton, 1993.

Tennessee Drainage Manual (TDOT Design Division Drainage Manual, Chapter 5, May, 2011) Watson, K. M., and Schopp, R.D., USGS, Scientific Investigations Report 2009–5167, "Methodology for Estimation of Flood Magnitude and Frequency for New Jersey Streams," Federal Highway Administration, 2009.

Underwater Evaluation and Repairs of Bridge Components, FHWA-DP-98-1, Final Report, Federal Highway Administration, Washington, D.C., November 1995.

Underwater Inspection and Evaluation of New Jersey Bridges, New Jersey Department of Transportation, Trenton, May 1994.

Underwater Inspection of Bridges, FHWA-DP-80-1 Final Report, Federal Highway Administration, November 1989.

"Un-gauged Sites." USGS Water Resources of the United States. Web. Oct. 27, 2011. http://water.usgs.gov/osw/streamstats/un-gauged.html

"USGS Pennsylvania Water Science Center." *Water Resources of Pennsylvania*. Stream Stats. Web. http://pa.water.usgs.gov/

"USGS Surface-Water Data for the Nation." USGS Water Data for the Nation. USGS. Web. Oct. 28, 2011. http://waterdata.usgs.gov/usa/nwis/sw

Wilson, M., et al., "Modeling Large-Scale Inundation of Amazonian Seasonally Flooded Wetlands," *Geophysical Research Letters*, 34, L15404, 2007.

References

Agostini, R., L. Cesario, F. Ferraialo, and A. Papetti, *Flexible Gabion and Reno Mattress Structures in River and Stream Training Works*, Officine Maccaferri, Bologna, Italy, 1988.

Aktan, A. E., Chuntavan, C., Lee, K. L., and Toksoy, T. "Structural Identification of a Steel-Stringer Bridge for Nondestructive Evaluation," *Transportation Research Record*, no. 1393, 1993, pp. 1–36.

Aktan, A. E., and Helmicki, A. J. "Instrumentation, Testing, and Monitoring of Reinforced Deck-on-Steel Girder Bridges: Mid-Term Project Report," Report no. UCCII94, FHWA/ODOT Contract no. 7665, College of Engineering, University of Cincinnati, OH, 1995.

Applied University Research (AUR) Inc., Bridge Scour Prevention, Blacksburg, VA, 2010.

Aref A. and Alampalli, S. "Dynamic and fatigue response of a truss bridge with fiber reinforced polymer deck." International Journal of Fatigue, 2006.

Ashmore, P. E., and Parker, G. "Confluence Scour in Coarse Braided Streams," *Water Resources Research*, 19, 1983.

Balázs, G. L. and Borosnyói, A. "Prestressing with CFRP Tendons," Proceedings of the International Conference, High Performance Materials in Bridges, United Engineering Foundation, Inc., Kona, Hawaii, July 29–August 3, 2001.

Briaud, J. L., Chen, H. C., Li, Y., Nurtjahyo, P., and Wang, J. *Complex Pier Scour and Contraction Scour in Cohesive Soils*, NCH,RP Report 24-15, Transportation Research Board, National Research Council, Washington, D.C., 2012.

Bridge Scour Evaluation Program, "Stage II—In-Depth Scour Evaluation," Phase 3, New Jersey Department of Transportation, June 2000.

Brown, S. A., and Clyde, E. S. "Design of Riprap Revetment," *Journal of Hydraulic Engineering* Circular 11 (HEC-11), Report no. FHWA-IP-89-016, Federal Highway Administration, U.S. Department of Transportation, Washington, D.C., 1989.

Brunner, G. W., "Computer and Physical Modeling of Bridge Scour," *ASCE Compendium of Papers on "Stream Stability and Scour at Highway Bridges,"* ASCE Water Resources Engineering Conferences 1991 to 1998, published by American Society of Civil Engineers, Reston, Virginia 1998.

Chajes, M.J., Gillespie, J.W., Mertz, D.R. , Shenton III, H.W. and Eckel II, D.A., (2000) "Delaware's First All-Composite Bridge", Proceedings of the ASCE Structures Congress, Philadelphia, PA, May 2000.

Cinotto, P. J., and White, K.E., (in cooperation with Pennsylvania DOT), "Procedures for Bridge Scour Assessments at Bridges in Pennsylvania," Open File Report 00-64, U.S. Department of the Interior, U.S. Geological Survey, Lemoyne, PA, 2000.

Earls, C.J., Shah, B.J., Thomas, S.J., "Compactness and Bracing Requirements for High Performance Steel Girders," ASCE Special Publication-High Performance Materials in Bridges, American Society of Civil Engineers, Reston, Virginia, 2003.

Ettema, R., "Flume Experiments on Abutment Scour, Hydroscience and Engineering," University of Iowa, vol. 1, *Proceedings of First International Conference on Scour of Foundations*, Texas A&M University, College Station, November 2002.

FHWA, Bridge Construction, Manual on Use of Self-Propelled Modular Transporters to Remove and Replace Bridges, Washington, D.C., June 2007.

Fotherby, L. M., "Alternatives to Riprap for Protection against Local Scour at Bridge Piers," *Transportation Research Record*, vol. 1420, Transportation Research Board, National Research Council, Washington, D.C., 1992.

Gao, D., Posada, L., and Nordin, C., "Pier Scour Equations Used in China," *ASCE Compendium of Papers on "Stream Stability and Scour at Highway Bridges,"* 1998.

Henneberg, M. F., and Strause, J.L., (in cooperation with Pennsylvania DOT), *Software User's Guide for Determining the Pennsylvania Scour-Critical Indicator Code and Streambed Scour Assessment Rating for Roadway Bridges*, Open-File Report 01-446, U.S. Department of the Interior, U.S. Geological Survey, New Cumberland, PA, 2002.

Khan, M. A., Zarriello, J. and Baig, S. A. "Stream Instability Study for New Jersey Bridges," *Proceedings of First International Conference on Foundation Engineering*, Texas A&M University, College Station, Texas, November, 2002.

Kibriya, T., Performance of Blended Cement in High Strength Self Compacting Concrete, ASCE Structures Congress, 2006.

Klaassen, G. J., and Vermeer, K. "Confluence Scour in Large Braided Rivers with Fine Bed Material," *Proceedings of International Conference on Fluvial Hydraulics*, Budapest, 1988.

Kosicki, A., and Johanesson, H. "The Challenges of Scour Prediction," *Proceedings of First International Conference on Scour of Foundations*, Texas A&M University, College Station, November 2002.

Kumar, Virender., Raju, K. and Vittal, N. "Reduction of Local Scour around Bridge Piers Using Slots and Collars," *Journal of Hydraulic Engineering*, 125(12), Federal Highway Administration, Washington, D.C., 1999.

Lagasse, P. F., Thompson, P. L. and Sabol, S. A. "Guarding against Scour," *Civil Engineering*, 65(6):56–59, 1995.

May, R. W. P., Ackers, J. C. and Kirby, A. M. *Manual on Scour at Bridges and Other Hydraulic Structures*, Construction Industry Research and Information Association, London, 2002.

May, R., "Comparison of General Scour Prediction Equations for River Crossings," *Proceedings of First International Conference on Scour of Foundations*, Texas A&M University, College Station, November 2002.

Maynord, S. T., "Gabion-Mattress Channel-Protection Design," *Journal of Hydraulic Engineering*, 121(7): 519–522, 1995.

McCorquodale, J. A., "Cable-Tied Concrete Block Erosion Protection," *Proceedings of Hydraulic Engineering Conference*, San Francisco, 1991.

Melville, B. W., and Hadfield, A., "Use of Sacrificial Piles as Pier Scour Countermeasures," *Journal of Hydraulic Engineering*, 125(11), 1999.

Melville, B. W., and Ettema, R. "Bridge Abutment Scour in Compound Channels," *ASCE Compendium of Papers on "Stream Stability and Scour at Highway Bridges,"* ASCE, Reston, VA, 1998.

Melville, B. W., and Coleman, S.E., *Bridge Scour*, Water Resource Publications, LLC, Highlands Ranch, CO, 2000.

Parker, G., Toro-Escobar, C., and Voigt, R.L., *Countermeasures to Protect Bridge Piers from Scour*, Users Guide and Final Report, NCHRP 24-7, National Cooperative Highway Research Program, Transportation Research Board, National Research Council, 1998.

Pennsylvania Department of Transportation, Bureau of Design, Bridge Management System (BMS), *Coding Manual*, 1993, office version: Pennsylvania Department of Transportation Publication 100A.

Predictions," *Proceedings of First International Conference on Scour of Foundations*, Texas A&M University, College Station, November 2002.

"Procedures for Estimating Scour at Bridge Abutments, Parts 1 and 2—Live Bed Scour and Clear Water Scour," Maryland State Highway Administration, *ASCE Compendium of Papers on "Stream Stability and Scour at Highway Bridges,"* 1998.

Racin, J. A., *Gabion Facilities along the Pacific Coast Highway*, Report no. FHWA-CA-TL-93-17, California Department of Transportation, Sacramento, 19912.

Reihsen, G., and Harrison, L. J. "Debris-Control Structures," *Journal of Hydraulic Engineering*, Circular no. 9, Federal Highway Administration, Washington, D.C., 1971.

Recording and Coding Guide for the Structure Inventory and Appraisal of the Nation's Bridges, Federal Highway Administration Report no. FHWA—PD-96-001, Washington, D.C., December 1995.

Recording and Coding Guide for the Structure Inventory and Appraisal of the Nation's Bridges, Federal Highway Administration, Report FHWA-ED-89-044, Washington, D.C., 1989, p. 115.

Revisions to the National Bridge Inspection Standards (NBIS), Federal Highway Administration, Washington, D.C., 1988.

Richardson, E. V., and Lagasse, P. F. "Instrumentation for Measuring Scour at Bridge Piers and Abutments," Final Report, Phase 3, Transportation Research Board, Washington, D.C., 1994.

Sarda, V. A., Protection of Bridge Abutments from Scour, *Journal of Indian Water Resources Society*, vol. 30, no.1, January 2010.

Shenton, III, H., Chajes, M., Johnson, W., Mertz, D., and Gillespie, J., "Continuous, Long-Term Monitoring of Two Advanced Polymer Composite Bridges," *Advanced Technology in Structural Engineering*, ASCE Library, 2000.

Smith, J.W. and Bright S., "Upgrading Orthotropic Bridge Decks with Fiber Reinforced Composites," ASCE Conference Proceedings, 2013.

Soil Erosion and Sediment Control Standards, New Jersey Department of Transportation, 1989.

Tedrow, J. C. F., *Soils of New Jersey*, Krieger, Malabar, FL, 1986.

U.S. Department of Agriculture, Forest Service, Bridge Scour Evaluation, Screening, Analysis and Countermeasures, Technical and Development Program.

U.S. Department of the Interior, U.S. Geological Survey, edited by Secretary, United States Department of the Interior, Washington, D.C.

Underwater Evaluation and Repairs of Bridge Components, Federal Highway Administration, Report no. FHWA-DP-98-1, Final Report, Washington, D.C., November 1995.

Underwater Inspection of Bridges, Federal Highway Administration, Report no. FHWA-DP-80-1, Final Report, Washington, D.C., November 1989.

Wells, A., "Asset Management—Listen and Learn," *Bridge Design and Engineering*, Third Quarter, 2001.

Weston, D.F., Stuffle, T.J., West, J., Shenton, H.W. III, and Chajes, M.J.,

Wire Mesh Gabions, U.S. Army Corps of Engineers, Report Cw-02541, Washington, D.C., 1991.

Accelerated Bridge Construction

Capers, H. "Accelerated Bridge Pier Construction in the U.S.: Seismic Implications," *Rapid Bridge Construction in New Jersey. What Was in It for Us?* Head of Bridge Design Department, New Jersey Department of Transportation Seminar, Trenton, 1999.

Capers, Harry, *Innovative Ideas for Designing New Jersey's Bridges*, Head of Bridge Design Department, New Jersey Department of Transportation Seminar, Trenton, 2000.

J. Muller. High Performance Steel Bridge Concepts. American Iron and Steel Institute, November 1996.

Khan, M. A., *Accelerated Bridge Construction, Best Practices and Techniques*, Elsevier, Butterworth-Heinemann, 2015.

Khan, M. A., and Dunne, R. W., "Recent Developments in ABC Technology," *FHWA Conference on Accelerated Bridge Construction*, Baltimore, MD, 2008.

Mistry, V., and Magnus, A., *Prefabricated Bridge Technology: Get In, Get Out and Stay Out*, Public Roads, FHWA, Nov/Dec 2006.

"Precast Slab System Workshop Summary Report," Minnesota DOT/FHWA, September 2005.

Prefabricated Bridges, American Association of State Highway and Transportation Officials, Technology Implementation Group, May 2002.

Prefabricated Bridge Elements and Systems, Accelerating Innovations, Federal Highway Administration, Washington, D.C., August, 2013.

New Materials and Technology

ACI 440.2R-02, "The Design and Construction of Externally Bonded FRP Systems for Strengthening Concrete Structures," ACI Committee 440, July 2002.

Guide for Design-Build Procurement, American Association of State Highway and Transportation Officials, Item Code 10-DPB-1, Washington, D.C., 2008.

Guide Specifications for Polymer Concrete Bridge Deck Overlays, American Association of State Highway and Transportation Officials, Task Force 34 Report, Washington, D.C., 1995.

Manual for Corrosion Protection of Concrete Components in Bridges, American Association of State Highway and Transportation Officials, Task Force 32 Report, Washington, D.C.

Manual on Uniform Traffic Control Devices (MUTCD), 3d ed., American Association of State Highway and Transportation Officials, Washington, D.C., 2003.

Talebi, M., Meehan, C., Cacciola, D., and Becker, M., "Rapid Replacement," American Society of Civil Engineers, *Civil Engineering Magazine*, April 2014.

The Use and State of the Practice of Fiber Reinforced Concrete, American Association of State Highway and Transportation Officials, AGC-ARTBA Joint Committee, Task Force 36. Washington, D.C., 2001.

Coating System

Attanayake, U., Liang, X., Ng, S., and Aktan, H. "Penetrating Sealants for Concrete Bridge Decks—Selection Procedure," *Journal of Bridge Engineering*, American Society of Civil Engineers, September 2006.

Guide Specifications for Coating Systems with Inorganic Zinc-Rich Primer, American Association of State Highway and Transportation Officials and National Steel Bridge Association Steel Bridge Collaboration, Washington, D.C., 2006.

Constructability

Schlafly, T., and Ericken, J. "Saying What You Mean—Weld Symbols," *Modern Steel Construction*, July 2007.

Welding Design Procedures and Inspection, U.S. Army Corps of Engineers, Washington, D.C., 2000.

Context-Sensitive Design

Barratt, S. D., "New York's First Transparent Noise Barrier," *Journal of Bridge Engineering*, American Society of Civil Engineers, July/August 2006.

D'Agostino, M. J., "Context Sensitive Design Precast Arch Bridges Boost Auto Research Park's Image," *Journal of Bridge Engineering*, Amercan Society of Civil Engineers, July/August 2006.

Analysis and Design

Bridge Manual, New York State Department of Transportation, Office of Structures, Albany, May 2011.

Bridge Design Manual, BDM 2007, Ohio Department of Transportation, Columbus, 2007.

Design Manual, Part 4, vol. 1, *Structures*, Pennsylvania Department of Transportation, Harrisburg, 2006.

FDOT Structures Manual Florida Department of Transportation, Tallahassee, FL, January, 2015.

International Handbook of Earthquake Engineering: Codes, Programs and Examples (IHEE), Kluwer Academic Publishers, The Netherlands, 1995.

Khan, M. A., and Kemp, K. O. "Elastic Full Composite Action in Slab-Beam Systems," *The Structural Engineer*, Institution of Structural Engineers, London, 1970.
Structural Standards Manual, Maryland State Highway Administration, Baltimore, MD, 2006.

Inspection

Bridge Inspection Reference Manual (BIRM), Publication No. FHWA NHI 12-049 February, 2012.
Hill, David, "Researchers Have High Hopes for Drone Use in Transportation," *Civil Engineering Journal*, American Society of Civil Engineers, April 2014.
Underwater Evaluation and Repair of Bridge Components, Federal Highway Administration, Demonstration Project, Washington, D.C., 1995.
Underwater Inspection of Bridges, FHWA Demonstration Project, Washington, D.C., 1989.

CHAPTER 5

Past Failures of Bridges and Those Vulnerable to Failure

In this chapter an investigation is made into the common causes of bridge failures and the diagnosis of the type of deficiencies in existing bridges that may lead to further failures. A review is given of the collapse of historical bridges from a variety of causes and those that failed from flood scour and hurricanes such as Floyd and Katrina and from ship and floating ice impacts. Lessons are learned from selected case studies. The types of bridges considered are highway bridges and bridges on nonnavigable and navigable rivers.

Forensic engineering applications are introduced. Recommendations are made for avoiding failures, and rehabilitation and underwater repair methods are described. These methods include the countermeasures and the innovative methods described in Chapters 3 and 4.

Some significant historic examples are described in other chapters. Over 500 failures of bridge structures in the United States between 1989 and 2000 were studied by the FHWA and the states affected by floods. The age of the failed bridges ranged from 1 year (during construction) to more than 150 years. Those bridges found to be scour-critical are now in various stages of remediation, ranging from monitoring to outright replacement based on funding and budget allocation, on a yearly basis.

5.1 How River Bridges Become Structurally Deficient

Bridges become structurally deficient from a variety of causes. Floods are one of the main causes because if the foundation moves or settles, the entire bridge is damaged and becomes out of alignment. The entire length of the highway needs to be shut down since detours for existing number of lanes are not usually available. Structurally deficient bridges are vulnerable to failure. A deficiency may be regarded as the first step leading to failure unless it is resolved in a timely manner. Heavy and fast-moving traffic and the environment (elements of nature) can act separately or combine to cause failures.

Examples of common environmental forces are rain, wind, snow, floods, daily and seasonal temperature changes, landslides, and earthquakes. Evaporation of river water and precipitation cause corrosion of the structural components. A moving live load may occur for 24 hours and for 365 days of the year. Reversals of live load deflection from asymmetric lane loads build up fatigue, residual stresses, and settlement of foundations.

Failures serve as our best guidelines in finding out the reason for the failure, what went wrong, and what to avoid to prevent similar failures in the future.

1. Cracks appear in the reinforced concrete deck. There are a variety of reasons as follows:
   - Debris prevents deck drainage.
   - Standing water promotes deck deterioration.
   - Rainwater and deicing chemicals penetrate through concrete and corrode steel reinforcement, causing deck spalling.
   - Speed, surface friction, roughness of concrete, and heavy truck suspension interact to amplify stress in the deck.
2. Steel or prestressed concrete girders can be damaged. Girders are subject to corrosion, water damage, metal fatigue, and reversible stresses from vibration. Debris-clogged deck joints prevent movement necessary to relieve superstructure stresses.
3. Bearings are subject to corrosion, misalignment, decay, and clogging from debris, preventing movement.
4. Substructure can suffer from surface cracks on walls, settlement, and damage due to the impact of floating debris.
5. Foundations can be damaged. Flood velocity can scour soil under foundations. Cracking of the substructure is caused by settling of the foundation and overload, when the soil bearing capacity is overestimated. Soil contents comprise clay of typical size < 0.001 mm, silt < 0.075 mm, sand < 4.75 mm, gravel < 100 mm, and rock > 100 mm.

5.2 Vulnerability to Failure of Bridges on Waterways

Bridges on waterways have the following problems: (1) There are restrictions from environmental agencies on placing piers in riverbeds, resulting in longer spans. (2) It is difficult to design and construct deep foundations in flowing water. (3) There are difficulties in maintaining the bridge substructure underwater and in regular painting of corroded girders.

Examples of Bridge Collapse over Rivers

The major physical causes of bridge failures include the following:

1. Extreme events or long-term fatigue and fracture cause bridge failures.
2. A maximum number of bridges have failed over waterways due to increased scour.
3. Bridge abutments and their approach embankments are the most common bridge components damaged during floods. Bridges with narrow waterway openings and erodible soils contribute to bridge collapse.
4. Most failures can be avoided through efficient monitoring and timely maintenance.

5. Foundation scour and soil erosion have contributed to bridge failures. A great majority of scour related bridge failures can be avoided by providing modern countermeasures at foundations. The health of the foundation is the most significant factor in the life of the structure.
6. Evaporation and condensation of river water on steel girders cause corrosion. The use of weathering steel will reduce maintenance and painting costs.

Typical Causes of Bridge Failures

- Foundation scour (Texas and New Jersey bridges)
- Excessive wind (Tacoma Narrows Bridge)
- Fatigue and fracture (numerous railway and highway bridges)
- Corrosion of concrete deck and steel girders (I-95 curved bridge)
- Collision damage due to limited vertical underclearance (bridge in New Jersey)
- Earthquake movements (bridges in California)
- Fire and excessive heat (I-95 bridge in northeast Philadelphia)

Bridge overload and lateral impact forces from trucks, barges, ships, and trains constitute 20 percent of the total bridge failures while substandard design, detailing, construction, material, and maintenance cause the remaining 80 percent of failures.

Much water has flown "both under and over the bridge" since the disaster at Schoharie Creek in New York State. It appears that you cannot always "cross the bridge when you come to it," as the well-known saying goes, especially when the bridge is submerged. Hence "prevention is better than cure."

A. In addition to the above causes, common structural damages can be summarized as
   1. Material failure
   2. Abutment cracking, movement, settlement of pier
   3. Skew effects
   4. Expansive soils
   5. Unknown foundations
   6. Defects in concrete arches
B. Modifying design criteria
C. Precautions and solutions
   1. Bridges are subject to displacement, movement, and sway of substructure due to floods.
   2. If design and reconstruction criteria are correctly implemented, many failures can be prevented.
   3. Some failures may occur due to experimentation with new types of materials or new systems such as undefined and unpredictable material properties in cast-in-place or precast construction.

4. Collapse can be reduced by using standard details and flexible and ductile moment resistance connections.
5. Effective inspection, structural evaluation, and timely rehabilitation is needed for modern bridges and their components to meet the expected life of 75 years or more estimated by AASHTO.

5.3 Analysis of River Bridge Failures

Common Causes of Bridge Failure

The most common causes of bridge failure include

- Structural and design deficiencies
- Construction and supervision mistakes
- Force majeure
- Scour
- Lack of maintenance or inspection

Flood and scour, with the major flood disaster in 1993, contributed to the frequency peak of bridge failures. Bridge overload and lateral impact forces from trucks, barges, ships, and trains constitutes 20 percent of all bridge failures.

- If the material strength is overestimated, then the factor of safety of the material is not high enough.
- If the load combinations are not adequate, then the factor of safety of the loads is not high enough.
- Analytical model was based on incorrect assumptions.

Events Leading to Failure

1. Structural deficiency (identified by inspections; action is required)
2. Appearance of minor cracks and signs of possible incremental collapse (immediate action required, bridge temporarily closed for investigation or partial lane closure)
3. Damage and unserviceability (bridge closed for major repairs or retrofit)
4. Collapse and loss of life (replacement required)

Probable Modes of Failure

Material failure modes are likely to be different for steel, concrete, and timber bridges. For steel composite bridges, plastic hinges form at the midspan or at the ends of flange plates. Tension yielding occurs in the bottom flange and in the web to the underside of the top flange, accompanied by cracking at the bottom surface of the slab. In prestressed concrete beams, collapse may occur due to principal tensile stress at anchorages or breakage of the corroded strands.

Modes of Failure for Concrete Bridges

1. Deficiencies in mix design or hot and cold weather concreting procedures
2. Corrosion of reinforcing bars or prestressing tendons
3. Cracking at anchorages due to high principal tensile stress
4. Loss of prestress forces
5. Lack of provision of thermal, shrinkage, and creep stress in design
6. Shear stress or principal tensile stress exceeded at girder supports
7. Foundation movement due to scour during floods, leading to settlement of pier or abutment
8. Excessive wind (Tacoma Narrows Bridge)
9. Fire and excessive heat (I-95 bridge in northeast Philadelphia)
10. Corrosion of concrete deck and steel girders (I-95 curved bridge)
11. Collision damage due to limited vertical underclearance (bridge in New Jersey)

Modes of Failure for Steel Bridges

1. Bending tension stress in a member is exceeded due to long-term fatigue.
2. Shear stress or principal tensile stress is exceeded at girder supports.
3. Failure of bolts or welds occurs at joints.
4. There is local buckling of compression members.
5. Thermal stress in members is increased due to malfunction of the bearings.
6. Foundation movement due to scour during floods leads to settlement of pier or abutment.

Inelastic Dynamic Progressive Collapse

- In steel truss structures, the truss is subjected to a slight quasi-static overload such that the failure is initiated by buckling of a member.
- Experimental results have shown that postbuckling and subsequent cycles of inelastic force deformation behavior of truss members give rise to adverse member force redistribution that leads to complete failure.

5.4 Typical Bridge Failures from Scour

Field experience has shown that following major causes contribute to the scour process:

- Flowing water, its duration and speed.
- The resilience against erosion of the soil in the riverbed, riverbanks, and supporting the bridge foundations. Cohesive soil refers to fine-grain soil, and cohesionless soil refers to coarse-grain soil.

- True cohesion is caused by chemical bond, for example, calcium carbonate or limestone. Apparent cohesion is caused by negative pore pressure.
- The bridge substructure geometry, opening sizes must allow free and unobstructed flow of floodwater
- Foundation design, whether shallow or deep.
- Provision of existing river training measures and countermeasures shielding foundations.

Flood and scour, with the major flood disaster in 1993, contributed to the frequency peak of bridge failures (almost 53 percent of all failures).

Historically, the most frequent causes of bridge failures are attributed to floods.

5.5 Lessons Learned from European Bridge Failures

See Tables 5.1 and 5.2.

Bridge Site	Country	Causes and Lessons Learned
Reuss River	Switzerland	Lateral channel migration and use of "prisms" as countermeasure to correct channel instability.
Oder River	Poland	During the 1997 and 1998 floods, hundreds of bridges were damaged or destroyed due to inadequate foundation design or inadequate waterway opening. More than 8000 bridges were in need of repair.
Salzach River	Salzburg, Austria	Installation of a temporary access dike for bridge repair caused unavoidable construction delays extending into the flood season. It led to a 12-m-deep scour hole at the main bridge pier under spread footing, which then required structural underpinning.

TABLE 5.1 Bridge Failures in Europe

Bridge Site	Country	Causes and Lessons Learned
Eider Barrier	Germany	Geotextile containers as filter layer. Use of fascine mat as toe protection. Extent of riprap mat. Specialized equipment.
Peena River	Germany	Use of geotextile container and riprap for pier protection.
Eastern Scheldt Barrier	Netherlands	Geotextile mat as filter layer. Individually placed riprap stone. Extent of riprap mat. Specialized equipment.
New Waterway Rotterdam	Netherlands	Riprap sizing based on turbulence intensity.

TABLE 5.2 Installation of Countermeasures on European Bridges

5.6 Bridge Failures from Ship Impact on Substructure or Superstructure

Table 5.3 shows the history of bridge failures due to accidents or impact from ships (human error) which occurred in Australia, Canada, Sweden, and the United States. Brief details of failures are listed. In the United States, more than 36,000 bridges are either scour-critical or scour-susceptible.

Bridges	Location	Year	Failure Details
Australia			
Tasman Bridge over Derwent River	Hobart, Tasmania	1975	Ship impact, inexperienced captain
Canada			
Fraser River Swing Bridge	New Westminster/ Vancouver	1975	Ship impact caused 120-m span to collapse—accident
Sweden			
Almö Bridge	Near Gothenburg	1980	Ship impact on steel arch due to lack of visibility in bad weather
United States			
Swing Bridge	Charlestown, Massachusetts	1945	Ship impact hit half-open swing bridge
John P. Grace Memorial Bridge	Cooper River, South Carolina	1946	Ship forced by wind into bridge deck
Bridge near Charleston	Cooper River, South Carolina	1965	Ship impact, error of ship captain
Chesapeake Bay Bridge	Annapolis, Maryland	1970	Military ship hit the bridge during stormy weather, five spans collapse, 11 other spans damaged
Sidney-Lanier Bridge	Brunswick, Georgia	1972	Ship impact, misunderstanding between captain and staff
Truss bridge	Near Mobile, Alabama	1993	Ship impact
Queen Isabella Causeway	South Padre Island, Texas	2001	Bridge struck by four barges and a tugboat
Interstate 40 Bridge over the Arkansas River	Webber Falls, Oklahoma	2002	Ship collides with one of the piers, bridge collapses on 150 m of length
Sunshine Skyway Bridge	Tampa Bay, Florida	1980	Ship impact, not enough care by captain in bad weather
Herbert C. Bonner Bridge	Oregon Inlet, North Carolina	1990	Ship impact, four piers damaged, five spans collapsed

TABLE 5.3 History of Bridge Failures due to Accidents or Impact from Ships (Human Error)

5.7 Additional List of Failures of River Bridges

Bjorn Akesson, in his book *Understanding Bridge Collapses,* has compiled a long list of major bridges located on wide rivers. Although the causes of the failures are mainly design and construction issues rather than foundation scour, it may be pointed out that the construction of bridge substructure in rivers is more difficult than that constructed on dry land. Land bridge piers and abutments are not subjected to moving water forces and soil erosion.

Following is a list of some river bridge failures (extending over 150 years), with the year of failure:

Nineteenth-century failures

Dee Bridge	1847
Ashtabula Bridge	1876
Tay Bridge	1879

Twentieth-century failures

Quebec Bridge	1907
Hasselt Bridge	1938
Sando Bridge	1939
Tacoma Narrows Bridge	1940
Peace River Bridge	1957
Second Narrows Bridge	1958
Kings Bridge	1962
Point Pleasant Bridge	1967
Fourth Danube Bridge	1969
Britannia Bridge	1970
Cleddau Bridge	1970
West Gate Bridge	1970
Rhine Bridge	1971
Zeulenroda Bridge	1973
Reichsbrucke	1976
Almö Bridge	1980

Twenty-first-century failures

Sgt. Aubrey Cosens VC Memorial Bridge 2003

Additional causes of failures are as follows:

Excessive wind (Tacoma Narrows Bridge)
Corrosion of concrete deck and steel girders (I-95 curved bridge)
Collision damage due to limited vertical underclearance (New Jersey bridge)
Fire and excessive heat (I-95 bridge in northeast Philadelphia)

List of Failures due to Additional Causes

- Schoharie Creek Bridge located on New York State Thruway in 1987
- U.S. 51 Bridge over Hatchie River in Tennessee in 1989
- Damages to bridges located on Mississippi River in 1993
- Interstate 5 northbound and southbound bridges over Los Gatos Creek in California in 1995
- Route 46 Bridge on Peckman's River in Passaic County, New Jersey, in 1998
- Ovilla Road Bridge located in Ellis County, Texas, in 2004

5.8 Case Studies of Well-Known Bridge Failures

Flood Damage to Key Iowa Highway I-29 Bridges and in Many States

The failure of important bridges was reported in the media by news agencies at the time of failure. But mainly the social aspects rather than deficiencies were discussed. The technical reasons are highlighted here. Flooding along the Missouri River has eroded bridges and roadbed on Interstate Highway 29 in southwest Iowa, between Kansas City, Missouri, and Fargo, North Dakota.

Prevention of bridge scour has now been a national priority for two decades. A typical example of bridge failure on rivers is underlined by the May 23, 2013, Interstate bridge collapse in Washington. The collapse disrupted the traffic between the United States and Canada, until repairs can be made or a temporary replacement put in place.

Case Studies of Interstate 5 Bridge

The bridge carrying Interstate 5 over the Skagit River in southwestern British Columbia in Canada and northwestern Washington in the United States collapsed. The four-lane highway bridge lies 62 miles north of Seattle and was constructed in 1955. It is one of over 78,000 bridges in Washington State highlighted as not being meeting required standards. The incident also reopens the issue of the safety of bridges in the United States.

Following the fatal collapse of an Interstate bridge in Minnesota the spotlight was thrown on the poor condition of many major structures. The ensuing investigations showed a worrisome large number of bridges in the United States to be functionally obsolete or structurally deficient. Some structures were repaired or replaced, but the state of many bridges in poor condition was ignored.

Case Study of the Tay Bridge Disaster

In December 1879, the central navigation spans of the Tay Bridge collapsed into the Firth of Tay at Dundee in Scotland. The collapse of the bridge, open only 19 months and passed as safe by the Board of Trade, sent shock waves through the Victorian engineering profession and general public.

FIGURE 5.1 Historical marker plaque.

The Historical Wilson Multiple-Arch Bridge Failure

The Wilson Bridge was built in Hagerstown in Washington County, Maryland. It carried traffic until seriously damaged by storm flooding in 1972. Figure 5.1 shows the progressive collapse of the historic Wilson Bridge occurred due to continued scour.

Please refer to "Maryland Historical Trust." *National Register of Historic Places: Wilson's Bridge*. Maryland Historical Trust. 2008-10-05

New Woodrow Wilson Bridge near Washington, D.C.

The Woodrow Wilson Memorial Bridge opened in 1961. Hydraulics Testing of Woodrow Wilson Bridge Designs. Bridge designs were carried out by Sterling Jones of FHWA to make it safe against scour. Testing was carried out at the flume at the FHWA Hydraulics Laboratory. It was reported in the FHWA journal *Public Roads*, March/April 2000. Scour at extremely wide piers has been an area of needed research for several years.

The new bridge shown in Figure 5.2 has fewer but much wider piers than the existing bridge. The piers are designed to have exposed pile foundations to be capped near the water surface. The soil borings indicate that there are some relatively erosion-resistant layers below the streambed in the vicinity of the proposed new bridge.

It connects Alexandria, Virginia, to Oxon Hill, Maryland, carrying both Interstate 495 and Interstate 95. By 1999 it was carrying 200,000 vehicles a day even though it was designed to handle a maximum of 75,000. Eventually the highways on either side of the bridge widened from six lanes to eight, but the bridge obviously couldn't expand, and so it created a terrible bottleneck.

Figure 5.2 Multiple-span Woodrow Wilson Bridge on Potomac River near Washington, D.C.

Failures of Important Bridges on Rivers

Failures of bridges on rivers receive a great deal of publicity due to the loss of billions of dollars and life and property. Such bridges are shown in Figures 5.3 to 5.12, and each one failed for specific reasons. Figure 5.12 shows an aerial view of the collapsed I-35W Bridge on the Mississippi River, which occurred recently.

The discussion of each bridge failure is not presented here.

Figure 5.3 Failure of Webber's Falls Bridge.

Failure of Peace River Bridge

A major river bridge failure due to soil scour occurred in October 1957. The Peace River Long Span Suspension Bridge (on the Alaska Highway) in Taylor, British Columbia, collapsed. The abutment foundation was not designed to resist the pulling tension forces from the cable anchors. Two years earlier, in 1955, one of the tower pier foundations was subjected to scour. Its foundation was already underpinned.

Failure of the Reichbrucke Bridge on the Danube River

Another major river bridge failure due to soil scour occurred in August 1976. The Reichbrucke Long Span Suspension Bridge (on the Danube River in Vienna, Austria) collapsed. Failure occurred due to fracturing of the river pier. Changes in temperature caused shrinkage and creep stress. The concrete pier was unreinforced. In 1936, the same pier was hit by a ship, causing permanent internal stress.

Case Study for Author's Replacement Design of Route 46 Bridge on Peckman's River

Due to Hurricane Floyd in 1997, overtopping of this bridge occurred. Much of the Route 46, New Jersey, area was fully flooded. Abutment settlement occurred and heavy cracking of approaches took place. Temporary pile bents were constructed to transfer the load from abutments. A replacement bridge was designed by the author, using integral abutments with a single row of piles (one of New Jersey's first integral abutment bridges on an active river).

Modern Approach in Planning New Route 46 Bridge Using HEC-18 Recommendations

Due to river meander over 50 years, the alignment of river thalweg became skew to the bridge opening. The wall of one abutment was subjected to impact from high velocity of flood. The author was the design engineer who replaced the Peckman's River Bridge, which was damaged by Hurricane Floyd. See Figures 5.4 and 5.5.

1. Direction of abutment skew is now made parallel to the direction of river flow to minimize scour.
2. Based on hydraulic analysis, the opening size is increased to minimize overtopping flood.
3. Use of integral abutments and integral approaches makes the bridge highly resistant to longitudinal forces.
4. Steel girders are replaced by prestressed spread box beams to prevent corrosion.
5. Deep sheet piling sections are used to minimize scour of abutment and wing walls.

In the United States, more than 36,000 bridges are either scour-critical or scour-susceptible. Some examples of recent bridge failures include these:

- Schoharie Creek bridge on New York State Thruway in 1987
- U.S. Route 51 Bridge over Hatchie River in Tennessee in 1989
- Bridges on Mississippi River in 1993
- Interstate 5 northbound and southbound bridges over Los Gatos Creek in California in 1995
- Route 46 Bridge on Peckman's River in Passaic County, New Jersey, in 1998
- Ovilla Road Bridge in Ellis County, Texas, in 2004

FIGURE 5.4 Collapse of approach slab on Route 46, Peckman's River Bridge, due to Hurricane Floyd.

FIGURE 5.5 Overtopping of the bridge deck resulted in flood debris trapped between the girders.

Figure 5.6 Collapse of a bridge in China.

Figure 5.7 The completed integral abutment bridge located on Peckman's River and important Route 46 in New Jersey designed by author. The original bridge was destroyed during Hurricane Floyd.

By analyzing the reasons for failures, a great deal can be learned in order to avoid the same mistakes and to design safe bridges. Case studies of other bridges that failed are provided in the text.

Note that with the advancement in technology, the failure rate of bridges has decreased. There are design codes available for river bridges. Design criteria for river bridges require hydrology, hydraulics, and scour analysis with provision of scour countermeasures. Construction equipment has advanced and quality control has improved. Modular bridges constructed in factory condition are being used widely. Typical examples of failures are shown in Figures 5.8 to 5.12.

FIGURE 5.8 Failure of Texas Queen Isabella Bridge.

FIGURE 5.9 Abandoned bridge due to overtopping floods.

Figure 5.10 Bridge foundation failure with column bent-type pier.

Figure 5.11 Failure of a hammerhead pier due to foundation settlement.

Figure 5.12 Recent failure of I-35W Bridge on Mississippi River.

5.9 Forensic Engineering Application to Bridge Failures

Engineers are answerable to the public for loss of life and damage resulting from oversight or mistakes. Technical and legal aspects of the investigation of scour failures as well as methods of conducting forensic engineering and expert witnessing need to be understood.

Various Phases of Investigation

The first steps following a collapse are critical. The structural forensic engineer, who is called in following a collapse, plays a crucial role in determining what those first steps should be. He or she is the most qualified to recognize perishable evidence and its potential value. The decisions made will directly affect the evidence upon which investigations will depend. Immediate issues that a structural forensic engineer may be faced with when arriving at a collapse site and in the ensuing days, include

- Safety
- Preserving perishable evidence
- Preserving samples
- Documentation
- Interviews
- Document gathering
- Preliminary evaluation

Case studies of failures have shown that a number of the factors that follow can contribute to the collapse of a bridge.

1. Precollapse deficiency—design defects such as
   - Incorrect assumptions
   - Error in data, incorrect analysis, and noncompliance with code guidelines
   - Incorrect connection details
   - Mistakes in drawings
   - Construction defects such as poor workmanship
   - Substandard materials
   - Inadequate concrete curing
   - Imperfections in steel
   - Lack of fit and lack of quality control
2. Underestimation—the effect of large number of small floods is to make the supporting soil media weaker and vulnerable to scour and foundation settlement. A major flood can do added damage.
3. Unexpected failures—in situations where considerable uncertainty exists in the prediction of accumulated damage or in the ability to reliably inspect critical details, preemptive retrofit strategies appear to be highly desirable.

Pile Design

The ultimate bearing capacity (UBC) of axially loaded piles must be limited to the compressive and/or tensile loads determined for reduced capacity for projected scour. Load redistribution must not be permitted when the axial pile capacity is reached; rather, axial capacity must be limited to the ultimate limit.

Currently only working loads are established by L-pile analysis and other programs. Lateral soil-pile response must be determined by concepts utilizing a coefficient of subgrade modulus provided or approved by the geotechnical engineer. Pile group effects must be considered. A dynamic screening tool is used for pile bents. An evaluation procedure developed by the Alabama DOT may be employed. It is a screening tool in macro and micro flood chart form and basically moves through the following precautionary sequence:

1. Make preliminary or general checks such as the bridge bent being located in water with scour possible.
2. Check bent piles for possible kick-out or plunging failure.
3. Check bent piles for buckling failure.
4. Check transverse bent for bridge centerline pushover failure, from combined gravity and floodwater loadings.

Substructure stabilization, repairs, and foundation improvement through underwater inspection need to be performed for all scour-critical bridges. Repair procedures are listed in the National Highway Institute's *Underwater Inspection Manual* (FHWA-NHI-10-027) published in June 2010.

Foundations can settle owing to scour and weak soil conditions. When a pier tilts, there's a potential for a bridge to collapse without warning. This is definitely a safety issue for travelers. The following types of countermeasures for risk reduction under flood conditions should be installed:

1. *River training measures*: Place guide banks to move scour away from the abutment foundation. Install spurs or bendways weirs at a bend that is migrating toward a bridge abutment. Spurs will redirect the flow away from the abutment.
2. *Hydraulic countermeasures*: Place armoring such as riprap around exposed foundation.
3. *Structural countermeasures*: Underpin the footings that were undermined by using grout or grout bags.
4. Effective monitoring (underwater inspection or instrumentation) is a bridge management tool.

Preventive Measures for Safety of Bridges

An in-depth study of bridge failures leads to following preventive measures:

1. Providing space for bearing inspection chambers.
2. Greater vendor and construction engineer participation in revising and developing design codes for countermeasures should be encouraged.
3. Studying failure mechanisms of different types of structural systems should be encouraged. Examples are using open web girders such as trusses which minimize the impact of flood flow.

4. Maintaining quality control and personnel safety during construction.
5. Introducing rehabilitation codes for old bridges, codes for rehabilitation of mixed old and new structural systems should be developed to enhance life and prevent early collapse. New techniques of repair and rehabilitation of substructure need to be developed and incorporated in the codes.

Diagnostic Design Deficient Modeling

The following are some of the unknowns a bridge designer will typically encounter:

1. Inability to define loads accurately, such as the magnitude and unpredictable level of stress distribution from settlement
2. Limited redundancy in the structural system
3. Inability to fully include plastic behavior of composite action between the concrete deck slab and repeated beams, arching action, creep, and shrinkage strain distribution in deck slab
4. Lack of information on fracture mechanics in general and lack of understanding of fracture of new materials, in particular
5. Inelastic behavior of connections and joints, splices, gusset plates, bolts, and welds
6. Complex behavior as a unified assembly of uncombined multiple components of mixed (old and new) materials and structural systems, resulting from rehabilitation or widening methods
7. Delamination and reduction in strength of concrete deck due to deicing salts (as observed from chain drag test)
8. Malfunction and locking of old bearing assemblies due to lack of maintenance; freezing of expansion bearings; large thermal forces causing compression and local buckling of truss members and flanges
9. Inability to prevent scour at pile top
10. Inability to fully incorporate different types of soil interaction at abutments
11. Lack of drainage behind abutments and pressure buildup behind abutments
12. Inadequacy of the Rankine, Coulomb, or Mononobe-Okabe theories for nonhomogeneous soil conditions for wing walls and stub abutments, resulting in unstable foundation design

Study of Failure Mechanism

Comparisons for 1977–1981, 1982–1988, and 1989–2000 revealed similar trends, with most failures occurring during the bridge's service life. Also, human-induced external events occurred frequently in all three periods.

Importance of Inspections

Regular inspection of transportation facilities has been critical to public safety over the past 20 years, and regulatory changes, identified problems, and continued growth of our population and commerce have resulted in a greater need to perform structural inspections of transportation facilities on a regular basis. Technological advances in information systems have a great impact on data collection and analysis.

A problem of growing damages to bridges was identified over time and it has resulted in increasing the minimum design stress. Older bridges were not designed to meet the new criterion. The AASHTO LRFD Code has recommended 14 load combinations, based on the probability for evaluating strength, serviceability, and fatigue. Extreme conditions such as scour are considered. Principal stresses due to bending and shear resulting from settlement of supports may exceed the allowable values and lead to cracking, formation of plastic hinges, and disintegration of elements. External actions need to be clearly defined in short- and long-term behavior. Excessive deformation such as deflection, rotation, curvature, vibration, side sway, and settlement needs to be kept to a minimum. Code guidelines and design criteria need to be refined based on failure study.

5.10 Recommendations for Avoiding Failures

Span Length

The length of the main span between centerlines of piers at the navigable channel must be based upon

- Coast Guard requirements
- The vessel's collision risk analysis
- Least-cost analysis
- Aesthetic considerations

Preventive and Diagnostic Design

A separate code for preventive and diagnostic design needs to be developed as distinct from AASHTO LRFD Specifications.

An accompanying rehabilitation construction code leading to selective reconstruction of potential damage to concrete or steel members or joints, based on fragility and vulnerability analysis, is also required.

- It should consider all failure modes in light of failure analysis.
- Advanced inspection techniques should be addressed.
- For effective maintenance there is a need for quality control in inspection methods similar to that of aircraft structures.
- At the time of construction, remote sensors at critical stress areas need to be installed and daily stress records monitored every 2 years.

5.11 Rehabilitation and Repair of Existing Structures

Chapter 8 presents detailed guidelines for planning new bridges to reduce or minimize the effects of scour. The important aspects of repair and retrofit of existing bridges are discussed.

Substructure Repairs Prior to Installing Structural Countermeasures

New bridges can be planned and designed to meet the latest criteria, making them safe from the beginning. Existing bridges may have smaller openings, creating

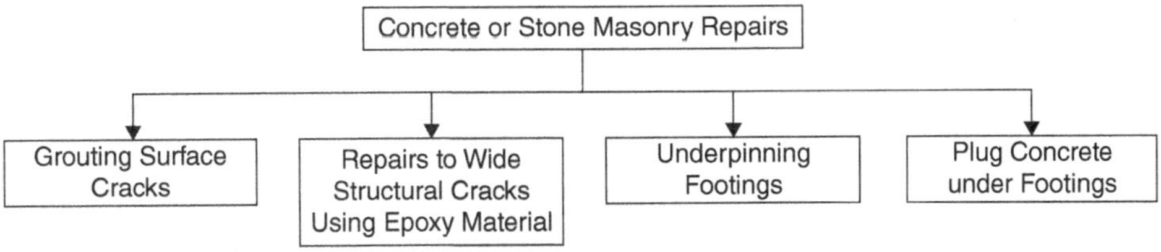

FIGURE 5.13 Types of concrete repairs.

higher velocity. These defects are difficult to fix. Rehabilitation methods and the ease of installation of countermeasures depend upon the construction techniques. Figure 5.13 shows the types of concrete repairs required for substructure and substructure components.

Maintenance of existing roads and bridges makes up the largest share, 44 percent, of New Jersey's Department of Transportation's capital program. According to Kate Slevin, Tri-State Transportation Campaign (TSTC) director, the spending on road and bridge capacity expansion projects makes up 11 percent of the capital program, the largest percentage in nearly a decade.

Use of Effective Monitoring Methods

1. *Underwater inspection or instrumentation as a bridge management tool*: These methods help flood watch and follow-up monitoring of scour-critical bridges.
2. *Use of modern techniques*: The use of wireless and remote sensors enables the movements of bridges to be monitored around the clock. This is most desirable in flood situations. Modern sensors, when installed on scour-critical bridges, minimize the possibility of sudden collapse and serve as a warning for a bridge to be closed. Modern techniques are further discussed in Chapter 9 on scour.
3. *Investigating unknown foundations*: There is a need for geotechnical procedures in evaluating unknown foundations.
4. *Scour safety evaluation*: The scour sufficiency rating needs to be evaluated. A scour analysis is carried out to evaluate scour depths based on FHWA Publication HEC-18. A scour report consists of a detailed field survey, substructure information, scour analysis results, hydraulics-related findings, and countermeasure recommendations. Harnessing the river to reduce flood velocities may also be done. Common countermeasures are riprap, gabion baskets, concrete blocks, and sheet piling.

Structural Planning and Solutions

Planning is a major area of concern. A mistake made in planning can be very difficult to overcome. Bad planning decisions will ultimately lead to bad results. The following planning principles shall be observed:

1. The bridge footprint must be located in a suitable site, ensuring stable foundations.

2. Sharp skew angles in the deck slab must be minimized to avoid stress concentration in the acute corners.
3. A structural system with redundancy and adequate construction materials should be chosen.
4. The optimum number of spans, continuity over supports, and equal spans need to be considered.
5. Provision for future widening of the highway below must be considered. This may require provision of a median, space for future lanes, an acceleration and deceleration lane, emergency shoulders, a sidewalk, and zebra crossing.
6. If a slab-beam bridge is selected, the number of girders and spacing shall be optimum. Girder spacing shall preferably be in the range of 6 to 12 ft. A higher degree of redundancy needs to be considered. In place of through girders that are nonredundant, a minimum of four girders are required. However, an odd number of girders is preferred for stage construction since the middle girder serves as a spine beam, reducing deck. Both fascia and interior girders need to be designed. If the fascia is lighter than the interior, good practice is to adopt the heavier interior girder design throughout. This is necessary for future widening; otherwise the fascia girder and part of the deck will need replacement.
7. Longitudinal joints shall be avoided. In place of transverse deck joints, integral abutments may be preferred. Deck joints usually have maintenance issues due to higher impact from vehicles.
8. Walls of the abutment and pier should be placed at 90° to the direction of traffic flow to avoid any skew. Minimum vertical clearance for new bridges shall be 16 ft 6 in. The horizontal distance to the face of the abutment and pier needs to be a minimum of 30 ft from the edge of the travel lane. This distance may be reduced if guardrails are provided in front of the abutment or pier.
9. Geotechnical borehole results must be applied in selecting shallow or deep foundations. Weaker soils require deep foundations, such as piles, drilled piers, or caissons. For firm soils with a bearing capacity of 2 ksf and above, shallow footings should be considered.
10. Traffic count is required to determine the number of lanes, with traffic projections for 20-year average daily truck traffic (ADTT). A provision for deck widening should be included.

Steps to Avoid Superstructure Failures

1. Durability requirements need to be addressed on a scientific basis.
2. The engineer needs to become familiar with the principles of related engineering disciplines, such as dynamic analysis, hydraulics, and materials science.
3. AASHTO LRFD code provisions and other relevant FHWA guidelines are to be followed:
   - Strength and serviceability load combinations for accelerated constructability that are not covered in the AASHTO LRFD code are presented in Chapter 4. Application of these loads will minimize the number of construction accidents.

- Strength and serviceability load combinations for ship collisions that are not covered in AASHTO LRFD code need to be addressed.
- Collapse can be reduced by using standard details and ductile moment resistance connections.
- The expected life of 75 years or more for modern bridges and their components may not always be achieved without regular inspection, structural evaluation, and timely rehabilitation.

4. If the modeling is deficient, a bridge designer will typically encounter some of the following unknowns:
   - Inability to define loads accurately, such as magnitude and unpredictable level of stress distribution from settlement
   - Limited redundancy in structural system
   - Inability to fully include plastic behavior of composite action between the concrete deck slab and repeated beams, arching action, creep, and shrinkage strain distribution in the deck slab
   - Lack of information on fracture mechanics in general and lack of understanding of fracture of new materials in particular
   - Inelastic behavior of connections and joints, splices, gusset plates, bolts, and welds
   - Complex behavior as a unified assembly of uncombined multiple components of mixed (old and new) materials and structural systems, resulting from rehabilitation or widening methods
   - Delamination and reduction in strength of concrete deck due to deicing salts (as observed from chain drag test)
   - Malfunction and locking of old bearing assemblies due to lack of maintenance; freezing of expansion bearings; large thermal forces causing compression and local buckling of truss members and flanges
   - Inability to prevent scour at pile top
   - Inability to fully incorporate different types of soil interaction at abutments
   - Lack of drainage behind abutments and pressure buildup behind abutments
   - Inadequacy of the Rankine, Coulomb, or Mononobe-Okabe theories for nonhomogeneous soil conditions for wing walls and stub abutments, resulting in unstable foundation design
5. Space for bearing inspection chambers should be provided.
6. Codes for rehabilitation of mixed (old and new) structural systems should be developed to enhance life and prevent early collapse.
7. Greater vendor and construction engineer participation in revising and developing design codes for countermeasures is encouraged.
8. Quality control and personnel safety during construction should be maintained.

Steps to Avoid Substructure Failures

1. The latest techniques of repair and rehabilitation of substructures need to be incorporated in codes.
2. Fragility analysis of reinforced concrete pier to include soil-structure interaction should be carried out. A procedure needs to be developed to evaluate effective soil strains as a function of dynamic shear forces generated at the planar surface of a riverbed due to flowing water. Uncertainties in soil properties can be included in the vulnerability assessment framework.

Steps to Avoid Foundation and Pile Failures

1. In the pile design, the ultimate bearing capacity (UBC) of axially loaded piles must be limited to the compressive and/or tensile loads determined for reduced capacity for projected scour.
2. Load redistribution must not be permitted when the axial pile capacity is reached; rather, axial capacity must be limited to the ultimate limit as established by L-pile analysis.
3. Lateral soil-pile response must be determined by concepts utilizing a coefficient of subgrade modulus provided or approved by the geotechnical engineer. Pile group effects must be considered.
4. Substructure stabilization, repairs, and foundation improvement through underwater inspection need to be carried out for all scour-critical bridges.
5. Use a dynamic screening tool for pile bents. An evaluation procedure developed by the Alabama DOT may be employed. It is a screening tool in macro- and microflood charts.
6. Make preliminary or general checks, such as the bridge bent being located in water with scour possible, including these:
   - Check bent piles for possible kick-out or plunging failure.
   - Check bent piles for buckling failure.
   - Check the bent for transverse to bridge centerline pushover failure from combined gravity and floodwater loadings.
   - Regarding foundation settlement from scour and weak soil conditions, when a pier tilts, there is potential for a bridge to collapse without warning. This definitely is a safety issue for travelers.
7. Install countermeasures for risk reduction as follows:
   - River training measures
   - Placement of guide banks to move scour away from the abutment foundation
   - Installation of spurs or bendways weirs at a bend that is migrating toward a bridge abutment with spurs that redirect the flow away from the abutment
   - For hydraulic countermeasures, placement of armoring such as riprap around exposed foundation

- For structural countermeasures, underpinning of footings that were undermined by using grout or grout bags

Repair versus Replacement Options

An investigation presented in the *Star Ledger* concluded that New Jersey should fix existing roads and bridges rather than build new ones. The same investment approach is applicable to the majority of states. In the past 20 years, many bridges have become deficient because of the increase in ADTT and the live load. The costs of construction of new bridges have gone up both over the rivers and over the intersections. Underwater repairs require a longer time and are difficult to fix in rivers which are deep.

5.12 Underwater Repairs

Table 5.4*a* and *b* shows practical solutions for problems created from floods and scour.

Foundation Items	Suggested Remedy
1. Debris accumulation	Clean debris
2. Undermining	Plug with concrete; use grout bags
3. Cracks in tremie concrete below water	Pressure-inject masonry cracks
4. Spalling in foundation	Use pressure grouting

TABLE 5.4*a* Underwater and Foundation Repairs

Substructure/Superstructure Items	Suggested Remedy
1. Spalls in abutment concrete	Surface repairs with approved patch material
2. Delaminations and cracks in substructure concrete above water	Pressure grouting
3. Voids in breast wall	Epoxy grouting with nonshrink grout
4. Mortar loss in masonry joints	Repointing of mortar
5. Broken stone masonry	Placement of stone and filling with mortar
6. Cracks in apron around piers	Repair of concrete apron or placement of riprap around piers
7. Corrosion of rebars	Clean, paint, and provide adequate cover; dowel anchor bars in concrete holes
8. Advance section loss of structural members	Strengthening of member

TABLE 5.4*b* Repairs above Foundations

5.13 Preventive Actions for Vessel Collision or Floating Ice Impacts

Bridge piers located on navigable rivers are likely to be hit in fog or in darkness, usually from barges or oceangoing ships. Damage to timber fenders which shield the piers may also be caused by floating ice at high velocities. The engineer must assemble the following information:

1. Characteristics of the waterway including:
   - A nautical chart of the waterway
   - Type and geometry of the bridge
   - Preliminary plan and elevation drawings depicting the number, size, and location of the proposed piers, navigation channel, width, depth, and geometry
   - Average current velocity across the waterway
2. Characteristics of the vessels and traffic including:
   - Ship, tug, and barge sizes (length, width, and height)
   - Number of passages for ships, tugs, and barges per year (prediction for 25 years)
   - Vessel displacements
   - Cargo displacements (deadweight tonnage)
   - Draft (depth below the waterline) of ships, tugs, and barges
   - The overall length and speed of tow
3. Accident reports
4. Bridge importance classification

The large number of impacts from ships is also a cause of concern for the shipping industry.

Design for Vessel Collision

The design of all bridges over navigable waters must be checked for possible vessel collision. Conduct a vessel risk analysis to determine the most economical method for protecting the bridge. The number of vessel passages and the vessel sizes are embedded as an integral part of the vessel collision risk analysis software.

1. The Florida DOT's MathCAD software for conducting vessel collision risk analysis may be used. The software computes the risk of collision for several vessel groups with every pier. When the loads and load factor probability are calculated, the overall length of each vessel group is used instead of the length overall (LOA) of a single design vessel.
2. Major widening spanning navigable waterways must be designed for vessel collision. Minor widening spanning navigable waterways will be considered on an individual basis for vessel collision design requirements.

Suggested Preventive Action against Ship Collision

1. Adequate lighting should be provided around bridge substructures to prevent accidents at night or in fog conditions.

2. Timber fenders should be provided around piers to resist the impact of ships on bridge substructures, and should be adequately maintained.
3. Coast Guard crews should be scheduled to monitor the bridges at night.
4. The training of ship captains and crew in navigation under bridges should be improved.

5.14 Conclusions

More bridges fail on waterways than those located on land due to scour. Such failures occur sooner than expected. "Failures are the pillars to success" and we should learn from failures as they suggest what to avoid and not making the same questions twice. The causes are scour, ship impacts, design and construction errors, etc.

Detailed lists of bridge failures have been presented. It is sometimes said that failures are the pillars to success. When the reasons for failures are known, preventive actions can be taken for the future.

It is important to study how bridges become structurally deficient. Functionally obsolete bridges are those where shoulders or median barriers are not provided but are required, where bicycle tracks are not provided but are now required, sight distance is not adequate, or the number of lanes needs to be increased.

After a failure a forensic engineering study is a legal requirement. Preventive actions such as underwater repairs become necessary. Recommendations for the protection of foundations against scour have been discussed in this chapter.

5.15 Bibliography

Akesson, B., *Understanding Bridge Collapses*, Taylor & Francis, London, 2008.

American Association of State Highway and Transportation Officials, *AASHTO LRFD Specifications for Design of Bridges*, 1998 with interims for 1999 and 2000.

American Association of State Highway and Transportation Officials, AASHTO Model Drainage Manual, AASHTO, Washington D.C., 2005.

Baig, S.M., Monaco, A.F., and Patel, J.C., Screening of Bridges in New Jersey for Scour, in Richardson, E.V., and Lagasse, P.F., eds. Stream Stability and Scour at Highway Bridges: Water Resources Engineering Division, ASCE, 1999.

Davis, S. R., J. E. Pagan-Ortiz, and Kelbaugh, L., "Streamlining the Permit Approval Process for the Installation of Scour Countermeasures at Bridges, Compendium of Stream Stability and Scour; papers presented at conferences sponsored by the Water Resources Engineering (Hydraulics) Division of the American Society of Civil Engineers, 1993.

Federal Highway Administration, *Evaluating Scour at Bridges*, Hydraulic Engineering Circular no. 18, 1996.

Federal Highway Administration, *Stream Stability at Highway Structures*, Hydraulic Engineering Circular no. 20, 1997.

Federal Highway Administration, *Countermeasures*, Hydraulic Engineering Circular no. 23, 1998.

Frassinelli, M., and Sciarrino, R., Advocacy Group, "NJ should fix existing roads, bridges rather than build new ones: Advocacy Group, "*The Star-Ledger*, May 17, 2011,

Horne, William A., (1998), "Evaluating Unknown Bridge Foundations: Is it Worth It?," *Proceedings of the 1998 International Water Resources Engineering Conference*, ASCE, Memphis, TN, Part 1, 145-150.

Melville, B., and Coleman, S., *Bridge Scour*, Water Resources Publications, Highlands Ranch, Colorado, 2000.

National Highway Institute, *Underwater Inspection Manual*, Report No. FHWA-NHI-10-027 Washington, D.C., June 2010.

Richardson, E. V., *History of Bridge Scour Research and Evaluations in the U.S.*, American Society of Civil Engineers, 1999.

Sterling Jones, J., "Hydraulics Testing of Wilson Bridge Designs," *FHWA Public Roads*, vol. 63, no. 5, March/April 2000.

U.S. National Transportation Safety Board, *Collapse of New York Thruway (I-90)* Bridge *over the Schoharie Creek near Amsterdam, NY*, Report no. NTSB/HAR-RR:2, Washington, D.C., 1989.

U.S. National Transportation Safety Board, *Collapse of the Northbound U.S. Route 51 Bridge Spans over the Hatchie River near Covington, TN*, Report NTSB/HAR-90-01, Washington, D.C., 1990.

PART 2

Hydrology, Hydraulics, Scour Analysis, and Inspection Methods

CHAPTER 6

Geotechnical and Hydraulic Issues at Scour-Critical Rivers and Bridges

6.1 Standard Procedures for Hydraulic Analysis and Design

The magnitudes of scour and erosion at bridges are dependent upon the geotechnical conditions, hydraulics, and hydrology. This chapter will address the geotechnical and hydraulic issues, while Chapter 7 will address hydrology.

In increasing order of causing scour depths, the following soil conditions are responsible:

1. Stiff soils and rock
2. Cohesive soils
3. Noncohesive soils

The ASCE report card on the performance of bridges has shown that with passage of time and climate changes, more bridges are becoming scour-critical. The factors causing higher scour are linked to geotechnical and hydraulic conditions that can be controlled, while hydrology (or floods) is a natural feature.

Hydraulic analysis in some cases will result in higher velocities and high water elevations. If the area of opening under the bridge is limited, or if there is aggradation and sediment deposit at the upstream or under the bridge or the alignment of flow is at skew angle with bridge geometry, then scour will be high.

AASHTO specifications for resisting scour describe data collection and modeling procedures for analysis. In this chapter, a number of case studies on scour-critical bridges are presented, and the effects of geotechnical conditions are emphasized. Standard methods of preparing hydraulic, geotechnical, and scour evaluation reports are described. In addition, the importance of investigating unknown foundations is discussed.

The following steps are standard requirements:

1. Geotechnical considerations
   a. Boreholes and augers
   b. Soil classification

2. Hydrologic and hydraulic analyses
   a. Bridge waterway
   b. Methods of hydrologic analysis
   c. Methods of hydraulic analysis
   d. Methods of scour analysis using HEC-18 and scour in cohesive soils
   e. Methods of countermeasures design, with the objectives of reducing stream instability using HEC-23 procedures. Emphasis is on using structural countermeasures and different types of deep foundations and scour of piles.
   f. Three levels of analysis are described, and scour code is assigned to scour-critical bridges.
3. Computer software
   a. Hydrology—TR55, PENNSTATE
   b. Hydraulics—HEC-RAS, WSPRO, and UNET

Type of Soil

The type of soil is the most important factor in offering resistance to erosion. Soil properties and response to floods under heavy live loads on bridges comprise a subject in themselves. Rock or stiff soils are most suitable, and sand is the most easily erodible soil. Bed material properties due to sediment transport may be different from those of the original soil in which the river is carved. An in-depth geotechnical investigation based on boreholes is required.

The intensity and duration of floods will affect the rate of scour in the soil. The following represents the duration of maximum scour depths in different soil conditions:

- Sand and gravel bed materials—in hours
- Cohesive bed materials—in days
- Glacial tills, poorly cemented sandstones and shales—in months
- Hard, dense, well-cemented sandstones and shales—in years
- Granites—in centuries

Special river conditions such as debris accumulation under bridges would reduce the available opening size. The formation of scour holes may further contribute to create soil erosion. The condition of any unknown foundation also needs to be determined.

In cohesive soils such as clay, both local scour and contraction scour magnitudes may be similar. However, scour takes place considerably later than in noncohesive sand. The clay content in soil increases cohesion, and relatively large forces are required to erode the riverbed. Higher pulsating drag and lift forces increase the dynamic action on aggregates until the bonds between aggregates are gradually destroyed. Professor Jean-Louis Briaud (Texas A&M University) has proposed the SRICOS method of scour measurement in cohesive soils.

Geological Subdivisions to Identify Soil Types

To determine the grain size distribution for a scour analysis, samples from the riverbed need to be tested in accordance with ASTM D422. Soil types may be broadly classified as follows:

1. **Noncohesive materials:**
   a. Gravels
   b. Sands
   c. Silt

Noncohesive sediments have a granular structure. Individual particles are susceptible to erosion when the applied fluid forces (drag and lift) are greater than the stabilizing forces due to gravity and cohesion with adjacent bed particles. The threshold of movement depends on

   a. Particle size
   b. Density
   c. Shape
   d. Packing
   e. Orientation of the bed material

2. **Cohesive materials:** Most fine-grained sediments possess some cohesion. Cohesive sediments typically require relatively large forces to detach the particles and initiate movement, but relatively small forces to transport the particles away. Cohesive materials are
   a. Silts
   b. Clays

Erodibility of Soil and Shear Stress in Cohesive Soils

Briaud has linked the rate of erodibility to the shear resistance of soil cohesion against critical shear stress caused by the flowing water and its speed. A large number of soils were tested, with clay being cohesive and sand being noncohesive, depending upon the D_{50} grain sizes. Figure 6.1 shows the comparison of critical shear stress between Briaud's experimental results as reported by Briaud et al and those obtained by other scientists.

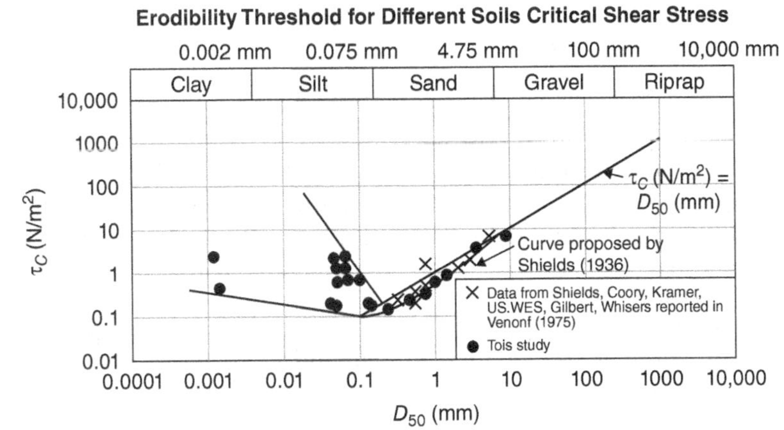

FIGURE 6.1 Critical shear resistance and erodibility thresholds of cohesive and noncohesive soils.

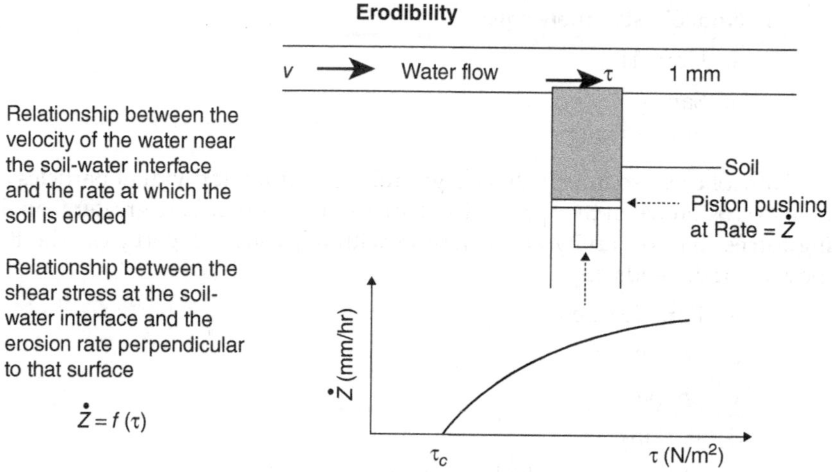

FIGURE 6.2 Definitions of the erosion rate of soils based on velocity (on *y* axis) and shear stress (on *x* axis) of flowing water. (*Courtesy of J. L. Briaud, Texas A&M.*)

Figure 6.2 shows that as velocity increases, the critical shear stress will also increase, causing greater erosion. Bridge abutments and foundations are subjected to heavy loads from truck traffic. The soil in the beds and banks of soil is submerged in water, and it is likely to result in low bearing capacity and is subject to settlement. Instead of shallow footings, deep foundations are preferred. However, spread footings may be used when rock is present below the river to place the spread footings.

Deep foundations using the following methods are preferred:

- Driven piles—either frictional or bearing types
- Augered cast-in-place (ACIP) piles and drilled displacement piles
- Drilled shafts
- Micropiles
- Helical piles and tiebacks
- Additional installation of cut-off walls at fissures

Commonly Used Software

In the geotechnical field, force is evidenced by drilling equipment that has greater power, weight, and winch and rotary capacity. The first hydraulic drilling rigs for mast movement were manufactured in the 1970s and did not exceed 55 tons, 268-hp engines, and 150 kN•m of torque.

The most widely used method today is to predrill one or more smaller-diameter holes and then enlarge the holes to their full diameters. This method causes less environment impact by reducing emissions, noise, and fuel consumption for deep foundations over rivers.

Alternative drilling systems have different power requirements for excavation, as follows:

1. Bored pile with Kelly bars
2. Full-face down-the-hole (DTH) hammer
3. Cluster DTH hammer
4. Large pilot hole and reaming
5. Small pilot hole and reaming

6.2 Application of AASHTO Specifications for Foundation Design

AASHTO Subsection 2.6.2, Site Data

A site-specific data collection plan shall include consideration of the following:

- Collection of aerial and/or ground survey data for appropriate distances upstream and downstream from the bridge for the main stream channel and its floodplain
- Estimation of roughness elements for the stream and the floodplain within the reach of the stream under study
- Sampling of streambed material to a depth sufficient to ascertain material characteristics for scour analysis
- Subsurface borings to required depth
- Factors affecting water stages during floods, including high water from streams, reservoirs, detention basins, tides, and flood control structures, and operating procedures
- Existing studies and reports, including those conducted in accordance with the provisions of the National Flood Insurance Program (NFIP) or other flood control programs

Available historical information on the behavior of the stream and the performance of the structure during past floods, includes observed scour, bank erosion, and structural damage due to debris or ice flows.

The guidelines given in AASHTO Subsection 2.6.4.4, Bridge Foundations, will be followed.

AASHTO Subsection 2.6.4.4.2, Bridge Scour

AASHTO Commentary C2.6.4.4.2 on bridge scour states, "The design flood for scour shall be determined on the basis of the Engineer's judgment of the hydrologic and hydraulic flow conditions at the site. The recommended procedure is to evaluate scour due to the specified flood flows and to design the foundation for the event expected to cause the deepest total scour."

The stability of abutments in areas of turbulent flow should be thoroughly investigated. Exposed embankment slopes should be protected with appropriate scour countermeasures.

Possible geomorphic response to regulation and changes in channel flow and floodplain alteration should be investigated.

Scour in Cohesive Soils

The clay content in soil increases cohesion, and relatively large forces are required to erode the riverbed. Higher pulsating drag and lift forces increase dynamic action on aggregates until the bonds between aggregates are gradually destroyed. Aggregates are carried away by the flow. Briaud has proposed SRICOS method of scour measurement in cohesive soils (2003). One of the earliest works on erodibility threshold was by Shields in 1936 since then other researchers have conducted studies using laboratory flumes.

> Scour analysis methods are different for cohesive and noncohesive soils. Bridge foundations supported by cohesive soils resist erosion for a much longer period than usually calculated, resulting in a longer life of bridge.

The bed material may be comprised of sediments (alluvial deposits) or other erodible materials. If bed materials are stratified, a conservative approach needs to be adopted regarding the risks of the scour breaking through the more resistant layer into the less resistant layer. Scour analysis of bridge piers and abutments in cohesive soils can be carried out on the basis of the NCHRP 516 report "Pier and Contraction Scour in Cohesive Soils" (Briaud et al. 2004).

6.3 Modeling for Hydraulic Analysis

From hydraulic analysis, the output required to compute the scour depth is mainly the velocity of flow on the upstream and downstream of the bridge and the water surface elevation.

The user must first develop a hydraulic model of the river reach containing the bridge to be analyzed. This model should include several cross sections downstream from the bridge, so that no user-defined downstream boundary condition affects the hydraulic results inside and just upstream of the bridge. The model should also include several cross sections upstream of the bridge, to evaluate the long-term effects of the bridge on the water surface profile upstream. The model geometry of the floodplain should consist of a minimum of four cross sections, two upstream of the bridge and two downstream, and the type of openings with dimensions. Photographs of the bridge taken from various angles to a known scale can be used in a known coordinate system such as NGVD or NAD83. U.S. Geological Survey (USGS) quad maps will show the surrounding terrain. Based on the hydrology of peak floods and a field survey of the bridge geometry (and known foundations) of existing bridges, it is possible to set up a hydraulic model. In addition, data for roughness coefficients or Manning's coefficients of the river soil and of the bridge will be required for the computer model.

Soil Profiles for Typical Scour-Critical Bridges

Case studies in each state need to be carried out for the types of soil. The soil profile for a particular bridge site is based on boring logs. To document sample soil profiles at different bridge sites, a detailed review of New Jersey DOT Phase 2 scour inspection reports was carried out. Table 6.1 shows soil profiles for selected scour-critical bridges

Bridge Name	County	Soil Type
U.S. 322 over Hospitality Brook	Atlantic	Brown gravel little F/M sand, trace silt
U.S. 322 over Big Ditch	Atlantic	Gray F/M sand, little silt
N.J. Route 4 and Hackensack River and Access Rd	Bergen	Silt and muck (fine sand and silt)
N.J. Route 70 over Friendship Creek	Burlington	Gray fine sand, trace of silt
Route 206 over Assiscunk Creek	Burlington	Brown F/M sand, trace of silt
Route 73 over Pennsauken Creek	Camden	Gray silt, little clay, small amount of fine sand (ML)
Route 49 over Mill Creek	Cape May	Brown F/M/C sand and gravel, trace of silt
Berkeley Avenue over Second River	Essex	Gray coarse/fine gravel with some coarse/ fine sand
I-295 northbound over Raccoon Creek	Gloucester	Brown F/M sand, little silt, trace of gravel
Route 29 over Swan Creek	Hunterdon	Medium to fine gravel and coarse to fine sand, trace of silt
Route 165 over Swan Creek	Hunterdon	Gravel and cobbles
U.S. Route 1 over Shipetaukin Creek	Mercer	Coarse to fine sand with varying proportions of silt and gravel
U.S. Route 130 over Rocky Brook	Mercer	Brown F/M sand, trace of silt
U.S. Route 9 over Deep Run Brook	Middlesex	Brown F/M sand, trace of silt, trace of gravel
U.S. Route 9 over Milford Brook	Monmouth	Brown F/M sand, trace of silt
U.S. Route 9 over Cedar Creek	Ocean	Brown gravel and F/M/C sand, trace of silt
Sicomac Road over Molly Ann's Brook	Passaic	Gray coarse/medium sand with little gravel
I-295 southbound over Oldmans Creek	Salem	Dark gray, hard, gravely fine to medium
U.S. Route 206 over Crusers Brook	Somerset	Medium to fine gravel and coarse to fine sand, trace of silt
New Jersey Route 15 over Paulins Kill Creek	Sussex	Poorly graded coarse to fine sand with some gravel, trace of silt
Route 173 over Pohatcong Creek	Warren	Brown sandy silt

TABLE 6.1 Soil Profiles at Selected Scour-Critical Bridge Sites in New Jersey

for scour studies. *These data are for informational purposes only and should not be used for the design.* Detailed geotechnical and borehole testing should be carried out to obtain site-specific information.

Soil Types Affecting Scour

The percentages of soil types with scour problems are listed in Table 6.2. It can be seen that sand foundations have 48 percent of scour problems while silt foundations do not display scour problems.

SeCdiment Type Percent	
Sand	48
Cohesive	19
Mixed	13
Gravel	10
Bedrock	5
Uncertain	5
Silt	0
Total	100

TABLE 6.2 Soil Types with Scour Problems

Design Policy and Coordination

Consideration of the effects of constructing a bridge across a waterway is key to ensuring the long-term stability of the structure. Confining the floodwater may cause excessive backwater or overtopping of the roadway, or it may induce excessive scour. These effects may result in damage to upstream land and improvements or endanger the bridge. Conversely, an excessively long bridge may cause flooding downstream or cost far more than can be justified by the benefits obtained. Somewhere between these extremes is the design that will be the most economical to the public over a long time.

The designer must evaluate existing upstream conditions as well as future upstream development in 25 years, if possible, in sizing a structure. The amount of information available to assist the designer in predicting future development is limited. Available information includes the following:

- **Responsibilities for drainage design:** These are divided between the bridge design section and the project development sections based primarily on the size of the drainage area.
- **Responsibilities for bridge design:** The bridge design department is responsible for the design of all locations requiring structures exceeding 20 ft^2 (1.86 m^2) of waterway opening for a single location. Bridge design will also be responsible for the design of pipe culverts, closed drainage systems, roadside ditches, and storm water management on "bridge only" projects.

The guidelines noted in this chapter are applicable to all structures and culverts. Refer to *AASHTO Drainage Manual 2014* for the design and construction of adjacent drainage ditches, pipe culverts, closed drainage systems, and erosion control near stream crossings.

Risk Analysis

If the designer is using a smaller design flood frequency, a risk analysis must be performed. This needs approval by the bridge design engineer. Where a risk analysis is needed, a complete hydraulic report should be prepared that gives consideration to each alternative under study. The lower level of study—risk assessment—should always be considered as the first course of action. Only if a detailed economic accounting of the risks and potential harm is needed should an extensive risk analysis be performed. A detailed analysis must be performed in situations involving substantial losses resulting in high encroachment costs. Good engineering judgment must always be used in determining the level of evaluation to be performed. HEC-17, *Design of Encroachments of Flood Plains Using Risk Analysis*, may provides valuable information.

Hydraulic and Geomorphic Factors Affecting Scour at New Jersey Bridges

The author investigated scour-critical bridges in New Jersey. Hydraulic and geomorphic factors appear to be the most important. Various papers in the ASCE Compendium (ASCE, 1999). The two bridges studied are Guinea Hollow Road and Water Street Bridges.

Basic parameters were as follows:

- *Upstream features*: Guinea Hollow Road Bridge is located downstream of a small dam and a footbridge. The presence of the dam has helped to reduce high velocities downstream of the bridge during flood events. However, the Water Street Bridge is located in a heavily forested area. The flow velocities are average.
- *Existing river training measures*: For both bridges, there is no evidence of dyke or spurs constructed along the outer bend to reduce the effects of long-term scour. Hence the scour has remained unmitigated throughout the long life of the structures.
- *Longitudinal slope*: The longitudinal slopes of the channel vary between 3 and 4 percent upstream for Guinea Hollow Road Bridge and between 2.4 and 3 percent downstream for Water Street Bridge. The slopes are not steep; and per Manning's equation for normal depth in an open channel, the critical or supercritical velocities are not affected.
- *Aggradation*: The channel beds are littered with stones and small boulders that provide partial armoring of the bed against scour holes. The coefficient of roughness is affected. Part of the Water Street Bridge opening is clogged by medium-size stones deposited by earlier floods, thereby reducing the effective channel area and increasing the flow velocity under the bridge.
- *Type of streambed*: Soil samples tested in the laboratory have shown the Guinea Hollow Road Bridge streambed as granular, with some large stones mixed with gravel and a substantial fraction of sand. The median size of aggregate D_{50} varies from 1 to 5 mm. The gravel layer has an armoring effect and leads to low degradation. For the Water Street Bridge, medium sand and traces of fine sand mixed with gravel are found with the median sizes of the aggregate D_{50} varying from 1 to 2 mm. The degradation of the streambed or banks has been significant.
- *Bridge skew:* For Guinea Hollow Bridge the skew of 20° in plan has resulted from a bend. The meander may continue to move laterally outward and downstream, eroding the left bank and increasing the scour depth in the future. For the Water Street Bridge the channel makes an abrupt change in direction upstream of the bridge, creating an increase in water velocity. The bridge skew of 45° has resulted from a bend or river meander. The meander may continue to move laterally outward and downstream, eroding the right bank and increasing scour in the future.
- *Depth below riverbed to top of footings*: Test borings have shown the depths of the bottom of spread footings as "shallow," making them scour-critical.

6.4 Performing Scour Analysis

Procedures for Scour Analysis

The methodology described here was developed for listed scour-critical bridges in New Jersey, Pennsylvania, Delaware, and Massachusetts. The highway agencies included the New Jersey Department of Transportation, Garden State Parkway, Pennsylvania Transportation Commission, Delaware Department of Transportation, U.S. Army Corps of Engineers, and the Massachusetts Bay Transportation Authority. Some of the notable bridges on which the author participated were Schuylkill River Bridge in Philadelphia, emergency replacement of Peckman's River Bridge on Route 46 in North Jersey and Lumberton Bridge on Rancocas Creek in South Jersey.

Data Collection and Processing of Data for Scour Assessment

Variables that require calculating the potential depth of scour and the susceptibility of existing foundations to scour damage need to be determined. These variables include the foundation details of timber crib or pile geotechnical data, survey of channel configurations at the bridge, upstream and downstream flood discharges, velocities, etc. A data collection memorandum need to be prepared to distinguish the amount of data presently available and the amount of additional data needed to complete the scour analysis.

Factors Affecting Stream Instability

These are the site-specific factors that directly contribute to soil erosion and are required for scour analysis:

1. *Geology and types of soils*: Geotechnical investigation is to be performed.
2. *Hydrology*: Meteorological data are required.
3. *Hydraulics*: Manning's coefficient is used to model the waterway for HEC-RAS software.
4. *Location and width of waterways*: Field survey is required.
5. *Longitudinal slope*: Field survey is required.
6. *Environmental conditions*: Both data for marine and plant life and New Jersey Department of Environmental Protection (DEP) permit approval are needed.
7. *Bridge characteristics*: Bridge characteristics, size of opening, site conditions need to be considered.
8. As for footing type and depth, unknown foundations require boreholes.

Office Data

An interdisciplinary team of licensed surveyors and structural, hydraulic, and geotechnical engineers need to organize this effort. A suggested data collection procedure will include the following:

- An inventory of data of a reliable nature, which is already available; preliminary scour evaluation report, if any
- Contract plans, as-built drawings, aerial survey, drainage area, and flood history
- Photographic documentation
- FEMA Flood Insurance Studies from the DEP

- Bridge Evaluation Survey and Underwater Inspection Report, if any, from the county or DOT
- Foundation reports and boring logs
- Existing H&H models, if any, from FEMA
- A list of remaining data needed to perform scour analysis and the methods, extent, and duration required for obtaining such data

Field Data Required at Bridge Site

The following field data are needed at the bridge site:

1. Geomorphic information
   a. Channel stability and site characteristics, meander, and channel cross sections upstream and downstream of bridge
   b. Bridge geometry using surveying, GPS (global positioning system) because location data use latitudes and longitudes
   c. Geotechnical data using state or USGS database
   d. Grain size analysis from laboratory tests
2. Hydrologic data
   a. Evaluation of catchment and drainage area
   b. Changes in demography
   c. Flood recurrence interval
   d. Computation of discharge and available gauge data
3. Hydraulic data
   a. Records of 50- and 100-year floods
   b. Setup of hydraulic model
   c. Size of bridge opening
   d. Evaluation of flood velocity and flood elevation

Field surveys are required to obtain a minimum of two cross sections upstream and two downstream of the bridge location. For rivers with no previous hydraulic models, six or more cross sections may be required. An underwater inspection report is required to ascertain footing conditions. For pile footings, both the total lengths of piles and the lengths exposed due to scour are determined. If the exposed length is high, the bridge will be classified as scour-critical. For small exposed pile lengths, nondestructive testing (NDT) techniques to determine crack distribution and depth of cracks in foundations are used. Visual observations, photographic documentation, evaluation of stream and waterway characteristics, and the roughness coefficients of the channel are required. Refer to AASHTO LRFD Specifications, Sec. 2.6.2, for additional requirements.

Methods for Investigating Conditions of Existing Footings

This topic is discussed in "Evaluating unknown bridge foundations, Is it worth it?" by W. A. Horne in Water Resources Engineering Conference in 1998.

Spread Footings Direct methods such as probes, augers, rotary drilling, or digging test pits adjacent to footings need to be used. To identify areas of loose sediment, organic debris, or scour holes, the soil may be probed adjacent to existing footings.

Deep Foundations Geotechnical analysis is performed to determine the foundation elevation and density of subsurface strata, by using techniques such as borings, standard penetration sampling, seismic refraction, and resistivity surveys. A geotechnical engineer should be consulted.

In addition, nondestructive, ground-penetrating radar, or acoustical emission techniques may be used subject to the project approval of New Jersey DOT Bureau of Structural Evaluation.

Scour Effects under Dams Dams retain huge volumes of flowing water. The water is under pressure, and a flow diagram may show some water percolating through soil under the dam. An example is installing cutoff walls on fissures at Arapuni Dam on Waikato River near Hamilton, New Zealand. The dam was built in 1928. Soil investigation revealed four fissures underlying the dam. It required installing cutoff walls across the fissures.

Geotechnical Considerations

Review Subsurface Information Given in the Geotechnical Report Evaluate historic scour-related conditions and potential scour holes at the bridge site. Concerning soil classification, laboratory tests for grain size samples are required. From streambed photographs, estimate the distribution of stones in the riverbed.

Data Collection

Data collection comprises the following tasks:

1. Determine the location and number of required channel cross sections, based on the channel characteristics to be surveyed for the HEC-RAS analyses.
2. Probe footings for abutments, pier, and wing walls.
3. Obtain soil samples for soil testing.
4. Determine the type and size of the stream.
5. Determine the bed material type and the classification of the roughness coefficient to be used in hydraulic analysis.
6. Determine the type of scour and the condition of the banks.
7. Plot the size and location of both scour holes and sediment deposits.

Information currently available and the recommended methods of obtaining additional required information for analysis are shown in Table 6.3, with the assigned applicable scour classification codes in Table 6.4.

Bridge No.	Hydrologic	Survey	Hydraulic	Geotechnical	Foundation Details
1000-041	FEMA studies available	Six channel sections required	Verification of existing HEC-2 analysis	Soil (samples obtained from field) properties required	Two test pits required

Table 6.3 Currently Available and Recommended Methods of Obtaining Information

Scour Code	Abutments	Piers	History
U	Type of foundation not known	Type of foundation not known	No history of scour problems –No scour at any pier –No movement, scour, or erosion at abutment –No channel lowering or lateral movement
9	Bridge foundations (including piles) well above floodwater elevations	Bridge foundations (including piles) well above floodwater elevations	No history of scour problems –No scour at any pier –No movement, scour, or erosion at abutment –No channel lowering or lateral movement
8	One of the following is true: –Pile depth greater than 40 ft –Adequate protection (class III riprap or greater; grouted riprap; gabions) in good condition –Spread footing on erosion-resistant rock (granite, basalt, gabbro quartzite, or gneiss) –When DA < 400 mi², Stiff clay (shear stress >2000 psi)	One of the following is true: –There are no piers –Pile depth is greater than 40 ft –Spread footing on erosion-resistant rock (granite, basalt, gabbro quartzite, or gneiss) –When DA < 400 mi², stiff clay (shear stress > 2000 psi)	No history of scour problems –No scour at any pier –No movement, scour, or erosion at abutment –No channel lowering or lateral movement
3			1. No history of uncorrected scour 2. No debris blocking over 10 percent flow XS

Table 6.4 Applicable Scour Codes

Levels of Analysis

The following types of analyses need to be carried out:

Level 1: Qualitative assessment of stream stability, including lateral stability, vertical stability, and determination of the profiles of rivers. (Refer to HEC-20.)

Level 2: Detailed quantitative analysis, including hydrologic, hydraulic, and scour analyses to assess scour vulnerability. (Refer to HEC-18.)

Level 3: Bridge scour design of stream instability countermeasures. (Refer to HEC-23.) Also refer to "Scour of Bridges" and "Bridge Scour in Tidal Waterways" for methods of scour analysis and the semiempirical formulas for scour depths at abutment and pier footings.

Scour Related to Soil Types

Per NCHRP 24-7, the percentages of foundation types with scour problems are listed in Table 6.2 shown earlier. It is seen that foundations on sand have 48 percent of scour problems while silt foundations do not display any scour problems.

Duration for Sour Depth

The intensity and duration of floods will affect the rate of scour in the soil. NCHRP 24-7 has reported the following important phenomenon of scour duration on different soil conditions.

Scour Component Identification for Abutments and Piers

The long-term effects of general scour such as degradation or aggradation may be assumed to be the same at piers and abutments. For scour at bridges located on bends and a confluence of rivers, the flow depths based on Maynard's equation will be equally applicable at abutments and piers. For scour at bridges located on a confluence of a river, the flow depths based on Ashmore and Parker's (1983) or Klaassen and Vermeer's (1988) equations will be applicable at both the abutment and the pier. Scour due to thalweg effects and migration of bed forms will be neglected since the magnitudes are small.

The results of the scour study will be compiled into a report and database for analysis of the dominant approaches and their effectiveness.

Literature Search The presentation of the literature search will be organized on the basis of

- Limitations of existing methods
- Review of the extent of erosion in New Jersey
- Geotechnical evaluation
- Frequency of floods, discharges, and hydrology
- Mathematical modeling approach
- Experimental modeling approach
- Correlation between theory and experiment
- Scaling and correction factors to be used for HEC-18 equations

Based on the results of the literature search, potential scour models are reviewed. The models will be classified according to their applicable use and the parameters utilized in their equations.

Evaluation shall include but not be limited to the following:

1. An evaluation of constructability will be performed.
2. No construction staging would be required for substructure rehabilitation work.

3. Regarding community impacts during construction, the impacts to emergency vehicle response, the tourist industry, traffic delays, pedestrians and bicyclists, and local business and noise should be considered. Construction cost estimates must be developed for each feasible alternative along with the anticipated construction schedule.
4. Maintenance protection of traffic schemes may be required for certain alternatives.
5. Right-of-way requirements and the environmental impact on the waterway and endangered species should be evaluated. Permit applications with technical documentation in support need to be prepared.

Procedures for performing this type of analysis are outlined in the 2001 HEC-18 report. These scour analysis equations utilize the hydraulic analysis data and in certain situations may give conservative results for scour depths. Further, significant changes to scour equations were made in the latest HEC-18, 5th edition (2012) as outlined in Chapter 4 of this book.

In addition to Section 6.3 of this chapter, the hydraulic modeling of the bridge should be based on the procedures outlined in Chapter 5 of the 2010 HEC-RAS *Hydraulic Reference Manual*. If observed data are available, the model should be calibrated to the fullest extent possible. Once the hydraulic model has been calibrated (if observed data are available), the modeler can enter the design events to be used for the scour analysis. In general, the design event for a scour analysis is usually the 100-year (1 percent chance) event. In addition to this event, it is recommended that a 500-year (0.2 percent chance) event also be used to evaluate the bridge foundation under a superflood condition.

The next step is to turn on the flow distribution option in the HEC-RAS software. This option allows for additional output showing the distribution of flow for multiple subdivisions of the left and right overbanks as well as the main channel. The output of the flow distribution option includes the following items for each flow slice: percentage of flow, flow area, wetted perimeter, conveyance, hydraulic depth, and average velocity. The user can control the number of slices in each flow element (left overbank, main channel, and right overbank), up to a maximum of 45 total slices. The flow distribution output is controlled from the Options menu of the Steady Flow.

Analysis Window Using HEC-RAS

The user must request the flow distribution output for the cross sections inside the bridge, the cross section just upstream of the bridge, and the approach section (cross section upstream of the bridge at such a distance that the flow lines are parallel and the flow has not yet begun to contract due to the bridge constriction). Flow distribution output can be requested at additional cross sections, but these are the only cross sections that will be used in the bridge scour computations. The flow distribution option must be turned on to get more detailed estimates of the depth and velocity at various locations within the cross section. Once the user has turned on this option, the profile computations must be performed again in order for the flow distribution output to be computed and included in the output file.

After the water surface profile calculations are performed for the design events and the flow distribution output is computed, the bridge scour can be evaluated. The total scour at a highway crossing is comprised of three components: long-term aggradation and degradation, contraction scour, and local scour at piers and abutments. The scour computations in the HEC-RAS software allow the user to compute contraction scour and local scour at piers and abutments. The current version of the HEC-RAS software does not allow the user to evaluate long-term aggradation and degradation. Long-term aggradation and degradation should be evaluated before the bridge scour analysis is carried out.

Entering Bridge Scour Data

The bridge scour computations are performed by opening the Hydraulic Design Functions window and selecting the Scour at Bridges function. Once this option is selected, the program will automatically go to the output file and get the computed output for the approach section, the section just upstream of the bridge, and the sections inside of the bridge. The Hydraulic Design window for Scour at Bridges will appear.

The Scour at Bridges window contains the input data, a graphic, and a window for summary results. Input data tabs are available for contraction scour, pier scour, and abutment scour. The user is required to enter only a minimal amount of input, and the computations can be performed. If the user does not agree with any of the data that the program has selected from the output file, the user can override it by entering her or his own values. This provides maximum flexibility in using the software.

Software for Design of Deep Foundations

Commonly used software for hydrologic and hydraulic analyses was summarized in the beginning of the chapter.

- HydroCAD Stormwater Modeling Software incorporates TR-20, TR-55, SBUH and rational hydrology, hydraulics, pond design etc. It is available with HydroCAD Software Solutions LLC, in New Hampshire.

The following software is available with ENSOFT Inc. (www.ensoftinc.com) of Austin, Texas:

- LPILE 6.0 (drilled shafts and driven piles under lateral loads, new p-y criteria (pile stress and lateral deflection using push over and non-linear pile buckling analysis), soil movements, nonlinear stiffness, and LRFD analysis)
- Group 8.0 (3D) (pile groups under combined loads, advanced 3D model-graphics-contours, group-reduction factors)
- SHAFT6, APILE Plus 5 (axial capacity and settlement of drilled shafts and piles)
- Dynamic Suite (DynaPile 2.0, DynaN 3.0, DynaMat 1.0): Novak's method is now available.
- GRLWEAP 2010 (wave equation analysis for pile driving)
- PyWall (flexible retaining walls)

Some of the above software products have been updated, and 2012 to 2014 versions are available. COM624P Software by Adama Engineering is also widely used for pile analysis and was developed in 1989. Laterally Loaded Pile Analysis, Version 2.0. Computer Program COM624P has been developed for use in the analysis of stresses and deflection of piles or drilled shafts under lateral loads. The technology on which the program is based is the widely used *p–y* curve method. The program solves the equations giving pile deflection, rotation, bending moment, and shear by using iterative procedures because of the nonlinear response of the soil.

Evaluating (Scoured) Unsupported Pile Lengths

Scour around piles leads to exposed pile lengths. Exposed pile bents at Peckman's River Bridge on Route 46 in New Jersey showed reduction in axial capacity. The higher the scour, the lower the axial capacity. After a peak flood has exposed pile lengths, it is necessary to compute the axial capacity, a reduced factor of safety, and the safety of the foundation. A commonly used pile program such as LPile may be used.

Use of Modern Underpinning Methods

A footing settlement is likely to occur owing to erosion of soil during a flood. Detailed subsurface information and an understanding of critical ground behavior are essential. The following methods can be used:

1. *Conventional pit method* consists of enlarging or deepening existing foundations by selectively removing soil from beneath the foundations and replacing it with concrete, soil reinforcement, or grout. The superstructure needs to be shored to prevent settlement. The disadvantage is that the superstructure deformation cannot be avoided.
2. *Soil treatment* changes the physical properties of the ground to make it more supportive and stronger. Ground treatment will strengthen the ground, which acts as a load transfer mechanism.
3. *Minipiles or micropiles* (as defined by FHWA) are 2- to 8-in-diameter steel piles. They transfer the load by relieving existing footings, which would function only as pile cap. This technology requires powerful drills applied from above the foundation.
4. *Jet grouting* serves as an excavation support and groundwater control system. High-velocity injection fluid erodes the soil and replaces the soil with an engineered grout, forming a cementitious product known as soilcrete, which has a compression strength exceeding 1000 psi. The technique has been used for reconnecting pile caps to decayed piles, for enlarging existing spread footings, and for creating deep footings in difficult access locations. Specialized equipment and experienced contractors are required.
5. *Compaction grouting* densifies and strengthens the loose soil by injecting a very viscous sand-cement grout. Very stiff, continuous grout columns can be formed directly beneath the structure. Specialized equipment and experienced contractors are required.
6. *Chemical grouting* is a soil treatment system that virtually glues the soil together to increase its strength. An injection system is used to give soil strengths greater than 300 psi.

Table 6.5 shows a comparative study of use of alternate countermeasures, highlighting their advantages and disadvantages.

Scour Type	Description	Advantages	Disadvantages	Remarks
1. Concrete Apron /Curtain Wall				
Contraction and local scour	Concrete walls precast or cast in place against the sides of footing	New wall can rest on hard strata/ rock	Local sheet pile degradation	Cofferdam required for construction is recommended
2. Piles Driven as Shields Adjacent to Bridge Foundations				
Scour from high flood velocities	Pile shields deflect flow	Helpful in dewatering	Scour can occur near sheet piling, construction is difficult, rust	Recommended for high-scour situations with riprap protection
3. Extended Footing Local Scour				
Local scour	Cast wider concrete slab footing to prevent settlement	Suitable for low scour depths Acts as curtain wall and apron on side of spread footing	Not suitable for masonry footings	Acts as curtain wall and apron on side of spread footing Recommended for concrete spread footings
4. Constructing Minipiles through Spread Footings				
Degradation	Driving minipiles through footings	Commonly used for footing strengthening	Expensive. Not suitable for old masonry footings	Not recommended for high-traffic-volume bridges
5. Underpinning				
Contraction scour Local scour	Lowering the bottom of footing elevation below scour depth Underpinning pit required	Commonly used for extensive repair or footing strengthening	Expensive Not suitable for old masonry footings Disturbance of streambed during construction	Not recommended for high-traffic-volume bridges
6. Use of Open Parapets or Railings				
Contraction scour and pressure flow	Increases flow area and prevents overtopping of floodwater	Effective for small openings or where vertical alignment is limited	Additional overflow downstream needs to be checked	Recommended only for overtopping flood situation
7. Relief Bridge				
Contraction and local scour	Approach bridge to increase size of waterway opening	Floodwater will be discharged rapidly	Expensive The key scour problem at main bridge may still remain unchanged	Not recommended since utilities need to be relocated

Table 6.5 Evaluation of Countermeasures

6.5 Scour of Piles, Pile Groups, and Caissons

Deep foundations such as piles and caissons are normally required in soft soils, where the bearing capacity of the soil is not sufficient to support a shallow footing. Such soils have greater scour potential.

Piles are designed as bearing piles or friction piles; in either case, they must be fully embedded in undisturbed soil at all times. If the surrounding soil is eroded or disturbed, the load bearing capacity of the pile is reduced, leading to settlement and collapse of the pile cap. Raking piles, which are driven closer to the river, are more susceptible to erosion than vertical piles. The following steps are required to minimize the erosion of raking piles:

1. When raking piles are used, they must not be placed at the edge of the river. Instead the span of bridge may be increased.
2. The top 5 to 10 ft of pile, which may be exposed, may be neglected in the pile design length. The exposed length of steel or concrete pile should be checked for local buckling.
3. The pile cap should be placed below the calculated scour depth. However, this results in taller and full-height abutments, increasing the cost. Shielding of piles by driving sheet piling would prevent erosion.
4. Interference of raking piles with sheet piles is likely to occur. It is possible to allow for sufficient clearance, but that would reduce the width of the waterway; making it difficult to obtain a stream encroachment permit from the DEP. Armoring or river training measures may be considered.
5. Caissons may be preferred over piles.
6. For long span bridges on harder ground and dense soils, drilled piers will be more resistant to erosion than caissons.
7. Integral abutment piles with a single row of piles would require greater protection by providing shielding in the form of sheeting left in place.

6.6 Contents of a Hydraulic Report and Guidelines

The minimum items required in a hydraulic report are the following:

- A location map
- A hydrologic and hydraulic narrative that provides site data, a hydraulic analysis, and flood profiles for the design year, the 100- and 500-year floods, and recommendations.

The report should be supplemented as needed with relevant information.

- Photographs as applicable
- USGS topographic maps
- Flood frequency curves
- Stage discharge curves for existing and proposed conditions
- A scour analysis

- Water surface profiles for existing and proposed conditions based on HEC-RAS, WSPRO, or other approved method; a 100-year flood boundary plan; stream cross sections; stream and flood surface profiles; and hydraulic assessment list

Stream Channel

The natural or altered condition of stream channels affects the flow characteristics. Any work being performed, proposed, or completed that modifies a stream channel changes the hydraulic efficiency of the stream and must be studied to determine its effect on the stream flow. The effect on peak flows at the structure site due to modification of a stream's hydraulic characteristics must be determined. The engineer should be aware of plans for channel modifications that might affect the stream hydraulics. Similarly, the effects of storm drainage systems and other water-related projects should be investigated. Any modifications that affect stream alignment should be kept to a minimum, particularly for straightening meandering streams.

Ice and Debris

The quantity and size of ice and debris carried by a stream should be investigated and recorded for use in the design of drainage structures. The times of occurrence of ice or debris in relation to the occurrence of flood peaks should be determined, and the effect of backwater from ice or debris jams or recorded flood heights should be considered in using stream flow records. The location of the constriction or other obstacle-causing jams, whether at the site or structure under study or downstream, should be investigated, and the feasibility of correcting the problem should be considered.

Under normal circumstances, 1-ft (0.3-m) freeboard is sufficient to permit passage of ice flow and debris. At locations where large pieces or quantities of debris are anticipated, the designer should consider increasing the freeboard.

6.7 Guidelines for Preparing Geotechnical Reports and Plans, Specifications

A set of review checklists and technical guidelines has been developed to aid engineers in their review of projects containing major and unusual geotechnical features. These features may involve any earthwork or foundation-related activities, such as the construction of cuts, fills, or retaining structures that deserve special attention because of their size, scope, complexity, or cost. It also provides a description of a voluntary program by which FHWA generalist engineers determine what type and size of projects may warrant review by an FHWA geotechnical specialist. The review checklists and technical guidelines are provided to assist generalist highway engineers in

- Reviewing both geotechnical reports and plan, specification, and estimate (PS&E)[1] packages
- Recognizing cost-saving opportunities

[1]For the purposes of this document, PS&E refers to a plan and specification review at any time during a project's development. Hence, the review may occur at a preliminary or partial stage of plan development.

- Identifying deficiencies or potential claim problems due to inadequate geotechnical investigation, analysis, or design
- Recognizing when to request additional technical assistance from a geotechnical specialist

For projects that are submitted to an FHWA geotechnical specialist, the checklists and technical guidelines are provided to aid FHWA generalist engineers in making a quick review of the geotechnical report and accompanying support data provided by the state, to ensure that the information provided by the state is complete enough to allow adequate technical review by the FHWA geotechnical specialist.

Plan, Specifications, and Estimate

Geotechnical Reports

For projects that will not be submitted to an FHWA geotechnical specialist for formal review (which will be the majority of projects handled by the FHWA division office), the checklists and technical guidelines are provided to assist generalist engineers in reviewing geotechnical reports and preliminary plan and specification packages. The following objectives need to be investigated and included in the geotechnical report:

- Recognizing cost-saving opportunities
- Spotting deficiencies or potential claim problems due to inadequate geotechnical investigations, analysis, or design
- Recognizing when to request technical assistance for an FHWA geotechnical specialist

Note that the checklists and technical guidelines also include coverage of structure foundations. These review checklists and technical guidelines have been developed to fill an existing need in this area.

The geotechnical report sets forth minimum geotechnical standards or criteria to show transportation agencies and consultants the basic geotechnical information that the FHWA recommends be provided in geotechnical reports and PS&E packages.

The geotechnical report is the tool used to communicate the site conditions and design and construction recommendations to the roadway design, bridge design, and construction personnel. Site investigations for transportation projects have the objective of providing specific information on subsurface soil, rock, and water conditions. Interpretation of the site investigation information by a geotechnical engineer results in design and construction recommendations that should be presented in the project geotechnical report.

The information contained in this report is referred to often during the design period, during the construction period, and frequently after completion of the project (resolving claims).

Preliminary and Final Geotechnical Reports: The state or its consultant should prepare *preliminary* geotechnical reports for submittal to the design team whenever this information will benefit the design process. Early submittal of geotechnical information and recommendations or engineering evaluation of preliminary data may be necessary to establish basic design concepts or design criteria. The development of a *final* geotechnical report will not normally be completed until the design has progressed to the point where specific recommendations can be made for all the geotechnical aspects of the work. Final alignment, grade, and geometry will usually have been selected prior to issuance of the final geotechnical report. While the geotechnical report content and format will vary by project size and highway agency, all geotechnical reports should contain certain *basic* essential information:

- Summary of all subsurface exploration data, including subsurface soil profile, exploration logs, laboratory or in situ test results, and groundwater information
- Interpretation and analysis of the subsurface data
- Specific engineering recommendations for design
- Discussion of conditions for solution of anticipated problems
- Recommended geotechnical special provisions

For brevity, the term *geotechnical report* will be used as a general term to cover all types of geotechnical reports, such as the foundation report, centerline soils report, landslide study report, etc.

It may not be always possible to establish strict criteria for all geotechnical information that should be obtained or geotechnical analysis that should be performed for a particular project, in such cases only general or minimum guidelines based on the geotechnical data may be established.

- *Selected projects for review by FHWA*: Provide definitions of both major and unusual features and guidelines as to select projects which may be appropriate for review by the FHWA geotechnical specialist.
- *Scope of subsurface investigation*: Presenting guidelines to minimum boring, sampling, and testing criteria for subsurface investigations that should be conducted for major or unusual geotechnical features.
- *Use of geotechnical analyses*: Presenting general guidelines on the major types of geotechnical engineering analyses normally required for embankments and cut slopes, structure foundations, and retaining structures for major soil types.
- *Additional support data*: Presenting a list of technical support data that should be provided for correction of soil and rock instabilities (landslides). The review and technical guidelines should cover as much information as the following geotechnical features would provide:
  - Centerline Cuts and Embankments
  - Embankments over soft ground
  - Landslide corrections
  - Retaining structures
  - Structure foundations (spread footings, piles, drilled shafts)
  - Ground improvement techniques
  - Material sites

Preparing preliminary reports: Reviews made during the preliminary stage of project development will commonly examine only the geotechnical report, since detailed plans and specifications may not yet be prepared. The practice of most highway agencies is to prepare a single geotechnical report that includes subsurface information, interpretations, and design and construction recommendations.

Preparing final reports based on review checklists: Review checklists should be prepared and used as the working document while the guidelines and the indicated sections of the National Highway Institute *Soils and Foundations Workshop Manual* should be used as references. Upon completion of the checklists, the reviewer should summarize the negative or missing information s and discuss these with the appropriate geotechnical engineers to determine if additional follow-up is appropriate.

The following checklists cover the major information and recommendations that should be addressed in project geotechnical reports and preliminary plans.

Geotechnical Report Review Checklists

Section A Site Investigation Information
Section B Centerline Cuts and Embankments
Section C Embankments over Soft Ground
Section D Landslide Corrections
Section E Retaining Structures
Section F Structure Foundations—Spread Footings
Section G Structure Foundations—Driven Piles
Section H Structure Foundations—Drilled Shafts
Section I Ground Improvement Techniques
Section J Material Sites

PS&E Review Checklists

Section A General
Section B Centerline Cuts and Embankments
Section C Embankments over Soft Ground
Section D Landslide Corrections
Section E Retaining Structures
Section F Structure Foundations—Spread Footings
Section G Structure Foundations—Driven Piles
Section H Structure Foundations—Drilled Shafts
Section I Ground Improvement Techniques
Section J Material Sites

The inclusion of certain checklist items in the geotechnical report is of vital importance. A negative response to any of these asterisked items is cause to contact the geotechnical engineer for clarification of the omission for important asterisked marked omissions.

Scourability of Rock

The following criteria represent the values to define rock quality and scourability of rock:

- The Rock quality designation (RQD) value is a modified computation of the percent of rock core recovery that reflects the relative frequency of discontinuities and the compressibility of the rock mass and may be used indirectly as a measure of scourability. The RQD is determined by measuring and summing all the pieces of sound rock 6 in (150 mm) and longer in a core run and dividing this by the total core run length. Rock with an RQD value of less than 50 percent should be assumed to be soil-like with regard to scour potential.

- The primary intact rock property for foundation design is unconfined compressive strength (ASTM Test D1702-14). The unconfined compressive strength provides an upper limit of the rock mass bearing capacity and an index value for rock classification. There is only a generalized correlation between unconfined compressive strength and scourability.
- AASHTO Test T104 is a laboratory test for the soundness of rock. A soaking procedure in magnesium and sodium sulfate solution is used. Generally the less sound the rock, the more scourable it will be.
- The Los Angeles abrasion test (AASHTO T96) is an empirical test to assess the abrasion of aggregates. In general, the less a material abrades during this test, the less it will scour. Loss percentages greater than 40 percent indicate scourable rock.

Scour Evaluation Report

According to the DELDOT Bridge Design Manual, the scour evaluation report must contain the following items:

- Table of contents
- Bridge description—bridge number, type, size, location, and National Bridge Inventory Record Item 113, scour coding
- Executive summary of hydrologic and hydraulic methods, scour results, conclusions, and any countermeasure recommendations required, with plan and profile views showing scour depths and limits
- Scour computations (including computer input and output)
- Bridge drawings, cross sections, soils information, test results, and other miscellaneous data
- References

Plan Presentation

The following information will be provided in the Project Notes on the plans:

- A note stating that the structure has been analyzed for the effects of scour in accordance with the procedures described in HEC-18, *Evaluating Scour at Bridges*
- Scour analysis design flow volume, frequency, velocity, and water surface elevation
- Scour analysis check flow volume, frequency, velocity, and water surface evaluation
- The calculated design scour depth
- The calculated check scour depth

6.8 Foundation Modification by Structural Countermeasures

Structural countermeasures are site-specific, and there is no single technique that would apply to different problems at the same time. They involve minor to major repairs and modifications to the bridge structure (foundation) to prevent failure from scour.

Typically, a bridge foundation is modified to increase its stability after scour has occurred or when a bridge is classified as scour-critical. These modifications can involve foundation strengthening, conversion from a simple span to continuous span configuration, or pier geometry modifications. Special provisions for technical specifications based on recommended guidelines may be required and should be used for preparing construction drawings and contract documents.

Structural countermeasures are more effective and have more direct application than armoring. With advances in structural applications, more sophisticated structural countermeasures are becoming available for adaptation by structural engineers.

1. Foundation strengthening includes supplementing the original foundation. This involves use of additional reinforcement and/or increasing the size of the original foundation.
2. These countermeasures are designed to prevent failure when a channel bed is lowered to an expected scour elevation, or to restore structural integrity after scour has occurred.
3. The design and construction of bridges with continuous spans provide redundancy against catastrophic failure that is due to substructure displacement as a result of scour. Retrofitting a simple span bridge with continuous spans can also serve as a countermeasure after scour has occurred or when a bridge is classified as scour-critical.
4. Pier geometry modifications are used to either reduce local scour at bridge piers or transfer scour to another location.

Structural countermeasures can also involve the use of armoring or no armoring. Structural repairs such as grouting holes and cracks are generally required. Most commonly used for existing bridges are the following:

1. Foundation shielding such as constructing concrete apron and curtain walls
2. Sheet piling local to foundations to act as shielding (sheeting left in place after construction is completed)
3. Soldier piles and precast panels
4. Minipiles driven through spread footings
5. Underpinning under the footing
6. Extended footings
7. Use of open parapet or railings to permit deck drainage of floodwater
8. Use of integral abutments

Recommended Structural Countermeasures

Depending on flood conditions and overall site conditions, the following types of structural countermeasures are recommended:

1. Concrete apron and curtain walls
2. Local sheet piles and soldier piles
3. Extended footings
4. Underpinning methods such as driving minipiles
5. Open parapets or railings
6. Use of integral abutments

The final selection must consider environmental considerations and cost.

Use of Modular Bridges for Rapid Replacement of Flood Damaged Bridges

Figure 6.3 shows assembled truss steel bridges manufactured by Acrow of Parsippany, New Jersey to replace a long span bridge in Pakistan.

Countermeasures for Footings on Soil and Rock

Theoretically, good-quality bedrock will not require any countermeasure. However, irrespective of erosion, the footing should never be exposed. A layer of insulation of 1-ft minimum is required to address antifrost, thermal changes, and vegetation growth. This may be in the form of soil with riprap or an alternate countermeasure at the top of a footing.

FIGURE 6.3 U.S. Ambassador to Pakistan (*right*) offering in aid an Acrow type modular bridge to Pakistani officials to replace a long span bridge lost due to floods.

6.9 Investigating Unknown Foundations

Unknown foundation may be defined as an existing buried foundation for which no drawings, including as-built drawings from the time of construction, are available. It may also be defined as a foundation whose type and elevation are unknown or a foundation whose type is known but whose bottom elevation is unknown. Such foundations include the following:

- Short one-span bridge abutments on land. The "dig and see" method is cost-effective for such foundations.
- Piers of multiple-span bridge on land or water.
- *Boring and NDT Methods*: Borings and NDT methods are investigate used for such foundations.

For scour protection design, the physical size and type of foundation need to be investigated. The following approaches are used to determine unknown foundations:

- Conventional methods such as probes, hand augers, power augers, rotary drilling methods, and digging pits adjacent to footing
- Deductive methods for deep foundations that rely on geotechnical analysis, obtaining samples of subsurface strata using the Standard Penetration Method (SPT) ASTM D-1586, Cone Penetrometers, to evaluate the density of layers
- NDT methods such as surface NDT and borehole NDT methods. Surface methods include
  - Pulse echo (compression wave echo from stiffness change)
  - Bending wave method
  - Ultraseismic method
  - Surface wave spectral analysis
  - Ground penetration radar or impulse radar
  - Borehole NDT methods include
  - Parallel seismic method (measurement of compression and shear wave emitted by foundation impact)
  - Pulse echo in borehole
  - Borehole radar method

Usually, the work is performed by specialist geotechnical contractors. In New Jersey, common methods are digging pits adjacent to footings or drilling boreholes through the approaches to abutments, backfill, and footings and through deck slab adjacent to piers.

The actual type of NDT method chosen should be project-specific depending on the access, foundation configuration, nature of subsurface soils, and skills and equipment of the practitioner, etc. The state of the art for NDT methods is still evolving and interpretation is still somewhat subjective based on the interpreter's judgment and experience. The project engineer and consultants should consider the latest information and resources in selecting an appropriate NDT approach. The following reference sources may be used for more information on unknown foundations:

Olson and Jalinos, (1998).

Olson and Aouad, (2000).

Nondestructive Evaluation of Unknown Subsurface Bridge Foundation Depths, (2001).

Elias, (1994)

Limitations of Bridges with Unknown Foundations

When there are no existing drawings available (generally for older bridges), neither the footing elevations nor the type of buried footing is known. Scour analysis becomes difficult, and such unknown footings need to be investigated. Either NDT methods (seismic echo, etc.) or the drilling of deep holes behind the abutments on approaches and through pier footings needs to be carried out. Because of aggradation from sediment transport during floods, some bridge footings may be buried deeper with layers of sand or silt, changing the *Q* value and velocity due to reduction in river cross section under the bridge. The StreamStats method does not have records of aggradation. The other limitations are as follows:

1. Gauged site information does not normally cover 50- and 100-year floods or records of flood frequency or recurrence. There are only 191 bridge sites where gauges are located. In calibrating regression equations, an additional 63 sites from Pennsylvania, New York, Delaware, and Maryland were used.
2. Revised velocities based on StreamStats are independent of geological conditions.
3. The *Q* values for the latest FEMA Flood Insurance Maps of New Jersey need to be compared with flows from StreamStats.
4. Errors in the prediction of flow estimates exceed 50 percent and can be either increased or decreased.
5. Extrapolated estimates of adjacent drainage area using low-flow regression equations will increase the error further.
6. In New Jersey, there is continuous increased urbanization, and the volume of water taken from rivers will increase. Regression equations based on population density need to change.

The Long-Term Bridge Performance (LTBP) program was created to identify, collect, and analyze research quality data on the most critical aspects of bridge performance. To complete a thorough investigation of bridge performance issues, the FHWA sponsored the FHWA Workshop to Identify Bridge Substructure Performance Issues in Orlando, Florida, from March 4 to 6, 2010. The workshop included participants from the FHWA, state transportation departments, academia, industry, and consultants. The workshop had three focal points: (1) identify bridge performance issues impacted by geotechnical factors, (2) identify data needs and data gaps related to the geotechnical performance issues, and (3) identify tools, technology development, and monitoring to address the data needs and data gaps. The report describes the results and recommendations of the workshop are presented in the larger perspective of designing and implementing the LTBP program.

R.J.S. et al presented a paper on, Scour, Offshore Structures, in Proceedings of ICSE-5 Conference, San Francisco, 2010; American Society of Civil Engineers, 2010.

The drive for the development of marine offshore renewables has led to focused requirements for scour hazard assessment relating to foundations. Foundation options including monopiles, multipile tripods and jackets, gravity bases, or suction piles are being considered for a variety of offshore renewable installations. There are scour assessment challenges in currents and waves, including scour experience at built foundations.

The Delaware DOT's Division of Planning

The projections include population, dwelling units, number of vehicles, and employment for each traffic zone in the county's system zones as well as the area type—city, urban, inner suburban, outer suburban, or rural.

Each county maintains zoning maps, which show the current zoning status of land within the county. The Department of Planning in each county prepares a comprehensive development plan that presents its concept of the optimum use of the land within the county. The plans include detailed maps that show the current and proposed use of all lands. These plans must be used with judgment since they do not necessarily conform to current zoning and the land may not be developed within the projected time frame.

Article 10 of the *New Castle County Unified Development Code* establishes criteria for structures in or near floodplains and floodways. Any structure to be located, relocated, constructed, reconstructed, extended, enlarged, or structurally altered within a designated floodplain is subject to the county code. The major items that must be included in the application procedures that affect structure designers are as follows:

- Site location and tax parcel number
- A brief description of the proposed work
- A plan of the site showing the exact size and location of the proposed construction
- Plan of any existing structures
- An engineering analysis of the impact on the floodplain utilizing HEC-RAS, WSPRO, or another acceptable backwater analysis model
- An accurate delineation of the floodplain area, including the location of any adjacent floodplain development or structures
- The location of any existing or proposed subdivision and land development
- Delineation of existing and proposed contours
- Information concerning the 100-year flood elevations and other applicable information such as the size of structures, location and elevation of streets
- Water supply and sanitary sewer facilities, soil types, and flood-proofing measures
- A document certified by a registered professional engineer: Certificate is for proposed construction being adequately designed to withstand the 100-year flood pressures, velocities, impact and uplift forces, and other hydrostatic, hydrodynamic, and buoyancy factors.

6.10 Types of Foundation Selection for New Bridges

For the design of new bridges, bridge foundations should be designed for potential scour by assuming that all streambed material in the computed scour prism has been removed and is not available for bearing or lateral support. Bridge foundations should be designed to withstand scour during floods equal to or less than the 100-year flood, and should be checked to ensure that they will not fail during a superflood (of the order of 500-year flood event). All foundations should have a minimum factor of safety of 1.0 under superflood conditions.

The procedure to compute the scour prism, which represents calculated scour conditions, should be based on procedures outlined in HEC-18. All foundations should be designed in accordance with the AASHTO LRFD Bridge Design Specifications. In cases of pile foundations, the piling should be designed for additional lateral restraint and column action because of unsupported pile length after scour. In areas where local scour is confined to the proximity of the footing, the lateral ground stresses on the pile length, which remains embedded, may not be significantly reduced from the pre–local scour conditions.

Shallow Foundation

1. Spread footings on soil, sand, and silt.
   a. Ensure that the top of the footing is below the sum of the long-term degradation, contraction scour, and lateral migration.
   b. Place the bottom of the footing below the total scour line.
   c. The top of the footing can act as a local scour arrester.
2. Spread footings on hard rock.
   a. Place the bottom of the footing directly on the cleaned rock surface. The rock surface should be highly scour-resistant massive rock formations (such as granite).
   b. Small embedment (keying) should be avoided since blasting to achieve keying frequently damages the subfooting rock structure and makes it more susceptible to scour.
   c. If footings on smooth massive rock surfaces require lateral constraint, steel dowels should be drilled and grouted into the rock below the footing.
3. Spread footings on erodible rock.
   a. Weathered or other potentially erodible rock formations need to be carefully assessed for scour. An engineering geologist familiar with the area geology should be consulted to determine if rock, soil, or other criteria should be used to calculate the support for the spread footing foundation. The decision should be based on an analysis of intact rock cores, including rock quality designations and local geology, as well as hydraulic data and anticipated structure life.
   b. An important consideration in the analysis would be to determine the existence of a high-quality rock formation below a thin, weathered zone. For deep deposits of weathered rock, the potential scour depth should be estimated and the footing base placed below that depth. Excavation into weathered rock should be made with care.

c. If blasting is required, light, closely spaced charges should be used to minimize the breakup beneath the planned footing level. Loose rock pieces should be removed and the zone filled with clean concrete. The final footing should be poured in contact with the sides of the excavation to the fully designed footing thickness to minimize water intrusion below the footing level.

d. Guidance on scourability of rock formations is given in the FHWA Memorandum "Scourability of Rock Formations," July 19, 1991.

4. Spread footings placed on tremie seals and supported on soil.

   a. Ensure that the top of the footing is below the sum of the long-term degradation, contraction scour, and lateral migration.

   b. Place the bottom of the footing below the total scour line.

Deep Foundations

1. *For deep foundations (drilled shaft and driven piling) with footings or caps*: Placement of the top of the footing or pile cap below the streambed to a depth that is equal to the estimated long-term degradation and contraction scour depth will minimize obstruction to flood flows and resulting local scour. Lower footing elevations may be desirable for pile-supported footings when the piles could be damaged by erosion and corrosion from exposure to river or tidal currents.

2. *Stub abutments on piling*: Stub abutments positioned in an embankment should be founded on piling that is driven below the elevation of the thalweg. Long-term degradation and contraction scour in the bridge waterway should be considered to ensure structural integrity in the event the thalweg shifts and the bed material around the piling scours to the thalweg elevation.

Table 6.6 gives a summary of analysis and design methods for resisting scour.

6.11 Conclusions

The three methods of analysis are hydrologic, hydraulic, and scour analysis. Standard software and the method of data preparation are described.

Modeling of hydraulic analysis will conform to the software selected. (AASHTO Section 2.6.2 is applicable.) Software used includes HEC-RAS replacing HEC-2 or WSPRO.

Scour analysis and countermeasure design are essential parts of foundation protection and as such can be classified under foundation design. Scour analysis and estimation of scour depths can be performed using spreadsheets developed for HEC-18 equations or HEC-RAS. For hydrologic analysis several software packages are available, but StreamStats is the latest. A hydraulics report in standard format will be prepared.

Geotechnical investigation needs to be performed by a geotechnical engineer. New foundations selected shall be shallow or deep. LPile or SHAFT6 or alternative software will be selected.

For unknown foundations buried in the soil, where drawings are not available, either NDT (seismic echo, etc.) methods or drilling deep holes behind the abutments on

Simplified Procedure	Risk Assessment Using Risk Decision Matrix	Physical Classification	Action Type
Geotechnical Sampling Using FHWA Procedures	High geotechnical risk	non-cohesive soil	Increased geotechnical sampling Use deep foundations and increased inspection Erosion monitoring
	Medium geotechnical risk	cohesive soil	Install armoring countermeasures and monitoring
	Low geotechnical risk	rock	Remove from scour-critical list
Hydraulic Analysis Using USACE HEC-RAS	High hydraulic risk	high flood velocities and elevations	Use office and field data for hydraulic model Increased inspection/erosion monitoring Install river training countermeasures
	Medium hydraulic risk	With large bridge opening	Install armoring countermeasures or monitor
	Low hydraulic risk	With low flood velocities and low elevations	Remove from scour-critical list
Hydrologic Analysis Using USGS StreamStats	High hydrologic risk	High *Q* values and high flood frequency	Increased inspection and erosion monitoring Install river training countermeasures
	Medium hydrologic risk	Average *Q* values	Install armoring countermeasures or monitor
	Low hydrologic risk	Low *Q* values	Remove from scour-critical list
Scour Analysis Based on FHWA HEC-18	For high risk, use alternate analysis	Use standard formulas based on type of scour	Input bridge data including hydraulic or hydrologic assessment
Countermeasures Design Using FHWA HEC-23	For high risk and high scour depths, use river training measures	Select countermeasures based on type of scour, contraction, or local scour	Apply durability and cost considerations
	For medium risk use armoring countermeasures	Select countermeasures based on type of scour, contraction or local scour	Same as above
	For low risk use nominal countermeasures	Minimum countermeasures required such as riprap	Same as above

Table 6.6 Summary of Methods to Resist Scour at Bridge Foundations

approaches and through pier footings needs to be carried out. For existing bridge foundations, modifications will be required. An underpinning method will be used. A geotechnical report will be prepared for which checklist items are given.

6.12 Bibliography

AASHTO LRFD Bridge Design Specifications, Customary US Units, 6th Edition, Washington, D.C., 2012.

AASHTO Drainage Manual, Washington, D.C., 2014. (Replaces AASHTO Model Drainage Manual, 2005.)

ASCE "Stream Stability and Scour at Highway Bridges: Compendium of Stream Stability and Scour Papers Presented at Conferences Sponsored by the Water Resources Engineering (Hydraulics) Division of the American Society of Civil Engineers (1991-1998), ASCE, 1999.

Ashmore, P.E. and Parker G., "Confluence Scour in Coarse Braided Streams," Water Resources Research, Vol. 19, 1983.

Briaud J.-L. et al, "Simplified Method for Estimating Scour at Bridges," Project Performed in Cooperation with TXDOT and FHWA; Report No. FHWA/TX-09/0-5505-1, June 2009.

Briaud J.-L., Chen H.-C., Li Y., Nurtjahyo P., Wang J., 2004, "The SRICOS-EFA Method for Complex Piers in Fine Grained Soils," *Journal of Geotechnical and Geoenvironmental Engineering*, vol. 130, no. 11, p1180-1191. ASCE, Reston, Virginia.

Briaud J.-L., Chen, C.H., Kwak K., Wang J., Xu J., 2004, "The SRICOS-EFA Computer Program for Bridge Scour," Keynote Lecture, Proceedings of the Second International Conference on Scour and Erosion, Singapore, World Scientific, Singapore.

DelDOT *Bridge Design Manual*, Hydrology and Hydraulics, May 2005.

Elias, V., *Strategies for Managing Unknown Bridge Foundations*, FHWA-RD-92-030, Federal Highway Administration, U.S. Department of Transportation, Washington, D.C., February 1994.

Ensoft Inc. (http://ensoft-inc.software.informer.com), Informer Technologies Inc., 2015.

FEMA, *Procedures for Coordinating Highway Encroachments on Floodplains with Federal Emergency Management Agency*, Chapter 432, Summary of Floodplain Requirements, June 2014.

FHWA "Checklist and Guidelines for Review of Geotechnical Reports and Preliminary Plans and Specifications," Publication No. FHWA ED-88-053, Washington, D.C., February, 2003.

HEC-RAS River Analysis System, Hydraulic Reference Manual, US Army Corps of Engineers, 2010.

Horne, W. A., "Evaluating Unknown Bridge Foundations, Is It Worth It?" *Proceedings of the International Water Resources Engineering Conference*, Memphis, TN, August 1998.

Klassen, G. J. and Vermeer, K., "Confluence Scour in Large Braided Rivers with Fine Bed Material," International Conference on Fluvial HHydraulics, Budapest, 1988.

NHI Course no. 132012, Soils and Foundations, Reference Manual, Volume 1 Publication no. FHWA NHI-06-088, Federal Highway Administration December 2006.

"Nondestructive Evaluation of Unknown Subsurface Bridge Foundation Depths," NCHRP Project 21-5(2), National Cooperative Highway Research Program, Washington, D.C., June 2001.

Olson, L. D., Jalinoos, and M. F. Aouad, *FHWA Geotechnical Engineering Notebook*, Geotechnical Guideline no. 16, Summary of Interim Report for NCHRP Project 21-05, "Determination of Unknown Subsurface Bridge Foundations." FHWA, August, 1998.

Olson, L. D., and Aouad, M. F., "Unknown Subsurface Bridge Foundation Testing," unpublished draft final report for NCHRP projects 21-05 and 21-05(2), National Cooperative Highway Research Program, Washington, D.C., September 2000.

Richardson, E. V., and Davis, S. R., *Evaluating Scour at Bridges*, 4th ed., Federal Highway Administration, Hydraulic Engineering Circular (HEC) no. 18, Publication FHWA-NHI 01-001, 2001.

Richardson, E. V., and Lagasse P. F. (eds.), "Abutment Scour in Compound Channels," *Stream Stability and Scour at Highway Bridges*, ASCE Compendium, Reston, VA.

Richardson, E. V., Simons, D. B. and Lagasse, P. F. "River Engineering for Highway Encroachments—Highways in the River Environment," FHWA Report NHI 01-004, Hydraulic Series No. 6, Washington, D.C., 2001.

Shields A., "Application of Similarity Principles and Turbulence Research to Bedload Movement, Mitt. Preuss Versuchsanst, Wasserbau Schiffbau, 1936.

Sturm, T.W., "Abutment Scour in Compound Channels," *Stream Stability and Scour at Highway Bridges*, Richardson, E.V. and Lagasse, P. F. (eds), ASCE, 1999.

U.S. Army Corps of Engineers, *HEC-RAS River Analysis System*, Hydrologic Engineering Center, Davis, CA, 1997.

Wagner, C. R., Mueller, D. S. Parola, A. C. Hagerty, D. J. and Benedict, S. T. "Scour at Contracted Bridges," NCHRP-TRB, Doc. 83 (Project 21-14), 2006.

Watson, K. M., and Schopp, R. D. *Methodology for Estimation of Flood Magnitude and Frequency for New Jersey Streams*, USGS, SI Report 2009–5167, pp. 1–51, 2006.

Whitehouse, R. J. S., Harris, J. M. T. R. Mundon; and J. Sutherland, Scour, Offshore Structures, Proceedings of ICSE-5 Conference, San Francisco, American Society of Civil Engineers, 2010.

Yarnell, D. L., *Bridge Piers as Channel Obstructions*, Technical Bulletin no. 442, U.S. Department of Agriculture, Washington, D.C., 1934.

Additional Publications

Sample Guide Specifications for Construction of Geosynthetic Reinforced Soil—Integrated Bridge System (GRS-IBS), Publication no. FHWA-HRT-12-051, FHWA, August, 2012.

Checklist and Guidelines for Review of Geotechnical Reports and Preliminary Plans and Specifications, Publication no. FHWA ED-88-053, February 2003.

CHAPTER 7

Hydrology, Floods, and Selection of Scour-Critical Bridges

7.1 Causes and Effects of Floods

In simple terms, rainfall and snowmelt are the physical causes of floods, and scour and soil erosion are the physical effects of floods. Hydrology and hydraulics are the sciences by which the volume of floodwater, its depth, and its velocity can be studied and computed. The new subject of scour analysis is based mainly on geotechnical considerations. The practice of flood control and scour countermeasures may vary from state to state. A survey was carried out by the author's team to determine the selection, planning, and design of countermeasures. Risk analysis and optimum costs of implementation are practical considerations requiring engineering judgment and experience. If there is code compliance, the chances of failure are greatly reduced. Hence there is a need to frame effective design code and specifications. In addition, research in the gray areas will be required.

Chapter 6 considered geotechnical and hydraulics issues. This chapter addresses the accompanying science of hydrology and the selection of scour-critical bridges. As this subject is relatively new, scour practices in the United States and European countries need to be considered for their relative merits.

7.2 Methods of Hydrologic Analysis

Hydrologic analysis is used to determine the rate of flow, runoff, or discharge that the drainage facility will be required to accommodate.

Documentation

Figure 7.1 shows an example of contraction scour that is due to a wide floodplain leading to road closure accompanied by local scour at the foundations. The design of highway facilities should be adequately documented. Frequently, it is necessary to refer to plans, specifications, and hydrologic analyses long after the actual construction has been completed.

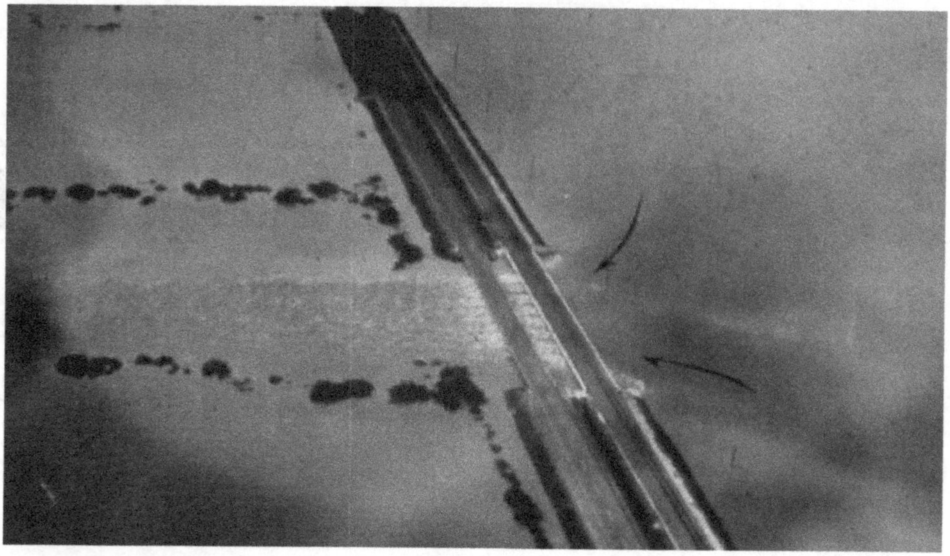

FIGURE 7.1 Arrows at right show contraction scour in a wide floodplain.

One of the primary reasons for documentation is to evaluate the hydraulic performance of structures after large floods to determine whether the structures performed as anticipated or to establish the cause of unexpected behavior. In the event of failure, it is essential that contributing factors be identified to avoid recurring damage.

The documentation of a hydrologic analysis is the compilation and preservation of all pertinent information on which the hydrologic decision was based. This might include drainage areas and other maps, field survey information, source references, photographs, hydrologic calculations, flood frequency analyses, stage discharge data, and flood history, including narratives from highway maintenance personnel and local residents who witnessed or had knowledge of an unusual event. The presence of hydrologic data on project plans ensures a permanency of record, serves as a reference in making plan reviews, and aids field engineers during construction.

Estimating Flood Runoff and Flood Magnitudes

In Delaware, there are seven methods of estimating flood magnitudes in sizing a waterway opening for a given structure:

1. Rational method
2. USGS method
3. Program TR-55
4. TR-20 method
5. Published reports
6. Recorded data
7. Frequency mixing method

Rational Method

This is the method used almost universally for computing urban runoff. The rational method is an empirical formula relating rainfall to runoff. It is also used to estimate bridge deck drainage from rainfall intensity, for the design of scuppers.

Discharge, as computed by this method, is related to frequency by assuming the discharge has the same frequency as the rainfall used. Because of the assumption that the rainfall is of equal intensity over the entire watershed, this formula should be used only for estimating runoff from small areas, up to 300 acres.

USGS Method

The Delaware DOT, for example, uses the equations in the current version of the U.S. Geological Survey (USGS) publication *Technique for Estimating Magnitude and Frequency of Floods in Delaware* to estimate flood runoff for drainage areas of 300 acres or greater. These equations are based on specific studies of the watersheds in the state and in adjacent states.

This method relies on data from stream flow gauging station records and combined statistically within a hydrologic homogeneous region, to produce flood frequency relationships applicable throughout the region.

The procedures in the USGS publication should be followed if the designer is using gauging station records and is evaluating these values for upstream or downstream sites. Reasonable estimates of flood runoff can be made by dividing the state into regions.

In the northern region in Delaware for example, only the size of the drainage area, basin development factor, and storage are considered in the equations. The other factors that affect runoff are considered in the constants and exponents. Land use is not considered in the flood runoff equations for the northern region.

In the southern region, basin relief, forest cover, and two soil categories must be considered, in addition to the drainage area and storage. Forest cover is a factor in the equations for southern region.

In areas where land use may change, the TR-55 method should be used. If the USGS method is used, the designer should consider the effects of possible changes in land use.

Program TR-55

TR-55 is a Soil Conservation Service (SCS) program that is applicable to small urban watersheds. The TR-55 program can generate and plot hydrographs, compute peak discharges, and perform detention pond storage estimates. It can account for hydrograph shift and attenuation due to reach routing.

The program uses the graphical peak discharge method to compute peak flows for different return events in a watershed. Refer to the TR-55 manual for additional limitations to ensure that the degree of error is small.

TR-20 Method

The TR-20 computer program (SCS) develops flood hydrographs from runoff. It also routes the flow through stream channels and reservoirs. Routed hydrographs are combined with those of tributaries. The program provides procedures for hydrograph separation by branching or diversion of flow and for adding base flow.

At any desired cross section or structure, peak discharges, their times of occurrence, water surface elevations and duration of flows can be computed. Complete discharge hydrographs and discharge hydrograph elevations can also be obtained.

The program provides for the analysis of up to nine different rainstorm distributions over a watershed, under various combinations of land treatment, floodwater-retarding structures, diversions, and channel modifications. Such analyses can be performed on a large number of subwatersheds, reaches, and structures in any one continuous run.

Recorded Data

This method of analyzing flood frequency relationships from actual stream flow data for a single gauging station enables the use of records of past events, to predict future occurrences. Consideration of future development must be included in these predictions.

The recorded data method assumes that there are no changes in the nature of the factors causing the peak magnitudes. The approximations in this assumption can be minimized by making efforts to determine the past conditions of the drainage area and also making some changes. Common changes consist of modifications as storage and land development. The user of hydrologic data must be familiar with:

- The procedures for evaluating stream flow data
- The techniques for preparing a flood frequency curve
- Proper interpretation of the curve

The data used for the computations of 50, 100 and 500 year discharge needs to be as accurate as possible since subsequent hydraulic analysis, scour analysis and countermeasures design are based on Q values. Exaggerated Q values will lead to expensive countermeasures. On the other hand, lower Q values will cause damage to the bridge in a future flood event.

Discrepancies in Data

The flood plain area is large and the physical parameters need to be recorded for discharge computations are are many. Various factors such as the length of runoff records, storm distribution, and parameters used in the regional analysis could account for some of the discrepancies.

If stream records are sufficiently long to give good flood frequency relationships, significant weight should be given to the stream record in estimating design floods.

When gauging station records are used, the designer should consult the USGS (http://www.usgs.gov/) publication *Techniques for Estimating Magnitude and Frequency of Floods in Delaware* and current USGS data.

Published Reports

Published reports may be used for comparison with the calculated runoff. The FEMA Flood Insurance Study contains runoff information for many streams in the United States. The reports document the methods used to determine runoff for each stream. Several reports have been prepared by the U.S. Army Corps of Engineers for floodplain information for many streams in certain counties. The reports include historical runoff data and in addition calculated runoff.

Design Flood Frequency *R*

The design frequencies for bridges and pipe culverts for each highway functional classification must be based on a risk analysis and must be approved by the bridge design engineer.

Frequency Mixing Method

Simultaneous flood in more than one adjacent streams may occur. Cumulative flood volume needs to be considered in the major river, when small streams are feeding the larger one and causing erosion.

Some states have greater frequency of floods due to higher intensity of rainfall. There may be situations, in which the hydraulic characteristics of the subject facility are influenced by a flood condition of a separate and independent drainage course. A small stream may outfall into a major river that itself is an outfall for a large and independently active watershed. These two waterways are not likely to peak at the same time. It may be assumed that there are two independent events:

- A storm event occurring on the small stream;
- A storm event applicable to the larger watershed.
- Two estimates must be made: one with the design storm applied to the main stream watershed and one with the tributary drainage area experiencing the design storm. The design should be based on the highest discharge.
- In coastal areas, the effects of tidal flows must be considered when the designer is evaluating the frequency mixing relationships.

7.3 USGS StreamStats Software

StreamStats is a Web application that provides

- Estimates of stream flow statistics
- Basin characteristics
- Other information for gauged and ungauged sites developed by the U.S. Geological Survey, using regression techniques

The method available on the USGS website in the public domain is summarized here for the benefit of accurate peak flood computation.

The methodology for computing flood magnitude should be based on latest topography surrounding the stream. For New Jersey streams, Kara Watson and Robert Schopp (2009) have developed a detailed StreamStats procedures.

StreamStats New Jersey

StreamStats is an interactive map-based USGS Web application available for public use. Users can access all the flood frequency statistics and basin characteristics published in the flood frequency report. It utilizes the updated regional regression equations to obtain flood frequency estimates for ungauged sites in New Jersey. The National Streamflow Statistics (NSS) program is used to solve regression equations.

StreamStats Benefits

The StreamStats web application is available for the public to determine the flood magnitude and frequency and basin characteristics for existing gauged sites and at ungauged sites.

- StreamStats is available for the majority of states in the United States.
- Published statistics are readily available.

- There is consistent information delivery.
- Little or no additional error is introduced.
- Estimates for ungauged sites take less than 5 min. (refer to Special Report 38 for New Jersey by Stankowski, 1974).

Need for Stream Flow Statistics

The output from statistics generated by StreamStats can be utilized for:

- Water resources planning, management, and permitting
- Floodplain mapping
- Engineering design of structures such as roads, bridges, dams, culverts, locks, and levees
- Design and permitting of facilities such as wastewater treatment plants, hydropower plants, and water supply reservoirs
- Lack of stream flow statistics where they are needed

Flood Frequency

- A total of 254 stream flow gauging stations in New Jersey and surrounding states were used in the study.
- Peak flow data were collected through the 2007 water year for most gauges.

Flood Frequency Regions

The state was divided into five geographic regions for the regression analysis as defined:

- **GP**, glaciated Piedmont
- **IN**, inner coastal
- **OC**, outer coastal
- **UP**, unglaciated Piedmont
- **VR**, valley and ridge

The regions follow physiographic province boundaries and take into account glaciated extent and coastal geology.

Equation Development

- Basin characteristics are calculated using ArcHydro tools.
- The ordinary least-squares (OLS) method is used for initial regression analysis.
- Four significant variables are chosen: drainage area DA, percent storage, 10-85 slope, and 1990 population density.
- The generalized least-squares (GLS) regression method is used for final regression analysis and equation development.
- GLS regression takes into account the correlation between peak flow estimates, as well as the variability and reliability of the variables.

Final Regional Regression Equations

Equations are calculated for 2-, 5-, 10-, 25-, 50-, 100-, and 500-year floods. All equations use the same variables: drainage area, percent storage, 10-85 slope, and 1990 population density. The constant K_{TR} varies by flood frequency region.

$$(\text{PopDen} + 1)^{0.093}\, Q_{100} = K_{TR}(\text{DA})^{0.713}(\text{Storage} + 1)^{-0.130}\,(\text{Slope})^{0.282}$$

The equations can be used to estimate instantaneous flood discharges for ungauged, unregulated, or slightly regulated nontidal streams in New Jersey.

Flood Frequency Equations

Table 7.1 shows the values for K_{TR}, the coefficient for the recurrence interval for the specified flood frequency region. K_{TR} is the coefficient preceding the drainage area.

Table 7.2 shows estimated percentages of error in flood frequency equations. Table 7.1 gives the values of K_{TR} for frequency intervals, and Table 7.2 shows a close comparison between the estimated and predicted values of the recurrence intervals of floods.

Flood Frequency Region					
Recurrence Interval (Years)	**OC**	**UP**	**GP**	**VR**	**IN**
2	8.9	37.8	20.6	13.8	17.5
5	18.1	75.3	40	28.5	37.6
10	26.5	108	56.7	42	56.2
25	39.7	159	82.2	63.6	86.2
50	51.5	204	105	83.2	114
100	65.1	256	130	106	146
500	105	408	205	176	241

TABLE 7.1 Sample Numerical Values K_{TR} for Frequency Intervals

Average standard errors			
Recurrence Interval (Years)	**Estimate (Percent)**	**Prediction (Percent)**	**Equivalent Years of Record (Years)**
2	49	50.3	1
5	49.4	50.9	2
10	50.5	52.2	3
25	52.6	54.5	4
50	54.7	56.8	5
100	57.1	59.5	5
500	63.9	66.3	6

TABLE 7.2 Estimated Percentages of Error in Flood Frequency Equations

Estimation at Ungauged Sites

- Use the regional regression equation.
- At ungauged sites, the accuracy of flood volume computation needs to be improved. The estimation is weighted by the percentage of the drainage basin within each region.

Estimation at an Ungauged Site Upstream or Downstream from a Gauged Site

- The ungauged site must be on the same stream as the gauged site.
- The drainage area of the ungauged site should be approximately between one-third and 3 times the drainage area of the gauged site.

Estimation at a Site between Gauged Sites

- Users can access all the flood frequency statistics and basin characteristics published in the flood frequency report.
- StreamStats utilizes the updated regional regression equations to obtain flood frequency estimates for ungauged sites in New Jersey.

StreamStats Results and Benefits

Tables 7.3 to 7.6 give examples of input and output data for the hydrologic analysis carried out by the author.[1] The benefits of the software were described above.

Development Strategy

- Program is based on the ArcGIS Hydro-Data Model and Tools (http://www.crwr.utexas.edu/giswr/)
- Standard GIS methods for measuring basin characteristics are built into the ArcGIS Hydro Tools.
- The National Streamflow Statistics Program (NSS) is used to solve regression equations (http://water.usgs.gov/software/NSS)

User Interface

- The user interface displays maps and enables site selection.
- It displays default data layers for selected scale ranges and allows addition and subtraction of map layers.
- It can zoom to a named geographic location or stream.
- It can print maps and download data shown in the map frame.

Flood Drainage Boundaries

Natural streams, canals, humanmade channels, and culverts are main sources of routing floodwater. Additional drainage facilities need to be provided for the disposal of

[1]The software is available on USGS website in the public domain. Tables 7.4 to 7.6 were printed out is similar to the computer output (courtesy of USGS to whom credit is given). The sample output tables from NJ StreamStats are likely to help the reader in using the hydrology program for New Jersey.

Units	
Drainage area	Square miles (mi^2)
Total stream length	Miles (mi)
Average basin elevation	Feet (ft)
Slope	Feet /mile (ft/mi)
Mean population density	People /square mile

TABLE 7.3*a* Data Used as Input in StreamStats Input Data

Water and wetland %
Forest %
Urban land use %
Wetland land use %

TABLE 7.3*b* Percentages of Total Area

Basin Characteristics Report	
New Jersey StreamStats **Date: Wed Jun 30 2010 11: 13:32 Mountain Daylight Time** **NAD27 Latitude: 41.1470 (41 08 49)** **NAD27 Longitude: –74.8834 (–74 53 00)** **NAD83 Latitude: 41.1471 (41 08 50)** **NAD83 Longitude: –74.8830 (–74 52 59)**	
Parameter	**Value**
Drainage area in square miles	57.8
Percent of area covered by water and wetland.	4.3
Mean population density, people per square mile, 2000 census block data.	42.4
10-85 slope in feet per mile based on perprocessed data	46.8
Total stream length in miles	108
Percent of area covered by forest	86.9
Percent of area covered by urban land use	1.24
Percent of area covered by wetland land use	3.24
Average basin elevation in feet	913

TABLE 7.4 Sample set of Data and Location

Peak Flows Region Basin Characteristics 100% Peak Valley and Ridge Region 2009 5167 (57.8 mi²)			
Parameter	**Value**	**Regression Equation Valid Range**	
		Min	**Max**
Drainge Area (Square miles)	57.8	0.87	763
Percent Storage (percent)	4.3	2.36	30.1
Stream Slope 10 and 85 Method (feet per mi)	46.8	2.56	268
Basin Population Density (persons per square mile)	42.4	35	1493

TABLE 7.5 Processing of Data by StreamStats

Peak Flows Region Streamflow Statistics					
Statistic	**Flow (ft^3/s)**	**Prediction Error (Percent)**	**Equivalent Years of Record**	**90-Percent Prediction Interval**	
				Minimum	**Maximum**
PK2	1130	50	1		
PK5	1970	51	2		
PK10	2670	52	3		
PK25	3710	55	4		
PK50	4580	57	5		
PK100	5590	60	5		
PK500	8330	66	6		

TABLE 7.6 Computation of Discharge with Estimates of Prediction Error

floodwater in urban areas. The floodplain boundaries and elevation modeling need to be determined.

A typical sustainable drainage system comprises the following:

- Basins (shallow landscape depressions that are dry most of the time), landscape depressions with shrub
- Swales (wide-based ditches)
- Filter drains (gravel filled trench drain)
- Bioretention basins (depressions with gravel and/or sand filtration layers)
- Wetland habitats

All major cities have best management practices (BMPs) for the urban drainage system and simultaneous water treatment. The Clean Water Act of the Environmental Protection Agency is applicable.

The United States is divided by USGS into successively smaller hydrologic units which are classified into four levels: regions, subregions, accounting units, and cataloging units. The hydrologic units are arranged or nested within one another, from the largest geographic area (regions) to the smallest geographic area (cataloging units).

Each hydrologic unit is identified by a unique hydrologic unit code (HUC) consisting of two to eight digits based on the four levels of classification in the hydrologic unit system. (Refer to USGS, *Water Resources of the United States*, National Water Information System.)

The geographic information system (GIS) stores, edits, analyzes, and displays geographic information. A GIS software is used for the floodplain volume computation which determines the boundary from the digital elevation model (DEM) and flow direction grid to USGS 14-digit hydrologic unit code (HUC 14) boundary lines.

StreamStats Statistics

StreamStats is a valuable tool to compute flows or check flows found by using other methods. It is based on the physical conditions in each state. It can predict the effect of changes in land use or population.

The hydrologic analysis method involves estimating flood flows using regression equations (developed earlier by experimentation). Both parametric field data and non-field data is prepared and used as input in the software.

The data input required for computation of parameters in NJ StreamStats is shown in Tables 7.3 (*a*) and (*b*).

StreamStats regression equations are developed differently for each hydrologic region in New Jersey. They are a function of drainage area, longitudinal slope of the main channel, population density, flood frequency, percentage of wetlands and lakes in the hydrologic region, and physical characteristics of the basin. Area-averaged estimates are based on the exceedance probability of unregulated sites.

Final Regional Regression Equations

$$(\text{PopDen} + 1)^{0.093}\, Q_{100} = K_{TR}(\text{DA})^{0.713}(\text{Storage} + 1)^{-0.130}\, (\text{Slope})^{0.282}$$

Equations are calculated for 2-, 5-, 10-, 25-, 50-, 100-, and 500-year floods. All equations use the same variables: drainage area, percent storage, 10-85 slope, and 1990 population density. Constant K_{TR} varies by flood frequency region.

Flood frequency equation values for K_{TR}, the coefficient for the recurrence interval for the specified flood frequency region.

User Interface The user interface displays maps and enables site selection. It also displays default data layers for selected scale ranges and allows the addition and subtraction of map layers. It can zoom to a named geographic location or stream. The interface can print maps and download data shown in the map frame.

The StreamStats New Jersey drainage boundaries are found as follows:

1. User selects a point on the stream.
2. GIS determines the boundary from digital elevation model (DEM) and the flow direction grid and HUC-14 boundary lines.
3. GIS accumulates all upstream areas and dissolves internal boundaries.

Limitations of StreamStats Method for New Jersey The StreamStats method does not apply to the following types of bridges:

- *Bridges on tidal rivers*: These are shown shaded on interactive maps. The **Q** icon gives a message that no data are available for tidal areas.
- *Parallel bridges*: On braided rivers, these are generally separated by 20 to 100 ft of trees. In river training measures with contraction problems, the river may be bifurcated and a second smaller span auxiliary bridge built.

Conclusions on StreamStats Applications

- The Stankowski method has been used so far for stage 2 and stage 3 reports for over 165 bridges in New Jersey. The StreamStats method, although more reliable than the Stankowski method, gives the same results in many cases. When it differs, a new HEC-RAS analysis needs to be carried out.
- If the velocities were to double by using regression equations, the increase in scour depth would be only about 25 percent, but this change could have a big impact both on qualitative scour analysis and in the selection and design of countermeasures.

Percentile errors are large (over 50 percent) when the regression equations are used. Besides assumptions in the regression equations, the approximate data input values for drainage area, longitudinal slope, population density, etc. used by the software are responsible for allowing abnormal error percentiles.

7.4 Case Studies of Peak Flood Computations Using Practical Examples

The StreamStats software developed by USGS was used. The required hydrologic data were input. The StreamStats sample output for the following bridge location in New Jersey is presented by the author:

Route 31 Bridge over Willoughby River, structure no. 1013152

Hydrologic Data Summary

Route 31 over Willoughby River **Structure no. 1013152**
40.67165–74.91518

- Our Field Inspection: Judge to be Category–Analysis Pending, visited but not classified.
- Constructed in 1930
- Stage 2 Flow Data: Q_{50} = 600 cfs Q_{100} = 760 cfs Q_{500} = 1180 cfs
  - Drainage Area N/A sq. mi.
- StreamStats Results: Q_{50} = 688 cfs Q_{100} = 826 cfs Q_{500} = 1180 cfs
  - Drainage Area 2.44 sq. mi.
- Stream gauge is located approximately 2 mi offstream, similar DA and location
- Q_{100} at gauge is 1930 cfs, with a Q of 2830, this gauge has seen a 100-year storm, along with its proximity, therefore the bridge is of low risk, and has seen a 100-year storm.

The Q_{50}, Q_{100}, and Q_{500} discharge values were computed for the USGS gage station. The highest historic peak flow values show the risk is low or moderate. The Q values were then used as data for HEC-RAS software which computed the velocities and peak floodwater surface elevations. HEC-RAS also performed scour analysis, giving scour depths at the foundations of the pier and abutments.

The StreamStats software data input procedure for the particular state must be followed as shown on the USGS website.

7.5 Scour Practice Survey in U.S. States

Ongoing screening and evaluation of the vulnerability of the nation's highway bridges to scour by state DOTs have identified more than 18,000 bridges that are considered scour-critical and in need of repair or replacement. Almost 100,000 bridges with unknown foundations have been identified. As state DOTs develop plans of action to address the effects of scour and stream instability, there is a need for innovative, effective, and economical countermeasures to be considered for the design of new bridges and for repairing existing bridges.

Limitations of HEC-18 Scour Analysis

For many years bridge designers in the United States have used FHWA Publication HEC-18 as a principal tool to determine scour depths. Increasingly, however, practitioners recognize that some of the circular's standard equations overpredict scour depth for certain geologic and hydraulic conditions. In an effort to improve scour design and evaluation methods, the New Jersey Institute of Technology (NJIT) team with the author recently conducted a survey of current scour practice of DOTs across the United States The 10-question survey queried agencies about their design standards, experiences with failures, monitoring programs, and countermeasure preferences, among other things. The joint research report "NJDOT Research Studies of Sixty New Jersey Scour-Critical Bridges" 2011, presents the results of the nationwide scour survey. Highlighted are the creative and diverse approaches by some states to either modify HEC-18 procedures or develop alternative scour prediction methods. The paper also discusses critical geologic, hydraulic, and hydrologic parameters for rational evaluation of scour depth, gleaned from both the survey and local experience with New Jersey bridges.

A summary of the study results comparing predicted and observed scour is presented in Table 7.7.

Scour Countermeasures Survey

These are the results of the survey:

- Riprap remains the preferred choice by more than a 2:1 ratio.
- Gabions, debris deflection and removal, and foundation strengthening were the next most applied countermeasure methods.
- A small minority of agencies report the use of articulated concrete blocks, concrete pavement, or other methods.

Those states that are currently using modified or alternate scour analysis methods also generously furnished supporting documentation. Other states have published formal design standards and/or rigorous scientific studies supporting their deviations

Highest Historical Peak Stream Flow for USGS 01396582	
Date	Streamflow (cfs)
1999-09-16	2830
2006-06-28	2350
1996-10-19	2280
2009-07-02	1930
1979-01-24	1870
2004-09-18	1860
1984-07-07	1730
2007-04-15	1720
1995-07-18	1660
2008-03-08	1590

Table 7.7a Complete Table of Peak Stream Flow for USGS Station 01396582

Method	Percent	Overpredictions		Underpredictions			Correlation Coefficient
		Avg (ft)	Max (ft)	Percent	Avg (ft)	Max (ft)	
Froehlich	96	10.8	33.2	4	2.2	3.9	0.00
Sturm	86	8.4	50.9	14	5.5	17.7	0.01
Maryland SHA	89	11.8	200.3	11	1.2	3.0	−0.09
Melville	86	4.3	21.3	14	1.4	3.2	0.08

Source: Modified from Lombard and Hodgkins (2008), USGS, Scientific Investigation Report 2008-5099, Prepared in Cooperation with Maine DOT.

Table 7.7b Summary of Predicted versus Observed Abutment Scour for Maine Study

from the standard methods in HEC-18. Selected examples of such modified or alternate scour analysis methods are listed and briefly described in Table 7.8. Reference links are also provided.

Case Study of New Jersey's Scour Program

In 1990, the New Jersey DOT launched a robust statewide scour evaluation program to assess the nearly 2400 existing state and county highway bridges over waterways. Based primarily upon underwater inspection reports, the stage 1 screening studies initially identified 313 state-owned bridges as potentially scour-susceptible. In-depth stage 2 scour evaluations were then carried out in four phases following the analysis procedures described in HEC-18. Upon completion of the stage 2 evaluations, a total of 165 state bridges were determined to be scour-critical.

The plan also prescribed a new real-time flood monitoring program for bridges on the state Watch List, to help safeguard the traveling public until corrective work was completed.

Q1: Have you had any bridges that have failed due to scour, including outright failures or preemptive replacements?		Yes = 69%	No = 31%				
Q2: In your experience, what are the most prevalent types of erosion that have caused failure and/or created potential danger of failure? (You may check more than one; the number of responses is shown.)	Local = 23	Meandering = 17	Contraction = 16	Debris = 15	Degradation = 14	Overtopping = 6	
Q3: Have you generated any summaries that compare field measurements with predicted scour (either published or unpublished)?	Yes = 17%		No = 83%				
Q4: Have you installed fixed instrumentation to measure scour at abutments or piers (either automated or semiautomated)?	Yes = 43%		No = 57%				
Q5: Have you made any field or laboratory measurements of erosion rates of soil or rock materials for the purpose of scour evaluation?	Yes = 17%		No = 83%				
Q6: In the light of your experience with different types of bridge scour, is there a need to modify current HEC-18 design procedures?	Yes = 79%		No = 21%				
Q7: What scour procedures and equations do you use in the design of new bridges?	Standard FHWA version		Modified/alternative version				
	75%	25%					
Q8: What scour procedures and equations do you use in the evaluation of existing bridges?	Standard FHWA version		Modified/alternative version				
	69%			31%			
Q9: Do you consider the effects of natural armoring in your scour computations?	Yes = 54%		No = 46%				
Q10: Does your agency have a preference for particular kinds of scour countermeasures? (You may check more than one; the number of responses is shown.)	Riprap = 32	Gabion = 15	Debris deflection/ removal = 11	Foundation streng-thening = 9	Articu-lated concrete blocks = 6	Other = 5	Concrete pavement = 3

Table 7.8 Summary of Scour Practice Survey Results

U.S. State	Method Description
Alabama	This USGS Scientific Investigations Report published in 2008 provides an alternative method to assess scour depth in the Black Prairie Belt soil, a consolidated, highly cohesive, organic clay within Alabama's Coastal Plain. Envelope curves are presented based on observations of clear-water contraction scour at 25 bridge sites. Related link: http://pubs.usgs.gov/sir/2007/5260/_
Illinois	The Illinois DOT permits reduction in scour depth computed by HEC-18 methods for bridges founded in cohesive soil or rock. Such reductions are graduated from 0 to 100 percent, depending on soil strength or degree of lithification of rock. Related link: http://www.dot.state.il.us/bridges/brmanuals.html
Maine	The USGS Water Resources Investigations Report (see URL below) collected and analyzed pier scour data for eight bridges across Maine over a 4-year period. Observed maximum scour depths were compared with predictions using the Colorado State University (CSU) equation in HEC-18. The relation performed well for rivers in Maine, and the Maine DOT currently uses it for evaluation of existing and new bridges. Related link: http://me.water.usgs.gov/reports/wrir02-4229.pdf
Pennsylvania	The Pennsylvania DOT scour design method recognizes the variable erosion behavior of geologic materials in scour design. It establishes three classifications: sound bedrock, erodible bedrock, and coarse soil (gravel, cobbles, and boulders); specific embedment depths and footing details are prescribed for each. Related link: ftp://ftp.dot.state.pa.us/public/PubsForms/Publications/PUB%2015M.pdf
South Carolina	A recently published USGS Report of Investigation extends the earlier 2006 USGS study described above in "Comparative Scour Studies." It recommends use of envelope design curves to estimate the scour depth. The curves are rigorously justified with field observations and laboratory data. The South Carolina DOT has incorporated the method into its latest scour design standards. Related links: http://pubs.usgs.gov/sir/2009/5099/ and http://www.scdot.org/doing/pdfs/requirements2009.pdf
Texas	This comprehensive study performed by Texas Transportation Institute summarizes a new method to assess a bridge for scour. It uses three levels of bridge scour assessment (BSA 1, 2, and 3) and erosion classification charts. Scour vulnerability is determined by comparing the predicted scour depth with the allowable scour depth of the foundation. The method is relatively simple to apply, and it overcomes some of the overconservatism in current methods. Related link: http://tti.tamu.edu/documents/0-5505-1.pdf

Table 7.9 Selected Examples of Modified or Alternative Scour Evaluation Methods

7.6 Identifying Scour-Critical Bridges on Flood-Prone Rivers

Bridges that are scour-critical and the hydrology of 50-, 100-, and 500-year floods are considered. For hydrology, flood discharge will be computed based on the rainfall intensity, catchment area, and soil conditions. A prioritization process for location on rivers is required, for identifying bridge importance based on the following factors:

- National security and military route
- Hospital route
- Those serving school busses
- Average daily truck traffic (ADTT)
- Site-specific conditions, such as overtopping of flood and opening blockage by debris

Structures that are scour-critical need to be identified by regular inspections, rating, and monitoring. The degree of scour criticality will be computed, using scour-critical bridge indicator (SCBI) software. A scour vulnerability index, giving relative weight to factors causing potential vulnerability to scour damage, will be computed. Bridges will be rated for scour as low, medium, high, and very high vulnerability. Factors to be considered are as follows:

- Scour history
- Debris accumulation
- River meander and angle of attack
- Longitudinal slope of river
- Hydraulic flow rate
- Streambed material
- Type of foundation
- Structural redundancy of substructure
- Structural redundancy of superstructure

For computation of the maximum flood discharge, a relevant hydrology method such as the StreamStats, rational, or Stankowski method for New Jersey will be selected. Alternative computation methods for scour depth and average flood velocity for 25-, 50-, and 100-year return periods will be discussed for H & H analysis. Factors affecting the magnitude of erosion and depth of scour are listed. Scour depths caused by aggradation, degradation, contraction, still water, live-bed and local scour, etc. will be computed based on the formula given in HEC-18 for riverine flow and in HEC-25 for tidal flow.

Figure 7.2*a* shows obstruction to flow from debris accumulation and vegetation.

Figure 7.2*b* shows riprap displaced after floods at a two-span New Jersey bridge. New Jersey DOT is not using riprap as a permanent countermeasure for scour-critical bridges.

As mentioned on Chapter 3, commercial computer software for solving HEC-18 scour analysis equations such as ABSCOUR and HEC-RAS is available, but an independent Excel spreadsheet also may be used. The results will be used for selecting and designing countermeasures.

Figure 7.2(*a*) Debris accumulation and vegetation growth at a single-span New Jersey bridge.

Figure 7.2(*b*) Riprap displaced after floods at a two-span New Jersey bridge.

Scour-critical bridges located on streams with high flood velocities have been the cause of major foundation settlements. Bridges with narrow waterway openings and erodible soils contribute to bridge collapse.

Underwater Inspection for Scour-Critical Bridges

Three methods are commonly used:

1. Wading inspection in shallow waters
2. Scuba diving in deep water
3. Hard-hat diving for rivers with high velocities

The levels of inspection are

- Visual
- Detailed inspection with partial cleaning
- Detailed inspection with nondestructive testing
- Use of special equipment such as fathometers

The National Bridge Inspection Standards (NBIS) have defined the criteria for selecting the type of inspection.

The two inspection forms are the waterway/channel form and the substructure scour form. Inspection of piers and abutments is carried out with special emphasis on piles and pile bent inspection. Priority repair categories and repair methods are outlined.

Figures 7.2*a* and *b* and 7.3 show several bridges subjected to varying degrees of scour after recent floods in the Northeast. It was a research project team of which author

FIGURE 7.3 Team members exploring riverbed for scour holes at a New Jersey bridge.

was a member. A special postflood report was prepared and recommendations made to repair or replace the bridges and remove the debris, to allow free flow of water and prevent further erosion of foundations.

Figures 7.4 to 7.7 also show several bridges subjected to varying degrees of scour after recent floods in the Northeast.

FIGURE 7.4 Pier footing exposed due to scour after riprap was washed.

FIGURE 7.5 Vertical cracks developed in pier wall and girders rusted due to floodwater.

FIGURE 7.6 Water flows at a sharp skew angle owing to meander.

FIGURE 7.7 Clogging of waterway is due to high velocity of water displacing the riprap.

FIGURE 7.8 A newly designed bridge with large size opening did not cause any overtopping.

Figure 7.8 shows a bridge designed to latest design code and FHWA circulars, which minimized scour effects.

Overtopping of Floodwater during Floods

Occasionally, floodwater may overtop a bridge deck. This may occur because of an unforeseen change in demography or unusual snowmelt or storms resulting in peak floods, the quantum of which was not allocated in the original hydrology analysis. This factor is generally beyond the control of a structural engineer since a large-span design or river training measures, in anticipation of unknown factors, is not economical. However, the following factors can be incorporated in the planning stage:

1. As an effective drainage provision, the longitudinal slope of profile of baseline geometry should be 2 percent or more for rapid disposal of water to the approaches and beyond. In the transverse direction, approximately 4 percent cross-slope needs to be provided. Drainage inlets and pipes need to be provided and connected directly to the public sewer system, instead of disposing of the water back into the river.
2. A sag curve on the deck should be avoided to prevent water accumulation.
3. Solid parapets, which trap water, should be avoided. Instead parapets with open rails should be used.
4. Bridge deck profile should be raised by a few feet, if possible. This may not be always possible since access to adjoining properties will be affected.

Figure 7.9 shows damage and corrosion to an old steel bridge resulting in bridge closure for repairs after floods.

FIGURE 7.9 A river steel truss bridge closed for repairs after floods.

Planning of Second Bridge over the Rio Guaíba in Brazil

A 1.9-km bridge will cost nearly $300 million and will help transportation projects in South America. Guaíba is located on the west side of the Guaíba, a lake that receives five rivers and carries them toward the Atlantic Ocean via the Lagoa dos Patos. It is also at the crossroads of two federal highways that link Brazil with Argentina and Uruguay, putting it at advantage for investments that seek to target the market of these Mercosul nations.

Continuity of Deck

Twin bridges, each catering to traffic flow in opposite directions, should be planned instead of a single bridge. The adverse scour or overtopping effects of one bridge will be less for the other bridge, due to cushioning of water overflow by the first bridge.

7.7 Planning of Structures in Relation to Hydrology Data

Some important guidelines for the planning and design of superstructures and substructures (piers and abutments) on the basis of HEC-23 guidance are described here.

Superstructure

1. The bridge superstructure elevation should be raised above the general elevation of the approach roadways wherever practicable. This provides for overtopping of approach embankments and relief from the hydraulic forces acting at the bridge. This is particularly important for streams carrying large amounts of debris, which could clog the waterway at the bridge.
2. Elevation of the lower chord of the bridge should be increased a minimum of 3 ft above the normal freeboard for the 100-year flood for streams that carry a large amount of debris.

3. Superstructures should be securely anchored to the substructure if buoyant or if debris and ice forces are probable. The superstructure should be shallow and open to minimize resistance to the flow where overtopping is likely.
4. Continuous span bridges withstand forces due to scour and resultant foundation movement better than simple span bridges. Continuous spans provide alternate load paths (redundancy) for unbalanced forces caused by settlement and/or rotation of the foundations. This type of structural design is recommended for bridges where there is a significant scour potential.
5. At some bridge sites, hydraulics and traffic conditions may necessitate consideration of a bridge that will be partially or even totally inundated during high flows. This consideration requires designing pressure flow through the bridge waterway.

Pier Substructures

1. Local scour holes at piers and abutments may overlap one another in some instances. If local scour holes do overlap, the scour depth becomes indeterminate and possibly deeper. The top width of a local scour hole on each side of a pier may range from 1.0 to 2.8 times the depth of local scour. A top width value of 2.0 times the depth of local scour on each side of a pier is suggested for practical applications.
2. For pile and drilled shaft-supported substructures subjected to scour, a reevaluation of the foundation design may require a change in the pile or shaft length, number, cross-sectional dimension, and type, based on the loading and performance requirements and site-specific conditions.
3. Pier foundations on floodplains should be designed to the same elevation as pier foundations in the stream channel, if there is likelihood that the channel will shift its location over the life of the bridge.
4. Align piers with the direction of flood flows. Assess the hydraulic advantages of round piers, particularly where there are complex flow patterns during flood events.
5. Streamline piers to decrease scour and minimize the potential for buildup of ice and debris. Use ice and debris deflectors where appropriate.
6. Evaluate the hazards of ice and debris buildup when considering use of multiple pile bents in stream channels. Where ice and debris buildup is a problem, consider the bent a solid pier for the purposes of estimating scour. Consider the use of other pier types where clogging of the waterway area could be a major problem.
7. Scour analyses of piers near abutments should consider the potential of larger velocities and skew angles from the flow coming around the abutment.

Abutment Substructures

1. Equations to calculate abutment scour in HEC-18 may tend to overestimate the scour depth because of lack of verification of field conditions. Recognizing this, the abutment scour equations may be used to develop insight into the scour potential at an abutment. Engineering judgment must then be used to determine

whether the abutment foundation should be designed to resist the computed local scour.

2. As an alternative, abutment foundations should be designed for the estimated long-term degradation and contraction scour. Riprap and/or guide banks should be used to protect the abutment with the use of this alternative.
3. Relief bridges, guide banks, and river training works should be used, where needed, to minimize the effects of adverse flow conditions at abutments.
4. Where ice buildup is likely to be a problem, the toe of spill-through slopes or vertical abutments should be back from the edge of the channel bank to facilitate the passage of ice.
5. Wherever possible, spill-through (sloping) abutments should be used. Scour at spill-through abutments is about 50 percent of that of vertical wall abutments.
6. When warranted, riprap, a guide bank 50 ft or longer, or other bank protection methods should be used on the downstream side of an abutment and approach embankment to protect them from erosion by the wake vortex.

Guidelines for new foundations suggest the use of riprap as a supplementary countermeasure for the protection of new foundations (piers and abutment). Riprap should be used with prior excavation and with a filter layer and only as a supplemental countermeasure.

7.8 Substructure Planning for New Bridges

In planning new bridges there is greater freedom in providing for resilience to soil erosion than for the existing bridges. The following steps need to be considered:

Bridge location: On straight segments is preferred. Avoid bends and downstream of dam Lateral meander of river is avoided. Increased scour if dam is breached.

Bridge may have sharp skew in plan: Place spurs on upstream side to retard flood flow, guide banks on upstream side to align flow in bridge opening,

Flow Direction: Align abutment and pier walls parallel to flow-Angle of attack is minimized.

Local scour is minimum.

Bridge plan has sharp skew: Monitoring is required to clean any debris which may be trapped between girders.

Abutment Location: Provide adequate setback of abutments to increase the width of bridge opening.

Advantage: As a result degradation is minimum.

Disadvantage: There may be a slight increase in bridge cost.

Advantage: Degradation is minimum.

Disadvantage: Increased bridge cost

Provide bed armoring or sheet piles.

Abutment Type: Use stub or integral in place of full height.

Advantage: Local scour is minimum. Use sheet piles,

Vertical or Sloping Wall Abutment: Use sloping wall in place of vertical wall

Advantage: Local scour is minimum.

Disadvantage: Increased cost of formwork

Abutment Foundation: Use deep piles in place of short piles

Minimum degradation, Loss of soil due to scour compensated by additional pile lengths.

Cost of piles increases.

Use sheet piles with riprap to protect scour of top of piles,

Wing Walls: Align with flow direction Local scour is minimum.

Sheet piles with riprap to protect scour of top of piles,

Pier Location: Place piers away from thalweg of river.

Advantage: Contraction scour is minimum. Pier height is reduced. Survey of river profile is required.

Use bed armoring or sheet piles.

Pier Type: Use pile bents or multiple columns with curtain wall to prevent debris deposit

Pier Shape: Use round or pointed shapes.

Advantage: Local scour is minimum since coefficient K_1 is lower.

Disadvantage: Increased formwork cost.

Pier Foundation: Use deep piles in place of short piles

Additional pile lengths compensate loss of soil due to scour.

Advantage: Minimum degradation

Disadvantage: Cost of piles increases

Use sheet piles with riprap to protect scour of top of piles,

Spread Footing on Soil: Place bottom of footing below total scour line. Footing settlement is avoided

Bed armoring such as riprap or gabion required

Spread Footing on Weathered Rock: Determine RQD to estimate rock erodibility

Allowance is made for scour of rock

Bed armoring such as riprap or gabion required,

Spread footing on rock: footing settlement is avoided

Embankment: Protect embankment at upstream and downstream of bridge with armoring

Advantage: Controlled floodplain width since bank erosion is avoided.

Contraction scour is minimum.

Disadvantage: Costly Protection of Sloping embankment.

Revetments required.

Length of Bridge: Bridge to span full floodplain width

Advantage: Opening area increased. Relief bridge is not required.

Contraction scour is minimum.

Disadvantage: Costs are high. Right-of-way issues to be resolved.

Use bed armoring or sheet piles.

Number of spans: Prefer multiple spans to single span

Advantage: Redundant load path. Safer bridge

Disadvantage: Foundation cost is higher for multiple spans.

Use bed armoring or sheet piles.

Length of Single span Increase of span length

Advantage: Opening area increased.

Contraction scour is minimum.

Disadvantage: Bridge cost increases Use bed armoring or sheet piles.

Parapet Wall: Use open spandrel. Avoid solid parapet.

Advantage: Increases flow during peak flood. Contraction scour is minimum.

Open Parapets and Railing: Avoid fence or use of solid parapet.

Increases flow during peak flood.

Contraction scour is minimum.

Deck Profile: Use of vertical curve

Advantage: Increased vertical clearance.

Disadvantage: Cost of approaches is increased

Girders: Use shallow depth Increased vertical clearance

Disadvantage: More girders required since girder spacing reduced.

Guidelines to Minimize Different Types of Scour of New Bridges

The following guidelines may be used to minimize different types of scour in the design phase of new bridges.

Contraction Scour

The following planning steps are desirable for reducing contraction scour:

1. Use larger openings to allow for debris height accumulation.
2. Use longer-span bridges, elevated decks, and crest vertical curves.
3. Use profiles for overtopping during floods and relief bridges.
4. Use reduced superstructure depths. Use open spandrel parapets.
5. Place piers away from the thalweg of a river.
6. Excavate waterways to remove debris from smaller floods.
7. Use guide banks on upstream side to align flow in bridge opening.
8. Use revetments on channel banks' fill slopes at bridge abutments.

Local Scour at Abutments

The following planning steps are desirable for reducing local scour:

1. Place foundations on sound rock.
2. Use deep piling.
3. Use stub abutments in lieu of full height.
4. Use sloping walls in place of vertical walls.
5. Use revetments (pervious rock or rigid concrete).

6. Use riprap on spill slopes.
7. Use guide banks at abutments.
8. Monitor and inspect after flood events.

Local Scour at Piers

The following planning steps are desirable for reducing local scour:

1. Place foundations in sound rock or below the total scour line.
2. Use deep piling as foundations.
3. Streamline pier noses (rounded shape).
4. Use pile bents or multiple columns with curtain walls to prevent debris deposit.
5. Use riprap as a temporary measure.
6. Cut cofferdams below contraction scour depths.
7. Monitor and inspect after flood events.

Aggradation

The following planning steps are desirable for reducing aggradation under the bridges:

1. Use debris basins.
2. Provide continual maintenance planning.

Degradation

The following planning steps are desirable for reducing degradation of bed elevation:

1. Use check dams or drop structures on small to medium streams.
2. Use channel lining.
3. Use deeper foundations.
4. Provide adequate setback of abutments.
5. Use rock and wire mattress for small channels.

River Meander

Locate bridges on straight reaches of streams between bends.

Braided Channels

1. Build one long bridge, more than one bridge, or a relief bridge.
2. Monitor and inspect after flood events.

Collecting Data for Bridges

See the evidence of postflood debris accumulation in Figure 7.10. This figure shows an example of construction of a pier on a scour hole in the case of twin bridges. Figure 7.11 shows a team member inspecting a scour-critical bridge in castern Pennsylvania. Figure 7.11 details a scour hole formed because of lateral movement of a main channel, exposing pier foundations. Figures 7.10 to 7.13 show debris accumulation and

Figure 7.10 Evidence of postflood debris accumulation in the Northeast. (Photo by author.)

Figure 7.11 Further evidence of postflood debris accumulation. (Photo by author.)

FIGURE 7.12 Flooding causing debris accumulation downstream of bridge. (Photo by author.)

FIGURE 7.13 Soil erosion at abutments of downstream twin bridge. (Photo by author.)

washing away of riprap given in Section 7.9. Figure 7.12 shows bridge failure due to settlement of foundation in a scour hole resulting from deficiencies in foundation design and soil conditions..

7.9 Planning Considerations for Superstructures Based on Hydrology and Hydraulics

Value Engineering (VE) for Lowering Project Costs

In Chapter 3, the importance of value engineering in the planning of low cost bridges was addressed. It is emphasized here also as a cost-effective planning tool. It requires research into the relative costs of alternative solutions. For example, a medium-span bridge can be constructed, using either concrete girders or steel girders. The following cost-related considerations need to be investigated:

- Span lengths to determine type of bridges
- Availability of precast concrete materials versus high-performance steel (HPS)
- Use of weathering steel versus painting for minimum corrosion
- Use of lightweight concrete versus normal strength for minimum creep and shrinkage losses
- Storage facilities at site for cast-in-place concentration compared to prefabrication
- Relative cost of indirect costs from adverse traffic impacts using conventional method versus accelerated bridge construction.

Figure 7.14 shows evidence of cracks in the abutment wall due to settlement.

Figure 7.14 Evidence of cracks in abutment wall due to settlement.

Value Engineering Goals

The primary goal is to reaffirm support for the selected scheme. The secondary goal is to suggest only improvements that are necessary to reaffirm this support or that will improve safety or efficiency.

- A thorough VE review takes no more than 3 months.
- It adds no time to the schedule.
- If concurrent with start-up of the designer's preliminary engineering activities, it takes less than 5 percent of the design budget.
- Peace of mind is worth the time and expense.

In the United States it can take 6 to 10 years to conceive, design, and award a project. Some typical reasons listed by Sung Park are

- Competition for limited funding
- Need for community support
- Environmental mandates
- Other federal and state mandates
- Bureaucracy and red tape
- Utilities and right of way

Economic Planning

Since major costs are involved in planning new bridges, the following factors should be considered in the preliminary planning of a new bridge construction.

1. A cost–benefit analysis that is based on the importance of a bridge to select the most appropriate foundation type should be considered.
2. Decide, if justifiable, on a design return period other than 100 years (such as 50 years) and a check return period of 500 years (superflood). In the case of embankments with levees, a 50-year flood may govern over a 100-year flood for peak flood velocity and water surface elevation.
3. Consider scour risk assessment based on hydraulic and hydrologic analysis.
4. Consider right-of-way and construction issues related to proposed scour countermeasures.

Hydraulic Planning

The goal of hydraulic scour reduction measures is to actively reduce scour around bridge foundations by improving flow conditions around the structure. Such hydraulic measures include

1. Location of a structure
2. Scour depth calculations using hydrologic and hydraulic analysis to design scour-resistant foundations
3. Streamlining of the structural elements

4. River training and deflectors to control stream instability, with the measures including guide banks, check dams, spurs, etc., and reduction of flood velocities or control of river migration
5. Avoiding the use of riprap for pier footings

Structural Planning

Structural measures involve designing bridge foundations that will not fail during design flood events. This is the most important, safest, and most reliable part of a countermeasure design for new bridges. The following factors should be considered in the structural planning of new bridges.

1. Locating bridges to avoid adverse flood flow patterns
2. Streamlining bridge elements to minimize obstructions to flow
3. Designing foundations to resist scour without relying on the use of riprap or other countermeasures
4. Designing deep foundations or foundations on rock, where practicable
5. For spread footings on soil, placing the footing to minimize the scour hazard

Scour Reduction Planning

The goal of scour reduction measures is to limit scour around the structure by shielding foundations by sheet piling and/or other suitable countermeasures. If countermeasures are applied to structures after they are constructed, their use is not of primary importance and should be considered as a supplement to structural planning.

Several important factors should be considered during the hydraulic design phase of the bridge to minimize the risk of scour.

Location-Related Factors

1. Avoid locating structures at a confluence of two or more channels.
2. Avoid locating structures at or near sharp bends; locations on straight reaches or gentle bends are preferred.
3. Locate bridge crossings at the head or apex of the alluvial fan.
4. Check channel stability using aerial and satellite photography, and historic maps, as changes in flow direction can increase scour significantly.
5. Consider the stability of the channel vertically and laterally.
6. Physical modeling should be considered for major structures on alluvial rivers or channels, major tidal crossings, barrages, and complex flow conditions that cannot be modeled using simple one-dimensional models.
7. Consider river morphology: channel widening, realignment, and changes in agricultural practices that may reduce scour.
8. Analyze the effect of existing nearby structures or effects of removal of old structures on flow conditions around the bridge

Hydraulic Design-Related Factors

1. Concerning the size of the waterway opening, consider construction cost versus scour risk.
2. The size and number of relief openings is a measure used for rivers with large waterway widths. Lack of relief openings may result in larger bridge crossing, increased water level on the floodplain, scour of embankments parallel to flow, scour of abutments from turbulence, and vortices generated by the interaction of main channel flow with returning floodplain flow. Determine the number of relief openings by hydraulic design. Consider scour countermeasures at relief openings.
3. Consider scour due to changes in flow direction at low flows and flood flows.
4. Consider the effect of operational requirements causing a concentration of flow on one side of the structure (e.g., opening of weir gate on one side).
5. Regarding the overtopping of approach embankments, a lower level of approach embankment than the bridge deck will avoid washout of the bridge.
6. Consider the effects of removal of downstream or upstream bed control.
7. Consider channel improvements.

Streamlining Structural Elements

1. Sloping ("spill-through") abutments cause significantly less scour than a vertical wall abutment.
2. Integral abutments with a single row of piles and semi-integral abutments, which are similar to stub abutments on piles, have a lower scour depth as calculated from standard scour analysis equations.
3. Angled wing walls (typically set at 30° to 75° to the longitudinal flow direction) or curved wing walls improve the hydraulic performance of vertical wall abutments. Under normal circumstances angled wing walls are adequate. Wing walls at 90° to the longitudinal flow direction are also acceptable if turbulence due to separation of flow is unlikely to be a significant problem.
4. Regarding pier groups, for bridges on scour-critical rivers, the number of piers selected should be as small as possible. This can be achieved by selecting longer spans. Minimum obstruction from the pier group during a peak flood event will prevent water overtopping the deck.
5. Regarding pier shapes, the best hydraulic performance is given by rectangular piers having a wedged-shaped nose (known as cutwaters).
6. Where the river may change its angle of approach over the life of the structure, a circular pier or a series of circular piles with a pile cap above the high-water elevation may be more appropriate.
7. Where debris accumulation is likely to be a problem, debris deflectors should be used.
8. Regarding the overall structure alignment and alignment of elements, where the bridge deck could become submerged by an extreme flood, it may be appropriate to streamline the underside of the bridge deck by rounding the upstream and downstream faces to facilitate passage of debris.

Comparison of Pier Scour Depths

A comparative study was made for the estimated scour depths at piers, using Melville's method and the widely used HEC-18 method. The difference was found to be on the order of 10 percent. The results for pier scour depth are 9.46 m by Melville's method and 8.58 m by the HEC-18 method.

Comparison of Abutment Scour Depths

The two results for the problem are 18.82 m using Melville's method and 17.69 m using the HEC-18 equations. Hence, the comparative study for abutment scour depths shows closer agreement and gives confidence in the use of HEC-18 equations.

Figure 7.15*a* to *c* shows severe scour exposing the foundations and partial collapse of substructure due to the increase in scour depths. Velocity of flow exceeded 10 ft/s.

FIGURE 7.15(*a*) Exposure of piles due to local scour.

FIGURE 7.15(*b*) Foundation scour at pier after floods.

Figure 7.15(c) Collapse of substructure due to increase in scour depth.

7.10 Formation of Scour Holes during Floods

Scour holes may evolve by progressive initiation, development, stabilization, and equilibrium. During the initiation phase, erosion is rapid and material is eroded to form scour hole. During the development phase, the depth of the scour hole increases. During the stabilization phase, the maximum length of hole occurs. In the equilibrium phase the dimensions of the hole are virtually fixed. These are usually created by the turbulence of flow and formation of eddy currents near the piers.

Debris accumulation upstream and downstream of a bridge also contributes to the formation of scour holes. In some cases, these holes may close over a long period by internal soil bed movement and aggradation. These are some important aspects of the scour hole formation process:

- They are more likely to be formed in soft, silty soils.
- The location and size of any scour holes must be determined. Sizes may typically vary from 2 to 5 ft in depth, while widths are wider and would depend upon the peak flood velocity and turbulence.
- Scour holes close to the footings may continue to erode and enlarge in area and may ultimately lead to foundation instability. Several small holes may enlarge to form a single large scour hole.
- Although they are filled up with water, scour holes need to be completely filled with a similar soil such as riverbed material or riprap. The DEP may not issue a permit for using concrete or a filling material due to chemical contamination and adverse effects on surrounding vegetation.

7.11 Conclusions

The causes and effects of sources of water resulting in scour and foundation settlement are addressed. Hydrologic analysis using StreamStats or a standard scour software such as the Melville method and ABSCOUR can also be used.

Methods of hydrologic analysis using the powerful USGS StreamStats software, which can address unique soil conditions in each state, are discussed. StreamStats examples for selected New Jersey bridges are presented.

In this chapter, scour practice is investigated using an scour practice inquiry proforma sent to many states. The responses helped to appreciate the recent developments in the new subject of countermeasures design.

Scour-critical bridges need to be identified as a result of intensive hydrology, although HEC-18 mandatory requirements provide nominal countermeasures at the foundations of every river bridge irrespective of the hydrology .and volume of flood.

Various steps for identifying scour-critical bridges are described. These include hydrologic and hydraulic analyses, designing scour-resistant foundations for new bridges, monitoring to identify scour issues, and use of monitoring measures as preventive countermeasures.

Planning considerations for new bridges and general procedures for the design of countermeasures are outlined. Case studies of scour-critical bridges in New Jersey are discussed.

Overtopping of floodwater during floods can be avoided by planning measures. Planning of substructures and bridges should consider hydrologic investigations. Other issues are debris accumulation and aggradation due to narrow waterway openings under the bridge.

Formation of scour holes at bridge sites and adjacent to abutments and piers is commonly seen due to vortex formation in soft soil.

7.12 Bibliography

American Association of State Highway and Transportation Officials, *AASHTO Load Resistance Factor Design (LRFD) Bridge Design Specifications*, Washington, D.C., 1998.

American Society of Civil Engineers, Stream Stability and Scour at Highway Bridges, Water Resources Engineering, Compendium of Papers presented at Conferences, 1991 to 1998, Richardson, E. V., and Lagasse, P. F., 1999.

DELDOT, *Bridge Design Manual*, Hydrology and Hydraulics, 2005.

FHWA, *Evaluating Scour at Bridges*, 4th ed., HEC-18, FHWA-IP-90-017, 2001.

FHWA, *Bridge Scour and Stream Instability Countermeasures*, HEC-23, Washington, D.C., July, 1997.

FHWA, *Federal Coding Guide*, 1994.

FHWA, *Determination of Unknown Subsurface Bridge Foundations*, Olson, L.D., Jalinoos, F. and Aouad, M.F., NCHRP Report no. 21-5, Geotechnical Guideline no. 16, 1998.

Hamill, Les, *Bridge Hydraulics*, E & FN Spon, London and New York, 1999.

Melville, B. W., and Coleman, S. E., *Bridge Scour*, Water Resources Publication LLC, Highlands Ranch, CO, 2000.

New Jersey Department of Transportation, *NJDOT Bridge Design Manual*, 1998.

Stein, S. M., Pearson, D., and Jones, S., *Economic Considerations in Designing Bridge Scour Countermeasures*, FHWA, 1998.

FHWA, *Design of Riprap Revetments SI*, HEC-11, 2000.
FHWA, *Evaluating Scour at Bridges*, HEC-18, 2010.
FHWA, *Stream Stability at Highway Structures*, 4th ed., HEC-20, 2012.
FHWA, *Bridge Scour and Stream Instability Countermeasures*, HEC-23, 2007.
FHWA, *Tidal Hydrology, Hydraulics, and Scour at Bridges*, 2d ed., HEC-25, 2008.
NCHRP Research Report 24-07, "Countermeasures to Protect Bridge Piers from Scour," Transportation Research Board, Washington, D.C., 2006.
Watson, Kata M., and Robert D. Schopp, *Methodology for Estimation of Flood Magnitude and Frequency for New Jersey Streams*, http://pubs.usgs.gov/sir/2009/5167/; for the StreamStats Web application, go to http://water.usgs.gov/osw/streamstats/

Mohiuddin Ali Khan Publications Related to Scour

Fuller, Derek, "Literature Review for Selection and Analysis of Scour Countermeasures for Pennsylvania Bridges," Master's thesis (committee members Bechara Abboud and Mohiuddin Ali Khan), Temple University, Philadelphia, PA, May 2011.
Khan, Mohiuddin Ali, "Bridge and Highway Structure Rehabilitation and Repair," Chapter 12, *Protection of Bridges against Extreme Events*, McGraw-Hill, New York, 2010.
Khan, Mohiuddin Ali, "Modern Bridge Engineering," Chapter 14, in Meyer Kutz (ed.), *Transportation Engineering Handbook*, McGraw-Hill, New York, 2011.
Khan, Mohiuddin A., and A. K. Agrawal, *Handbook of Scour Countermeasures Designs*,
Lombard, Pamela L., and Hodgkins Glenn A.," Comparison of Observed and Predicted Abutment Scour at Selected Bridges in Maine," USGS Scientific Investigation Report 2008-5099, (prepared in cooperation with Maine DOT).
NJDOT 2007, FHWA-NJ-2005-027, NJDOT, Bureau of Research, Trenton, NJ, 2007.
Khan, M. A., and Khalid Mukhtar Siddiqui, "Management and Improvement of Roads through Inspections," The Institution of Engineers Pakistan, 30th International Conference, Karachi, Pakistan, 1988.
Khan, M. Ali, John Schuring, Golub, E., Young, M.R. and Aboobaker, N. "Review of Bridge Scour Practice in U.S.," Proceedings of Fifth International Conference on Soil Erosion (ICSE-5), San Francisco, CA, 2010.
Khan, Mohiuddin Ali, Joseph Zarriello, and Salim Baig, "Stream Instability Study for New Jersey Bridges," First International Conference on Scour of Foundations (ICSF-1), Texas A&M, College Station, TX, 2003.
NJDOT Bridge Design Manual, Section 45 New Jersey Seismic Code and Section 46 Scour at Bridges, developed by Mohiuddin Ali Khan, chief bridge engineer, STV Inc., Trenton, NJ, 2002.

Mohiuddin Ali Khan Research Reports

NJDOT Research Project on Water Quality and Banking, jointly with RBA Group Inc., Consulting Engineers, Parsippany, NJ and City University of New York, 2006.
NJDOT Research Studies of Sixty New Jersey Scour-Critical Bridges; Hydrologic, Hydraulic, Scour Analysis, Countermeasures Selection and Design, Report submitted jointly by the author with New Jersey Institute of Technology (NJIT), Newark, NJ, 2011.
NJDOT Stage 2 Studies of Twelve Hunterdon County Scour-Critical Bridges; Hydrologic, Hydraulic, Scour Analysis, Countermeasures Selection and Design, STV Inc., Trenton, NJ, 2002.

PennDOT Studies of Four District 6 Scour-Critical Bridges; Hydrologic, Hydraulic, Scour Analysis, Countermeasures Selection and Design, jointly with Temple University, Philadelphia, PA, 2012.

USGS, *Water Resources of the U.S.*, National Water Information System (http://water.usgs.gov/GIS/huc.html).

Mohiuddin Ali Khan Seminars Presented to Practicing Engineers

ASCE Philadelphia Section, "Scour Analysis and Countermeasures Design Course," presented by Peggy Johnson, Nazhat Abubaker, and Mohiuddin Ali Khan, Philadelphia, PA, 2008.

SEI of ASCE Philadelphia Section, "Bridge Failures and Remediation," Mohiuddin Ali Khan, Philadelphia, PA, 2010.

Maryland Structural Engineering Institute (MDSEI), "Scour Analysis Course," presented by Mohiuddin Ali Khan, October 25, 2007, Marriott Hotel, Towson, MD.

CHAPTER 8

Estimating Scour Depths and Selecting Applicable Countermeasures

8.1 Need for Inspection, Scour Studies, and Remediation

This chapter supplements Chapters 3 to 7 by providing further information on the importance of scour analysis and countermeasures selection.

The objectives are to prevent collapse and settlement of bridges on rivers. In this chapter, the hydraulic assessment checklist and software details are included, in addition to geotechnical and hydraulic factors listed in Chapter 6.

The concepts of vulnerability index and essential aspects of scour technology with case studies of bridges that the author studied are presented. These include identification of scour-critical bridges, hydraulics, scour analysis, countermeasure selection and design, and continued monitoring. Soil scour studies and remediation are required.

Bridge rating based on NBIS can be evaluated regularly when bridge inspection records are well documented. If cross sections of the stream are maintained over many years, long-term scour can be calculated from the streambed elevation changes. Projections of scour based on a long-term trends can be made. Long-term scour is based on the concept that long-term streambed elevation changes can take place over the time scale of several years during the life of a bridge, due to aggradation or degradation.

Aggradation involves the deposition of material eroded from the channel or watershed upstream of the bridge. A common countermeasure used for aggradation is channel improvements by dredging or cleaning. Degradation involves the lowering or scouring of the longitudinal profile of a channel. Countermeasures for degradation are channel lining with concrete pavement, increasing a bridge opening width, or planting vegetation.

Scour Depth Computation Methodology

The following information is presented here:

1. Magnitude and depth of erosion depend upon discharge volume and velocity. Rivers, rivulets, streams, brooks, and channels are all subjected to scour to

varying degrees. A channel, for example, may have a small discharge but a high velocity. Hence scour is present in all types of rivers, narrow or wide.

2. As discussed in Chapter 6, the two main issues for computing erosion and scour depth are hydraulics and soil science, that is, the interaction between water and soil. Nonerodible rock will not be subjected to scour.
3. The location of the bridge also influences the formation of scour at its foundation. Bridges located on a straight, meandering, or sloping thalweg, at a confluence or downstream of a dam, will all have different degrees of scour.
4. The size of the opening or degree of obstruction from abutment, pier, and foundations will influence the velocity of water.
5. The catchment area of the river, its source of supply, population density, storms, and melting of glaciers will influence the volume of discharge and erosion.
6. To correct some of the conservatism in computing large values of scour depths, as a result of assumptions, it is proposed to use scaling factors, based on engineering judgment, as discussed below. Pennsylvania Design Manual (DM4) (Section 7.2.2) recommends such an approach: "For piers a scour depth reduction factor of up to 50% may be used for local scour when multi-layered riprap scour protection and appropriate inspection frequency are provided."
7. Flowcharts given in the Scour at Bridges Section of the *NJDOT Bridge Design Manual* may be followed in developing a comprehensive scour analysis. They are based on standard procedures from HEC-18 and other publications. These diagrams are included in Section 39 (original Section 46) of the *NJDOT Bridge Design Manual* and were developed by the author under Section 46 for the LRFD version.
8. AASHTO has published Guide Specifications for Bridges Vulnerable to Coastal Storms, ship collision, bridge construction and bridge security.

8.2 Factors Affecting Scour

The following is a summary of the factors affecting scour in addition to those described in Chapter 3.

1. River configuration types
    a. Straight
    b. Braided or multichannel
    c. Meandering
2. River bottom types
    a. Aggrading
    b. Stable
    c. Degrading

Soil types determine the grain size distribution for a scour analysis, and samples from the riverbed need to be tested in accordance with ASTM D422. Soil types may be broadly classified as noncohesive materials, such as gravels, sands, and silts.

Noncohesive Materials

The threshold of movement depends on the particle size, density, shape, packing, and orientation of bed material.

Cohesive Materials

Most fine-grained sediments possess some cohesion. The clay content is of great importance. Cohesive sediments typically require relatively large forces to detach the particles and initiate movement, but relatively small forces to transport the particles away. Cohesive materials are silts and clays.

Table 8.1 shows geotechnical factors in the rate of scour and long-term performance. After an innovative approach is tried, the following educational options are available for further research and advancement of the subject.

An Innovative Approach on Scour Presented in Conference Papers

Technical papers may be presented on innovative projects in the useful sessions at national and international conferences. Examples are:

- ASCE Water Resources Division Specialty Conference
- AASHTO Bridge conferences and meetings
- International Bridge Conference, Pittsburgh, PA
- TRB Committee presentations, Washington, D.C.
- International Society for Soil Mechanics and Geotechnical Engineering (ISSMGE)

Activities to disseminate information include the following:

- Incorporate new technology into FHWA planned update of HEC-18, HEC-20, and HEC-23 and in National Highway Institute (NHI) planned revisions to Stream Stability and Scour at Bridges courses (NHI nos. 13046 and 13047) and Highways in the River Environment course (NHI no. 13010).

Observed Scour in Rivers	Typical Duration of Maximum Scour Depth	Sediment Type	Percent of Scour
Sand	In hours	Sand	48
Cohesive materials	In days	Cohesive materials	19
Mixed soil	In hours	Mixed soil	13
Gravel bed	In hours	Gravel	10
Limestone	In months	Bedrock	5
Shale and sandstones	In years	Silt	0
Uncertain	In months	Uncertain	5
Granite	In centuries	Granite	0

TABLE 8.1 Geotechnical Factors and Soil Types in the Prediction of Scour

- Provide support for and invite selected hydraulic engineers from countries visited during the scanning review to an appropriate conference in the United States for continued dialog. FHWA International Programs and the NHI schedule a Stream Stability and Scour course in cooperation with the H. R. Wallingford Laboratories in the United Kingdom.
- Develop a decision-making frame worker: In 2001, the AASHTO Technology Implementation Group chose prefabricated bridge elements and systems as one of the innovative technologies that promised the highest payoff. According to Tang and Ralls (2005), AASHTO provisions on the ABC method led to the Innovative Bridge Research and Deployment, Highways for Life, and Federal Bridge Program.
- Seek support for a study of AASHTO specifications for countermeasures and work with AASHTO on recommendations for improvements.
- Coordinate with the Association of General Contractors regarding techniques, equipment, quality control, specifications, etc., for riprap and other countermeasure placement.

8.3 Rating of Scour-Critical Bridges Using the Pennsylvania SCBI Method

AASHTO has updated the guidelines on bridge evaluation. In Pennsylvania, for bridges located on rivers, the Web-based scour-critical bridge indicator (SCBI) code, developed by the U.S. Geological Survey (USGS), is used. The SCBI code indicates the vulnerability of the bridge to future scour. The SCBI code is based on the FHWA code (NBI item 113) (FHWA, 1989) and the Pennsylvania DOT's interpretation of the FHWA code (Spangler, 1999). The SCBI code contains a whole number between 9 and 2. Each code number has one or more cases. The scour assessment rating (SAR) is computed from select collected and compiled structure components as well as hydrologic and hydraulic data. Agency personnel assign the final scour-critical bridge indicator code and an SAR between 0 and 100 based on their review of all data. The SCBI code and SAR calculator use various factors from the field or office.

Scour Valuations to Determine the SCBI Code for Individual Subunits and the Bridge

The calculator allows inspection personnel in Pennsylvania to determine the overall bridge rating under these circumstances:

1. Review of bridge reports identifying undetermined historical data or revised field data.
2. When site conditions change.
3. When new scenarios have developed.
4. When new bridges are constructed.

The complete original procedures for determining the SCBI code can be found in Cinotto and White (2000). The SCBI code algorithm used by the Web-based SCBI code and SAR calculator was modified from Cinotto and White (2000) to eliminate the comparisons of USGS and Pennsylvania DOT data.

Scour study will be based on scour vulnerability analysis, which is a 1988 mandatory requirement of FHWA for bridges on waterways.

- The scour sufficiency rating needs to be evaluated.
- A scour analysis is carried out to evaluate scour depths based on FHWA Publication HEC-18.
- Scour depths for long-term scour, contraction scour, and local scour are assumed to develop independently.

Both boring information and grain size analysis will be needed for accurate determination of scour. The collection and processing of geomorphic, hydrologic, and hydraulic data for assessment of scour at bridges require borehole information for soil characteristics.

Figure 8.1 demonstrates how wire baskets prevent riprap from being displaced. Some effective alternatives are partially grouted riprap, cable-tied blocks, and sheet piles, although other types of countermeasures may be suitable under different conditions, environmental requirements, durability, availability, and cost. A value engineering study may also be carried out.

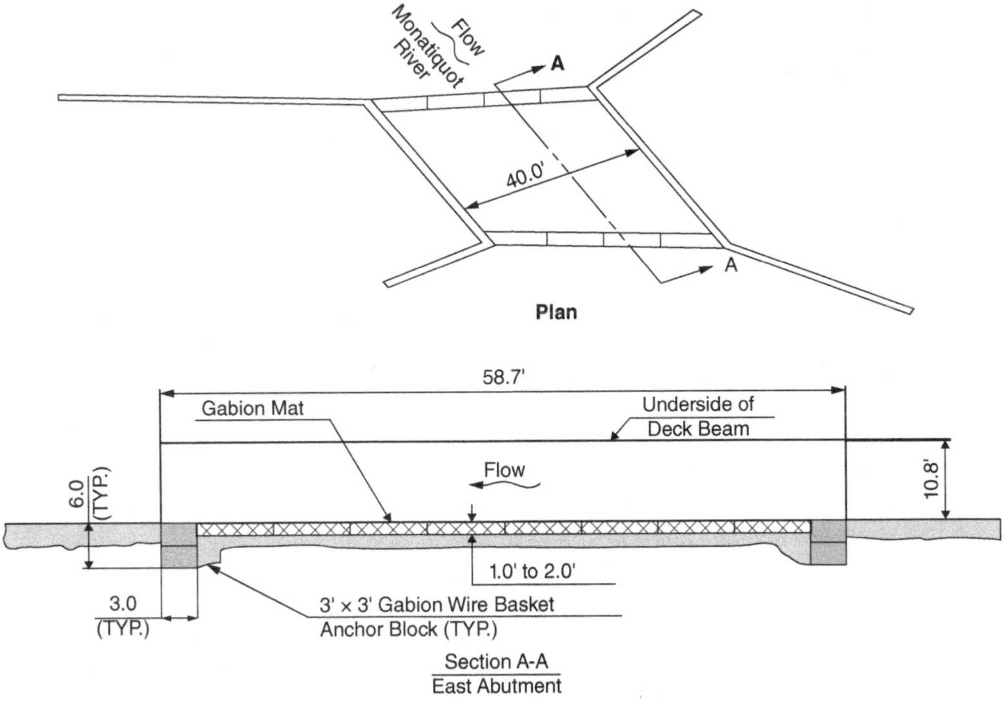

FIGURE 8.1 Use of gabion baskets by author on a scour-critical bridge.

Essentials of a Scour Report The scour report should consist of the following:

- A detailed field survey
- Substructure information
- Scour analysis results
- Hydraulics-related findings
- Countermeasure recommendations

Performing a Scour Analysis

The following should be conducted in the overall scour analysis of a bridge:

Level 1 Qualitative assessment of stream stability, including lateral stability, vertical stability, and determination of the profiles and plan formations of streams and rivers. (Refer to HEC-20.)

Level 2 More detailed quantitative analysis, including hydrologic, hydraulic, and scour analysis to assess scour vulnerability. (Refer to HEC-18.)

Level 3 Bridge scour design of stream instability countermeasures. (Refer to HEC-23.)

If the determined estimated scour depth that must be addressed in the design process seems excessive, an evaluation of the site should be made. The evaluation should include a flood history assessment as well as any evidence of scour.

The evaluation can be used as a basis to recommend a reduction of the determined scour depth. A percentile reduction recommendation may be submitted to the manager of the structural engineering department for approval.

If 500-year discharge is not available, use $Q_{500} = 1.7 \times Q_{100}$ in developing a comprehensive scour analysis.

Changes in the bed level of a stream affect highway structures and may be described by three types of actions: general scour (contraction scour), local scour, and degradation or aggradation of the stream channel. Scour and degradation are discussed in this section.

Scour Analysis Procedure

Scour analysis should be performed according to the FHWA publication HEC-18, *Evaluating Scour at Bridges*. Computer software HY-9, Scour at Bridges, should be used to check the manual calculations. Any countermeasures required should be designed using the methods in the following sources:

- HEC-18, *Evaluating Scour at Bridges*
- HEC-11, *Design of Riprap Revetment*
- FHWA-HI-90-016, *Highways in the River Environment*, 2009HEC-20, *Stream Stability at Highway Structures*
- HEC-23, *Bridge Scour and Stream Instability Countermeasures*

The preferred minimum riprap size must conform to the requirements of R-5 stone size. Larger riprap may be specified if needed. The riprap in the channel should be covered with a minimum of 1 ft of natural streambed channel. A low-flow channel should be formed at that point, if applicable.

In recent years, several studies have compared the field scour observed at bridge sites with the scour values predicted by various equations. The studies reflect the ever-increasing concern that current methods for estimating scour depth are principally based on laboratory experiments and do not necessarily correlate well with field conditions. These agencies are seeking more realistic procedures to estimate scour depth, since resources for construction and repair are chronically limited and bridges need to be better prioritized so that funds are expended where they are truly needed. Three recent comparative studies of bridge scour are summarized in this section. All studies were rigorous, and in total they comprise more than 200 bridges located in five states.

Investigation of Scour Depth at Bridge Piers Using BRI-STARS Model (Talebbeydokhti and Aghbolagh, 2006)

The BRI-STARS (*bri*dge *s*tream *t*ube model for *a*lluvial *r*iver *s*imulation) program was used to investigate the scour depth around bridge piers in some of the major river systems in Iran. Model calibration was performed by collecting different field data. Field data are cataloged in three categories: bridges having riverbeds formed by fine material (i.e., silt and clay), bridges having riverbeds formed by sand material, and bridges having riverbeds formed by gravel or cobble materials.

Verification involved application and comparison of the model to different known real-life cases. The results of this comparison determine the validity and quality of the calibration. Verification was performed with some field data in Fars Province. The scour depths of various bridges located in Iran during significant floods were calculated. The bridges are located in three different places with different bed materials: the Ghadir Bridge on the Zayanderood River, the Khan Bridge River, and the Chame Sohrabkhani Bridge on Kor.

Results show that for wide piers, the computed scour depth is greater than the measured depth. In gravel bed streams, the computed scour depth is also greater than the measured scour depth; the reason is the formation of the armor layer on the bed of the channel. Once this layer is eroded, the computed scour depth is close to the measured one.

8.4 Use of Hydraulic Assessment Checklist

The objectives of a hydraulic design are to identify the stream forces that may cause harm to the bridge or roadway system and to provide a safe level of service acceptable to the needs of the traveling public, without unreasonable effect on adjacent property. See Figure 8.2.

Completion of the Hydraulic Assessment Checklist

The checklist documents the design of the structure and provides guidance to the designer to ensure that all items are completed. The designer is responsible for completing the checklist. It is completed in the following phases:

- Part I: General site data are listed.
- Part II: The existing structure is to be completed in the office in preparation for the field inspection.
- Part III: Field inspection provides guidelines to ensure that all items are completed during the field inspection.
- Part IV: The proposed structure documents the design of the proposed structure and the level of evaluation.

Figure 8.2 The author (*left*) awarding a certificate of appreciation for a research paper to a professor from Turkey at a hydrology session that he chaired at the OMICS International Conference on Hydrology in Raleigh, North Carolina, in 2013.

Scour Mitigation by Applying FHWA Scour Technology

Hydraulic Engineering Circulars (HECs) are used. Based on the latest research, FHWA has published HEC-11, HEC-18, HEC-22 (Urban Drainage Design Manual), HEC-23, and HEC-25. Details are given in the Bibliography. These are guidelines for scour analysis and design of countermeasures and are revised every few years.

Studies by Lombard and Hodgkins (2008) This insightful study was recently completed by the USGS in cooperation with the Maine DOT. The investigators analyzed 50 bridges that were distributed geographically throughout the state. The median age of the bridges was 66 years, and all were single-span on nontidal waterways. Field surveys were conducted to determine the channel geometry and characteristics as well as to measure the observed abutment scour, which ranged from 0 to 6.8 ft. The average actual observed scour across all the sites was less than 1 ft. Skew angles of the abutments and embankments in relation to the channel showed wide variation, ranging from 0° to 50°. The four scour estimation methods (reported earlier in Table 7.7) applied to the bridges in the Maine study were the Froehlich/Hire method, the Sturm method, the Maryland SHA method, and the Melville method.

Progress in Foundation Protection and Scour Remediation For bridges located on waterways and subjected to recurring floods, the options for taking action are limited. There is no cost–benefit analysis. It is simply the "do or die" approach and fixing the issues as soon as possible, before the next flood strikes.

In recent years, some progress has been made in protecting bridge foundations by adequate structural planning and by implementing suitable countermeasures. The publications of Hydraulic Engineering Circulars such as HEC-18, HEC-20, HEC-23, and HEC-25 by FHWA have contributed to the progress to some extent, and any spread of bridge failures may have been reduced as a result.

Small openings under bridge decks causing obstruction to floodwater flow, soil erosion, and scour of foundations remain a leading cause of bridge failures in the United States. It will be expensive to divert the floodwater in channels through pressure flow by using large-diameter pipes over long distances.

8.5 Vulnerability Index for Scour-Critical Bridges

Since storms often encompass large geographical areas, it may not always be possible to monitor all Watch List bridges simultaneously with available department resources. Thus, the plan of action prescribes a system to rate and rank each bridge according to its potential vulnerability to scour damage. The vulnerability index focuses only on factors specifically related to scour, such as foundation type, existing scour problems, streambed material, and history of debris.

Every bridge over a waterway should be evaluated for its vulnerability to scour in order to determine the appropriate protective measures. Most waterways can be expected to experience scour over a bridge's service life (which could approach 100 years). The need to ensure public safety and to minimize the adverse effects stemming from bridge closures requires the best effort to improve the state of practice of designing and maintaining bridge foundations to resist the effects of scour. Current information on this subject has been assembled in HEC-18, *Evaluating Scour at Bridges*.

Scour evaluations of new and existing bridges should be conducted by an interdisciplinary team composed of hydraulic, geotechnical, and structural engineers. Bridges over waterways—both tidal and nontidal—with scourable beds should withstand the effects of scour from a superflood (500-year flood) without failing. Hydraulic studies should include estimates of scour at bridge piers and evaluation of abutment stability. Bridge foundations should be designed to withstand the effects of scour for the worst conditions resulting from floods. The design flood for scour will be the worst condition from the Q_{100} storm or the overtopping storm. If the overtopping flood is less than Q_{100}, then overtopping is the design flood; if overtopping is greater than Q_{100}, then Q_{100} is the design flood. The substructure must withstand scour resulting from the Q_{500} storm. The geotechnical analysis of bridge foundations should be performed on the basis that all streambed material in the scour prism above the total scour line for the design flood (for scour) has been removed and is not available for bearing or lateral support. In addition, the factor of safety should be greater than 1.0 for conditions of scour resulting from the superflood.

Because of the limited accuracy of the procedures for predicting abutment scour, scour protection, such as riprap, may be provided for abutments instead of designing the abutment to resist scour. The AASHTO Specifications contain requirements for designing bridges to resist scour. Particular attention is directed to Sections 2.6.4.4.2 and 3.7.5.

Table 8.2 shows a list of available hydraulic engineering software by FHWA and others. Some vendors have also developed their own designs such as for ACBs, cable-tied blocks, and gabion baskets.

FHWA Hydraulics Engineering Software		
Software	**Title**	**Year**
HEC-26	HEC-26 Spreadsheet	2011
Toolbox	FHWA Hydraulic Toolbox	2013
FESWMS	FESWMS (Version 3.1.5)	2003
FESWMS	FESWMS User's Manual	2003
Hydraulics Software by Others		
Software	**Title**	**Year**
CAESAR	Cataloging And Expert evaluation of Scour risk And River stability at bridge sites	2001
CHL	Coastal & Hydraulics Laboratory USACE	
HEC	Hydrologic Engineering Center USACE	
HEC-RAS tools	Instructions for extracting HEC-RAS cross sections from MicroStation (pdf, 5 mB)	2013
NSS	National Streamflow Statistics Program	
PEAKFQWin	Flood Frequency Analysis (USGS)	2007
SAMwin	SAMwin Hydraulic Design Package for Channels	2002
SMS	Surface-Water Modeling System (SMS) (download and registration instructions) (pdf, 0.7 mB)	2001
StreamStats	StreamStats	
USGS	Water Resources Applications Software USGS	
WMS	Watershed Modeling System (WMS) (download and registration instructions)	

FHWA Hydraulics Software Archive (http://www.fhwa.dot.gov/engineering/hydraulics/software/softwarearchive.cfm)

Table 8.2 List of Hydraulic Engineering Software Developed by FHWA and Others

River Geomorphology In the United States, HEC-20 stresses the need to take a river system geomorphic approach to channel instability problems. European countries recognize the value of a geomorphic analysis in bridge and countermeasure design.

Scour Prediction The investigation of improved scour prediction techniques is an important aspect of scour prediction. Scour at very wide piers needs to be fully investigated to develop techniques to estimate the time rate of scour. In Switzerland, the Netherlands, and the United Kingdom, the estimation of turbulence intensity in relation to the structure (bridge, storm surge barrier, etc.) is considered to be a key factor for estimating scour potential and designing scour countermeasures. In estimating scour at a bridge pier, researchers in the Netherlands consider the interaction between various scour components (long-term degradation, general or contraction scour, and local scour) when calculating total scour. The interaction between scour holes on adjacent substructure elements needs to be investigated.

Soil Type	Scaling Factor $\zeta < 1.0$
Weak soil	0.90
Sandy soil	0.85
Weathering rock	0.80

TABLE 8.3 Scaling Factor ζ as Correction Factor for the Design Experience Based on Geologic Conditions

Scaling Down Sizes of Armoring

With the view of obtaining an economical solution, certain scaling factors may be used in optimizing an armoring. These factors are approximate but are based on engineering judgment and observations of past performance of the type of countermeasure. Use of scaling factors provides the following advantages:

1. It reduces the volume of required armoring and associated labor charges.
2. It minimizes the adverse impacts of armoring on the river's natural environment.

In the following situations, the size of armoring may be scaled down by the method presented herein.

a. When the scour depth is determined by use of a 500-year storm, the determined value could be interpreted to be a conservative value. As discussed below, the determined scour value may be divided by using a maximum factor of safety of 2 or by a scale factor with maximum value of 2.

b. HEC-18 procedures for scour analysis are based on extreme hydraulic conditions, such as designing for 100-year floods and checking for 500 years. This approach may be satisfactory for planning of new bridges for a useful life of 100 years. But when the bridge is earmarked for replacement in the near future, this may become over conservative. Scaling factors in this section may be used in such situations to reduce the fiscal impact.

c. Deep excavation that is required for anchorage of armoring (equal to the computed scour depth) next to bridge footings may cause settlement and may be expensive due to the high cost of steel or timber cofferdams. It may also disturb the river environment. For such situations, the required armoring may be scaled down. The influence of scaling factors may be linear or non-linear.

The following scaling factors are proposed based on experience for modification of computed armoring sizes for economic design:

$$\text{Cumulative scaling factor (SF)} = \zeta\eta\xi\chi\psi$$

The cumulative scaling factor should not be less than 0.5. Nominal countermeasures will be provided for all river bridges even when the computed scour depth is low. Maximum scaling factor = 1.0; that is, armoring thickness provided = depth of local scour. Minimum scaling factor = 0.5 for weathering rock. For good-quality rock, no armoring is required due to minimal erosion.

Application	Width	Condition	Scaling Factor η
Bridge opening size	Small	< 30 ft	0.90
	Medium	> 30 but ≤ 50 ft	0.85
	Large	> 50 but ≤ 70 ft	0.80
River training measures	Very effective	Listed in HEC-23 matrix	Scaling factor ξ 0.80
	Medium	–do-	0.85
	None	N/A	0.90
Underwater inspections	For regular inspections	Every year and after floods	Scaling factor χ 0.80
	For routine inspections	Every 2 years but no postflood inspection	0.85
	For limited inspections	Frequency > 5 years	0.90
Remaining bridge life assessed	Major repairs or replacement	< 10 years	Scaling factor ψ 0.80
		>10 but ≤ 15 years	0.85
		> 15 years	0.90

TABLE 8.4 Scaling Factor η

8.6 Scour Case Studies by the Author

It will be noted that riprap alone is not recommended as a permanent countermeasure, but as emergency shielding only for a period of 5 years or longer, after regular evaluation from underwater bridge inspection reports. Riprap can, however, be used as secondary local armoring, in conjunction with primary structural countermeasures or river training measures.

Case studies of two of the eleven New Jersey scour-critical bridges investigated by the author are presented here. These bridges are located in the same watershed in Hunterdon County. The studies show that the first bridge can be salvaged by foundation repairs, while the second bridge would need replacement.

The Stankowski method of hydrologic studies as applicable to New Jersey rivers and HEC-RAS hydraulic analysis methods are used. Contraction and local scour depths are computed using Excel spreadsheets, which were developed based on the latest HEC-18 methods. Conceptual design of remedial measures, using an HEC-23 countermeasures matrix, is presented. Various techniques to evaluate the unknown foundations are summarized.

The factors that affect stream stability include the location of the bridge, either close to a river bend or at a skew angle to the direction of flow. The factors that affect foundation scour include the use of shallow foundations (placing the footing elevation above the computed scour depth) or any deficiencies in planning, such as absence of river training measures and lack of foundation armoring. The methodology for scour studies is in accordance with FHWA procedures and represents the current state of the art in United States. Stream stability and bridge scour were recognized as design criteria in the AASHTO LRFD Bridge Design Specifications 1998, Section 2.6.

Field Survey Results for Two Bridges in New Jersey

New Jersey DOT completed an in-depth study to evaluate the condition of more than 200 bridges whose foundations are subjected to continued undermining from floods.

Typical flow diagrams for new and replacement bridges, bridges requiring widening, and existing structurally deficient bridges are shown in Scour at Bridges Section of NJDOT Bridge Design Manual. Flow diagrams show the detailed procedure for evaluating scour. To perform all aspects of scour analysis, a data collection procedure is required, as follows:

1. An inventory of all reliable field data and data available in the design office should be compiled.
2. A list of remaining data, which are necessary to perform scour analysis, and the method, extent, and duration of obtaining such data should be prepared.

No gauge data for peak flood discharges were available. A field survey of the two bridges, by a team of licensed structural, hydraulic, and geotechnical engineers was carried out during March and April of 2001.

Foundation Details

No as-built plans were available for these bridges. The team performed the following tasks:

1. Recorded visual observations for the following items: scour holes; shift in thalweg of the main channel, causing increased scour, cracks, and open mortar joints in stone masonry abutments; deposit of stones and debris from earlier floods; and evidence of river training measures or scour countermeasures in place
2. Investigated the need to drill test borings to determine the depth of rock and the type of bearing material
3. Prepared photographic documentation
4. Obtained river cross sections
5. Evaluated stream and waterway characteristics
6. Evaluated roughness coefficients.
7. Prepared data collection items including the purpose and methodology, as shown in Table 8.5

Item	Purpose	Data Collection	Methodology for Obtaining Data
Field surveys	Waterway opening	Yes	Observations and measurements
Underwater surveys	Scour holes	Yes	Probing
River cross sections	HEC-RAS model	Yes	Surveying instruments
Soil parameters	Roughness coefficients	Inspection and photos	Characteristics of banks and bed
Foundation details	Size and scour damage	Inspection and photos	Probing and exposed footings
Test boring	Depth of footing and rock	Boring log	Drilling and coring

Table 8.5 Methodology for Obtaining Data

Field inspectors should have adequate training for collecting data in the field and researching data related to the bridge in the office. Table 8.4 shows the methodology for obtaining data.

Various Methods for Investigating Conditions of Unknown Foundations

The plan of action for scour of unknown foundations is described in HEC-18, Appendix K. As discussed in FHWA, Determination of Unknown Subsurface Bridge Foundations, and the NCHRP Report, the FHWA procedures for investigating unknown foundations may be summarized as follows:

1. Direct methods are employed such as probes, augers, or rotary drilling adjacent to footings. Probes were carried out to locate the top of existing footings.
2. Test pits were not dug because of the danger of undermining existing footings.
3. To identify areas of scour holes, probing of the soil of the riverbed was carried out adjacent to existing footings.
4. For **nondestructive** tests, ground-penetrating radar or acoustical emission techniques were considered. Because of the unknown type of soil and stone masonry supported on stone substrata, this technique could not determine an accurate depth of footing.
5. Drilling through the heel of abutment footing, obtaining borehole logs, and analyzing the results is a reliable method and was approved by the New Jersey DOT. Accordingly, a test boring to obtain existing foundation information such as footing elevations, bedrock elevation, and quality of rock was drilled at one abutment.

Methods of Hydrologic and Hydraulic (H&H) Analysis

The 50-, 100-, and 500-year storms were calculated for the watershed based on the Stankowski regional equation (reported earlier) found in Stankowski, 1974. This is the preferred method for determining storm flows in New Jersey. The Stankowski regional equation is based on a thorough study of relationships between flood data and meteorological characteristics for 103 sites in the state. The parameters for the Stankowski regional equation are drainage area, channel slope, surface storage index, and index of human-made impervious cover.

An HEC-RAS analysis was performed using HEC-RAS version 3.0 software (US Army Corps of Engineers). The input data consisted of eight recently surveyed cross sections, five upstream and three downstream. A FEMA study based on the HEC-2 study is available only for Guinea Hollow Road Bridge. However, the cross sections were determined to be outdated (more than 20 years old) and inaccurate as a result of sedimentation and erosion over many years.

Methods for Scour Analyses

Scour analysis was performed in accordance with HEC-18 and the *NJDOT Design Manual*, Section 46. The intent of HEC-18 is to establish methods for estimating various scour components for use in conjunction with engineering judgment, to evaluate the potential depth of total scour. Modes of bed transport were live bed and clear water. (See HEC-18, Melville and Coleman, and the NJDOT *Bridge Design Manual*.)

Long-Term Scour

Field observations and past inspection reports do not reveal evidence of continuing or significant long-term aggradation or degradation. Any long-term scour was neglected.

Contraction Scour

This occurs mainly from the contraction of flow area, which results in increased velocity and shear stress on channel bed. The computation of critical velocity was based on Laursen's equation (Equation 5.1, page 5.2, HEC-18). A comparison was made with the mean velocity value obtained from HEC-RAS analysis. A modified Laursen equation (Equation 5.2, page 5.10, HEC-18) for live bed mode determined the depth of contraction scour at the two abutments. Another modified Laursen equation (Equation 5.4, page 5.13, HEC-18) for clear water mode was used.

Local Scour

This occurs from vortices, which are caused by flow obstructions and are cyclic. Based on the ratio of projected abutment length to flow depth (L/y_1), either Froehlich's equation (Equation 7.1, page 7.8, HEC-18) or Hire's equation (Equation 7.9, page 7.9, HEC-18) for live bed scour was used to estimate the potential depth of local scour at each abutment. Resulting scour depths estimated by Froehlich's equation seem excessive. However, HEC-18 states that the equation is to be used to aid in the placement of countermeasures. Therefore calculated depths will be treated as an estimate of the potential depth of scour.

Local scour occurs due to the acceleration of flow around the pier and due to the formation of flow vortices. HEC-18 equations are applicable. However, comparison of pier scour depths from alternate methods need to be performed.

Countermeasure Evaluation for Bridge Foundations

Based on detailed field investigations including test boring and hydrologic, hydraulic, and scour studies, it is found that the depth of theoretical total scour is below the footings. The following measures need to be implemented, in order to preserve the long-term structural integrity of the bridge from damage by scour:

Structural Countermeasures

The following steps are required:

1. Strengthen the existing abutment footing by providing a reinforced concrete curtain wall of cross section with 1.25-ft width and extending to the bedrock.
2. Provide two vertical layers of #4 diameter reinforcing bars with spacing in both directions of 12 in.

Hydraulic (Armoring) Countermeasures

For existing bridges, the installation of riprap, steel sheeting, or cast-in-place concrete paving, to armor the exposed faces of the existing footing was investigated to arrive at the most cost-effective method. Because of the confined space under the bridge, driving steel sheeting would not be practical. Hence, the HEC-23 method was used of providing riprap adjacent to the abutment footing and placing the top of the riprap level with riverbed elevation (Equation 8.1, page 8.3, HEC-23). Also the countermeasures directive in HEC-18 (Appendix J, page J.3, HEC-18) was followed. Since the riverbed elevation varies, use a minimum depth of riprap of 3.0 ft below the riverbed. A permit from the New Jersey DEP is required to install armoring.

Monitoring as an Effective Initial Countermeasure

Visual inspection on a regular 2-year cycle and flood watch need to be carried out. The monitoring efforts need to be increased during high-flow events.

Countermeasures for Channel and Banks at the Wing Walls

- *Channel improvements by means of channel lining*: To prevent degradation, armoring of approach banks by revetment materials is recommended. Common materials used for the shielding of foundations, such as standard riprap, stones enclosed in wire nets, gabion mattresses, and artificial riprap (concrete bags or toskanes), are recommended.
- *Dredging*: Removing debris thereby increases the waterway width and reduces velocities.
- *Scour holes*: Scour holes should be filled with with dense material, heavy stones, or concrete blocks.

8.7 Case Study of Countermeasure Evaluation

Bridge Footings

Long-Term Countermeasures

These may be summarized as follows:

- Reconstructing the cast-in-place apron footing has not worked in the past because footings were placed at shallow depth. It has resulted in undermining and complete exposure of the east footing.
- Riprap placement requires a 3-ft minimum depth below the riverbed. The bottom of the east abutment footing is at the same level as the riverbed. Hence any riprap will be washed away during the flood and will not act as armor.
- Underpinning the east abutment footing will be expensive and will require bridge closure.

Interim Countermeasures

Replacement of the bridge will take several years, and interim countermeasures are required.

- Placement of grout bags on geotextile class C layers, adjacent to the east abutment, appears to be a cost-effective method, with an estimated cost of $10,000. Grout bags will measure approximately 4 ft × 4 ft × 1 ft and will not protrude above the riverbed elevation. Placement of riprap adjacent to the west abutment has an estimated cost of $20,000. The cost may change over long periods and at different sites.

Channel and Banks at the Wing Walls

- Both channel lining to prevent degradation and armoring of approach banks by revetment materials are recommended. An estimated length of 30 to 50 ft at each bank beyond the wing wall needs to be shielded. The cost is estimated at $20,000.

- Dredging to remove silting and to increase waterway width is recommended for an estimated cost of $5000.
- Filling up any additional scour holes located upstream of the bridge with dense material, heavy stones, or concrete blocks is recommended. The estimated cost is $1000.

Approximate Cost of Proposed Countermeasures

1. Repairs prior to implementing countermeasures (including removal of debris from channel bed and filling of scour holes) = $5000.
2. Filling the gap between the bottom of the footing and the top of the soil next to the east abutment with concrete for an approximate 20-ft length (including building temporary cofferdam and dewatering) = $5000.
3. Placement of grout bags at east abutment (including textile layer) = $8000.
4. Providing riprap at west abutment = $20,000.
5. Providing gabions at banks = $29,000.

Total = $67,000.

Summary of Results for Water Street Bridge

Two factors indicate that the bridge is potentially susceptible to scour:

- Evidence of undermining at the east abutment footing
- Evidence of large scour holes close to the abutments

The total estimated scour depth is below the estimated bottom of the footing elevations. Test boring at the east abutment has shown that there is good-quality rock within 2 ft below the bottom of the footing, thereby reducing the scour depth considerably.

There appears to be a deficiency in the original design—placing the footing at a shallow depth. The east abutment footing has become fully exposed, thereby making remedial measures ineffective.

Conclusions of Scour Study for Water Street Bridge

The abutment walls are located at a high skew to the direction of the floodwater, and shallow foundations are used. Absence of river training measures and lack of foundation armoring have aggravated the scour depth. The estimated cost for dredging, underpinning the east abutment footing, and countermeasures for both abutments will approach the cost of a replacement structure. Hence replacing the 63-year-old bridge appears to be a better choice in economic terms. Thus interim repairs and replacement of the bridge are recommended. Also, as an interim measure, the continued monitoring of the footings and the main channel in addition to biennial inspections is recommended.

Design Issues for Bridge and Highway Structure Foundation in Rivers

Foundation design requires guidelines on river training measures that may be used without armoring countermeasure guidelines on structural countermeasures for existing bridges. In some states, riprap is considered to be a temporary countermeasure. Foundation design and construction in fast-flowing rivers require special considerations. Most bridge collapses occur either from a single peak flood or from a series of repeated smaller floods, thereby causing soil settlement and resulting foundations failures. Shallow foundations are more vulnerable than deep foundations because of the depth of scour. An in-depth geotechnical report will be required to select shallow or deep foundations.

Countermeasures for Abutments, Piers, Arches, and Culverts

Footing Types and Scour at Abutments

Structures located on streams that are subjected to scour include highway bridges, railway bridges, movable bridges, swing bridges, arch bridges, and culverts. In practice, there are a large variety of scour-critical bridges with significant variations in the types of footings. The type and size of countermeasure required will vary for each type of footing.

Bridge abutments and their approach embankments are the most commonly damaged bridge components during floods. The type of footing, whether shallow or deep, has an influence on the scour depth and the type of countermeasure required. Two typical approaches for protecting bridge abutments from scour are:

1. Mechanically stabilize the abutment slopes with riprap, gabions, cable-tied blocks, or grout filled bags.
2. Align the upstream flow by using guide banks, dikes, or spurs, or in-channel devices such as vanes and bendways weirs.

Recommended Countermeasures for Abutments

Based on a detailed survey, the following countermeasures are recommended for application to bridges in New Jersey:

1. Local scour countermeasures
   a. Riprap
   b. Gabions
   c. Grout and cement bags
   d. Cable-tied blocks
2. River control countermeasures
   a. Guide banks
   b. Spur dikes
3. Scour at piers

Recommended Pier Scour Countermeasures

Figure 8.3, riprap detailing at a pier, is shown above. The following pier countermeasures are recommended for applications to New Jersey bridges:

1. Armoring countermeasures
   a. Riprap
   b. Gabions and Reno mattresses

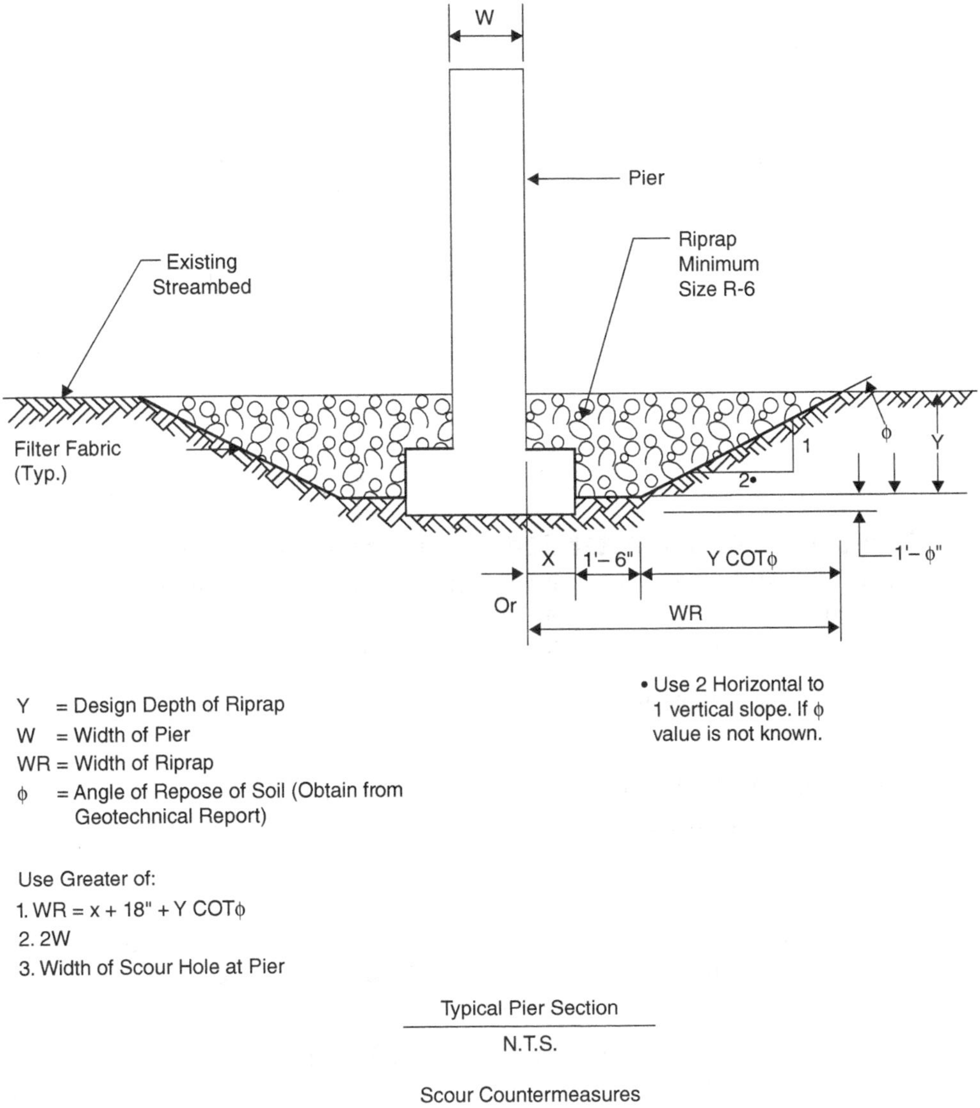

Figure 8.3 Typical pier section countermeasures on page 39-14 of NJDOT Bridge Design Manual.

   c. Articulated concrete blocks and cable-tied blocks
   d. Concrete armor units
2. Flow-altering countermeasures
   a. Upstream sheet piles
   b. Flow-deflecting vanes or plates
3. Structural countermeasures
   a. Structural repairs using tremie concrete
   b. Grout bags

c. Casting concrete aprons
d. Shielding by sacrificial piles
e. Sheet piles

Guidelines on applications of each of these are detailed in this handbook.

NOTE: *Flow-altering countermeasures alone are not considered effective for local scour at bridge piers. To increase their effectiveness during high-flow conditions, they must be used in combination with armoring.*

Construction Drawings

For scour countermeasures construction, the drawings must be based on detailed designs.

8.8 Applications of Computer Software for Scour Analysis

For detailed design approved commercial software may be used. A spreadsheet may be developed and used in lieu of software. The following software may be used for detailed analysis and design of scour and scour countermeasures.

1. Hydrology programs

   a. TR55

   b. PENNSTATE program

   c. USGS methods for different states. The Stankowski method is being used in New Jersey.

2. Hydraulics programs

   The following programs are available to determine the flood intensity, water elevations, and scour potential; to check FEMA compliance; and to help size the proposed structures over waterways.

   a. HEC-2 was developed by the U.S. Army Corps of Engineers and used by FEMA in the past for most of their studies. It has been replaced by the HEC-RAS computer program, which computes the water surface profiles and velocities using the stream cross sections, Manning's variable, and the design and basic flows.

   b. WSPRO is a computer program developed by FHWA that computes the water surface profiles and velocities using stream cross sections, Manning's variable, and design (Q_{50}) and basic (Q_{100}) flows.

   c. HEC-RAS is a U.S. Army Corps of Engineers program. This program can handle variable flows and has a WSPRO subroutine in the water profile routine. In addition, this program computes the possible scour depths at the substructures. Figure 8.4 shows a flow chart for scour analysis developed by the author. HEC-18 scour analysis procedure is followed.

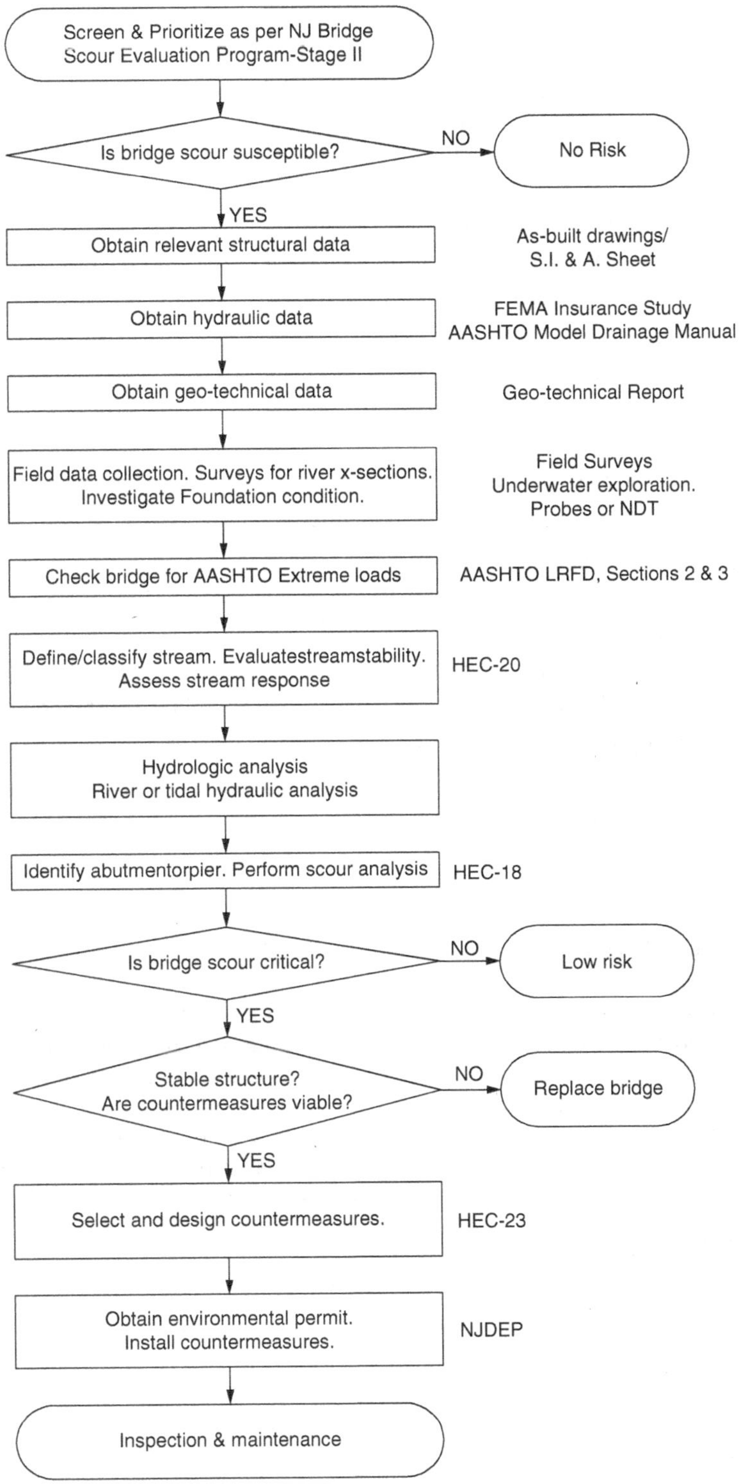

Figure 8.4 Flowchart for scour analysis of existing bridges and bridges to be widened.

d. HYDRAIN is a conglomeration of several programs developed as a pool fund project by several states and FHWA. It includes
   (i) HYDRO, which is a hydrology program
   (ii) HYDRA, which simulates the hydrology and hydraulics on storm drain or sanitary pipe network
   (iii) HY8, which simulates hydraulic analysis or design for culverts, reservoir routing, and energy dissipaters
   (iv) HYCHL, which analyzes and designs channel and rip-rap linings
   (v) NFF, which interactively calculates USGS regression equation flows
e. BRI-STARS is a pseudo two-dimensional hydraulic program that (through the use of stream tubes) provides a time- and flow-dependent, two-dimensional sediment routing (aggradation and degradation) in a bridge cross section.
f. UNET. Unlike the riverine flow, this program deals with tidal flow.
g. DYNET is a two-dimensional hydraulic program for tidal flow.

3. Scour analysis
   a. In-house Excel spreadsheets are based on HEC-18 equations developed by the author.
   b. A subroutine in the HEC-RAS program does scour analysis. However, in some cases high values of scour depth are obtained for contraction or local scour.
4. Rating software
   a. In-house Excel spreadsheets are based on *AASHTO Manual for Condition Evaluation of Bridges.*
   b. A similar approach was taken to the SCBI calculator for scour-critical indicator code as developed by USGS.

Scour Watch Software

The software allows bridge engineers to identify, predict, and record severe floods. Scour watch identifies the occurrence of a flood and processes scour monitoring data from equipment such as SHIFLO. The results obtained from the software would expedite maintenance and facilitate urgent action. Actions required when scour depth is above, within, or below the footing are shown.

Construction drawings for scour countermeasures must be based on detailed designs using approved commercial software. A spreadsheet may be developed and used in lieu of software. The following software may be used for detailed scour.

1. Hydrology
   a. TR55
   b. PENNSTATE program
2. Hydraulics
   a. HEC-RAS
   b. WSPRO
   c. UNET

3. Scour analysis
   a. HEC-RAS
   b. Excel spreadsheets based on HEC-18 equations

8.9 Scour Depth—A Concern for the Bridge Engineer

The undermining of pier or abutment by scouring to a level that exposes the footing can lead to collapse. It results in costly repairs, disruption of traffic, and possible death of passengers traveling on bridge.

There are a number of alternative methods to HEC-18. Research shows that the programs based on Melville equations for scour depth and Maryland DOT ABSCOUR programs are the most investigated.

Influence of Hydraulics on Erosion and Scour Depth

The following factors affect scour depth

- Return flow from the floodplain into the main channel
- Increased main-channel velocity due to the contraction between abutments
- Impact with the flow striking the leading abutment and pier corner
- Vortex formation due to obstruction
- Shear stress between flow of water and riverbed and banks

Procedures Used for the Implementation of Scour Countermeasures

In addition to HEC-18 methods, Melville's equations and ABSCOUR software methods, the following approaches are applicable:

- Both empirical and theoretical mathematical equations developed to estimate scour depth
- Scour mitigation by taking measures in designing a new bridge and countermeasures for an older bridge
- Peripheral and bank hardening by countermeasures: riprap, gabion baskets, cable-tied blocks, and geo-bags
- Flow-altering methods including spur dykes, guide banks, parallel walls, and collars
- The sheeting may initially serve as a cofferdam during construction in place of stream diversion. It may then be left in place as permanent sheeting.

Optimization of Countermeasures

There is a need for optimization of countermeasures relative to durability and cost considerations. For New Jersey, recent trends have been to focus on increased monitoring by inspections and instrumentation, use of armoring such as gabion mattresses and baskets, using sheeting left in place as permanent sheet piles, and channel improvements. However, greater use of these and other types of armoring and structural countermeasures needs to be considered.

Selection of a Countermeasure

A selected countermeasure needs to satisfy cost, environmental, and constructability requirements in addition to the functional requirement at a site. The selection of a countermeasure may depend on several site considerations:

1. For new bridges, effective measures such as optimum location of bridge, planning of deep foundations, or adopting adequate sizes of openings should serve a primary role. Countermeasures such as river training or structural countermeasures and armoring may serve a secondary role.
2. For existing bridges, if the scour depth and magnitude of erosion warrant the use of special countermeasures (such as river training or structural countermeasures), the accompanying use of armoring may serve in a secondary role. If the scour depth and magnitude of erosion are not high, armoring alone may be sufficient as a primary countermeasure.
3. If precast concrete segments (ACB or cable-tied blocks) are selected, they should be tested in a laboratory to determine if the use of any admixtures in the concrete mix could be harmful to marine life. The vendor may provide a certificate to this effect.
4. Since armoring countermeasures function as gravity forces on soil, use of high-density aggregates should be specified. This will increase the deadweight of precast blocks and decrease the volume of armoring by 15 to 25 percent.
5. The success rate of a similar countermeasure used in other states under similar types of flow conditions should be reviewed during the selection of a countermeasure. Riprap may be used as a temporary measure at bridges. Average rankings of different scour countermeasures by 35 state DOTs are presented in NCHRP 24-7 (Parker et al. 1998).

European Policy on Selecting the Types of Countermeasures

In addition to the information on riprap presented in Chapters 1 and 3, the following information is supplemental for the most commonly used countermeasures.

Riprap Is the Most Popular Countermeasure The use of riprap, that is, armor stone in combination with a geotextile or granular filter, is by far the most common scour and stream instability countermeasure in all countries visited in Europe by a select panel of experts from the United States. The team of experts included delegates from leading organizations in hydraulics and scour mitigation. These included AASHTO, FHWA, NCHRP, NHI, SHA, DOTs, private organizations, TRB, and universities. The DOTs included Caltrans, Illinois, Maryland, Minnesota, Oregon, and South Carolina. The universities represented were Texas A&M and Louisville, and the private sector was represented by Ayres Associates.

The following research organizations were visited and scour practice in their country studied:

Laboratory of hydraulics, hydrology and glaciology (VAW), Zurich, Switzerland
Swiss Federal Institute (ETH)
Federal Waterways and Engineering and Research Institute (BAW), Karlsruhe, Germany

Federal Highway Research Institute (BAST) Cologne, Germany
German Association for water resources and Land Improvement (DVWK), Germany
Netherlands Center for Civil Engineering Research and Codes (CUR)
Highways Agency (HA), England
National Rivers Authority (NRA), England.

Its availability, economy, ease of installation, and flexibility are considered highly desirable characteristics in all four countries visited namely England, Germany, Netherlands and Switzerland. As a result, considerable effort has been devoted to techniques for determining the size, gradation, layer thickness and horizontal extent, filters, and placement techniques and equipment for revetment and coastal applications. Riprap is considered to be an effective and permanent countermeasure against channel instability and scour, including local scour at bridge piers.

Generally, riprap is sized using the Hudson formula (coastal applications), Shields diagram, or methods developed in New Zealand, the Netherlands, the United Kingdom, or the United States. Research on erosion has common themes with the European countries. These include maintenance of an aging infrastructure and a similar conservative deterministic approach (with worst conditions of loads with safety factors) and a probabilistic statistical analysis (probability of occurrence of loads and load combinations).

At the BAW wave tank testing of prototype-scale partially grouted riprap was carried out. In general, the objective is to increase the stability of the riprap without sacrificing all the flexibility. Contractors in Germany have developed techniques and equipment to achieve the desired grout coverage (fill about 40 percent of the voids at the surface) and the right penetration (decreasing grout fill with depth into the riprap matrix and no grout in contact with the geotextile filter). With the correct slurry mix (recipe), partial grouting can be achieved underwater with minimal environmental impact. The topics discussed by the U.S. Team covered the following aspects: Environmental policy, scour prediction, inspection and monitoring, riverbed degradation, recommendations and implementation plan. An in depth study of the use of alternative countermeasures in Europe was made. The types of countermeasures included riprap, partially grouted riprap, precast concrete blocks, sacrificial piles and debris deflectors. BAW engineers believe that partial grouting, if done correctly, will ensure that the riprap retains sufficient flexibility while adding to stability. A number of references on partially grouted riprap from the Permanent International Association of Navigation Congresses (PIANC) were obtained in Germany.

Filters As in the United States, a properly designed geotextile or granular filter is considered essential to the success of riprap and most other countermeasures on sand or fine-grained material. In Germany and the Netherlands, a significant investment has been made in the development and testing of geosynthetic materials, and innovative installation techniques have been developed that could find application for bridge pier and abutment countermeasures in the United States.

River Training River training and stabilization techniques against lateral channel migration in the major navigable waterways of Europe are similar to those employed by the U.S. Army Corps of Engineers on navigable waterways in the United States (e.g., the Mississippi, Ohio, and Missouri river systems). Groins and jetties projecting roughly perpendicular to the riverbank, dikes placed parallel to the riverbank, or revetment placed on the riverbank are the most common river training works in Europe.

Riverbed Degradation NCHRP published an important study Research Results Digest 378 dated August, 2012 "Evaluation of Scour Research".2 projects on pier and abutment contraction scour (NCHRP 175 and 181) and NCHRP 177 on geomorphic scour were published.

Alternative Countermeasures Among the areas of particular interest are alternative countermeasures such as flow-altering devices or alternatives to riprap (particularly for the pier scour problem). The following paragraphs outline some of the panel's observations on alternative countermeasures in relation to U.S. practice.

In 1987, the Swiss experienced a near catastrophic failure of a major highway bridge when the Reuss River migrated laterally and undermined the foundation of a bridge pier. The countermeasure system developed by the VAW/ETH laboratory included large concrete groins to correct and prevent further channel migration. The riverbank between the groins was protected by very large precast concrete prisms, triangular in cross section, placed individually as revetment. In lieu of smaller interlocking armor units that would be costly to fabricate, the decision was made to cast much larger prisms with a simple shape and use the mass of the prisms to protect against riverbank scour. The precast, hollow prisms were filled with concrete after they were placed in their final position. The groin field and prism revetment were then covered with a layer of natural stone for aesthetic and environmental reasons. The economics of the trade-offs between smaller, high-cost interlocking shapes for artificial riprap and simpler shapes with greater mass are worth further consideration.

In general, proprietary products such as interlocking block, articulating cable-tied block, and articulating grout-filled mattresses for revetment and channel bed protection are not considered as effective as riprap in Europe. The need for adequate toe protection and anchoring was emphasized. Block and mattress manufacturers in the United States and Europe are developing design criteria based on full-scale laboratory testing of specific products. Such tests should provide the necessary guidance for the successful design and installation of proprietary products for revetment and channel protection.

Recent laboratory testing by NCHRP, FHWA, and others in the United States indicates that when articulating mat products are used as a pier scour countermeasure, the joint between the mattress and the pier must be protected to prevent scour under the mat. The panel encountered two approaches to solving this problem that justify further evaluation. In Germany, reference was made to a proprietary system for installing a collar and tying the geotextile filter underlying a block mattress to the bridge pier, using a pneumatic tie. This approach appears feasible for circular piers. In the Netherlands, the recommended approach is to place granular filter material to a depth of 1 m below the streambed for 5 m around the pier. The geotextile filter and block mat placed on the streambed overlap this granular filter layer, and the remaining gap between the mat and the pier is filled with riprap. Successful field installations have apparently been made using this technique.

Use of Debris Deflectors Protecting bridges from the accumulation of debris and predicting the increase in scour at a bridge caused by debris are a problem worldwide. The panel did not encounter any applications of "debris deflectors" or other devices at a bridge during the scanning review. The Swiss were, however, experimenting with the design of large "trash racks" at sedimentation basins to catch vegetative debris before it moved downstream to a bridge. At the University of Nottingham in the United Kingdom, a substantial effort has been made to develop software to aid in the prediction of scour when debris accumulates at a bridge.

Potential Total Scour

Case Study of Guinea Hollow Road Bridge in New Jersey Computed potential total scour is substantial for 50- and 100-year floods, and the scour depth exceeds the depth of footing, making the bridge scour-critical.

Hydraulic and Geomorphic Factors Affecting Scour

A FEMA study for this bridge was available and was used for comparing the results from a new HEC-RAS analysis.

- *Upstream features:* The bridge is located downstream of a small dam and a footbridge. The dam has helped to check high velocities during floods.
- *Direction of river flow:* The bridge is located on a well-developed bend. In plan it has a skew angle of 20° with the waterway. The bend causes the shifting of thalweg and nonuniform velocity distribution as well as scour of the bed and bank at the outside of the bend and deposition at the inside of the bend.
- *River training measures:* There is no evidence of any dyke or spurs constructed on the upstream side of the bend to reduce the effects of long-term scour. Hence the scour has remained unmitigated throughout the 100-year life of the structure.
- *Longitudinal slope:* The longitudinal slope of the channel is 4 percent upstream and 2.4 percent downstream. The slopes are not steep, and in keeping with Manning's equation for flow in an open channel, slopes will not affect the velocities.
- *Channel cross-section:* The channel bed is littered with stones and small boulders. This leads to a reduction in channel area, thereby increasing the velocity of flow and subsequently increasing scour at the bridge abutments.
- *Streambed description:* Soil samples tested in the laboratory have shown the streambed as granular, with some large stones mixed with gravel and a substantial fraction of sand. The median size of aggregate (D_{50}) varies from 1 to 5 mm. The gravel layer has an armoring effect and leads to low degradation.
- *Depth of foundations:* Existing foundations are supported on spread footings. The shallow depth of bottom of footings makes it scour-critical.

Table 8.6 shows author's application of HEC-18 for Guinea Hollow Road Project in New Jersey. Note that the bottom of footing elevation has been determined by test boring, elevations were referenced to National Geodetic Vertical Datum (NGVD), 1929, and scour below the bedrock was neglected for 50-, 100-, and 500-year floods.

Flood Discharge Frequency	Long-Term Scour, ft	Contraction Scour, ft	Local Scour, ft	Total Scour, ft	Net Scour to Bedrock, ft	Bottom of Footing Elevation, ft
50-year	0	4.30	14.95	19.25	7.0	439.0
100-year	0	4.59	15.33	19.92	7.0	439.0
500-year	0	5.49	16.07	21.56	7.0	439.0

Contraction, local, and total scour depths for 500 years are considered.

TABLE 8.6 Scour Depth Summary for South Abutment for the Case Study

Recommended Bridge Inventory Coding

Based upon the in-depth (stage II) bridge scour evaluation (earlier stage in Structural Inventory and Appraisal), SI&A coding for the case study bridge may be revised.

Table 8.7 shows Bridge Inventory Codes.

Proposed Countermeasures and Approximate Cost

Placement of riprap adjacent to the apron on the north and south abutments appears to be the most cost-effective method. The details of the work are as follows:

1. Repairs prior to implementing countermeasures (including filling missing mortar joints, removing debris from channel bed, and filling scour holes and gaps between the bottom of footing and the top of the soil next to the south abutment with concrete) = $5000
2. Construction of concrete aprons (including cofferdams and dewatering) = $12,000
3. Provision of riprap at abutments and wing walls = $29,000
4. Provision of gabions at banks adjacent to wing walls = $29,000

Approximate total cost = $75,000.

Coding Guides	Description	Stage I Coding	Stage II Coding
Item 61	Channel and channel protection (stream stability, channel condition, slope protection)	6	6
Item 71	Waterway adequacy (overtopping flood frequency)	9	9
Item 113	Scour-critical bridges (abutments are rated as unstable due to scour)	4	3*

*Computed scour calculations for stage II, phase 3 show that the bridge is scour-critical and scour depth is below the bottom of the spread footing base. Hence the stage I SI&A rating is revised to 3 based on scour analysis.

TABLE 8.7 Stage 1 and Stage 2 Coding Guides

Summary of Results for Guinea Hollow Road Bridge

Spreadsheets were developed based on applicable methods and formulas. The total estimated scour depth is below the estimated bottom of footing elevations). The following additional factors indicate that the bridge is scour-critical:

- Evidence of undermining at both abutment footings
- Evidence of poor-quality masonry walls with missing mortar in joints
- Evidence of large scour holes close to the abutments

Conclusions of Scour Study for Guinea Hollow Road Bridge

The abutment walls are located skew to the direction of the floodwater, and the footing elevation is placed above the computed scour depth. There is an absence of river training measures and lack of foundation armoring. Strengthening the footing and installation of suitable armoring, such as riprap, are recommended. Continued monitoring of the footings and of the main channel during routine biennial inspections will be required.

8.10 Precautions to Minimize Flood Damage

In addition to the protection of bridges and highways, the following recommendations will be helpful for toll plazas, office buildings and overall safety:

Awareness of Flood Occurrence

1. Set up a workable emergency plan in advance
2. Take necessary steps to prevent the release of dangerous chemicals stored in the vicinity
3. Locate gas main and electrical shut-offs,
4. Anchor fuel tanks, if any
5. Postpone any visits away from home
6. Maintain insurance policy, become familiar with damages from the floods
7. Establish emergency communication method (Alert System,voice mail, phone tree, e-mail messaging etc.);
8. Obtain Crisis Management Training
9. Update disaster recovery kits.
10. Install flood vents and flood proof barriers

Vigilance During the Flood

1. Stay tuned to local media & community messaging
2. Raise elevators to the 2nd floor and turn off when required, safely
3. Keep cell phones and chargers with you & emergency kits
4. Unplug electrical items, etc.

After the Flood Subsides

1. Learn from news reports, whether the community's water supply is safe to drink.
2. Avoid using floodwaters; water may be contaminated by oil, gasoline, or raw sewage.
3. Water may be electrically charged from underground or downed power lines.
4. Be aware of areas where flood waters have receded.
5. Roads may have weakened and could collapse under the weight of a car.
6. Clean and disinfect everything that got wet and may contain chemicals.
7. Contact your insurance agent for claims.
8. During evacuation have a central point of contact for all relatives and friends.
9. Following the flood, notify all critical people of next steps, based on damage
10. Check for any foundation settlement or subsidence to ensure safety.

8.11 Conclusions

This chapter deals with methods of hydraulic and scour analyses using HEC-18 procedures and selection of countermeasures using HEC-23. The need for inspections and floodwatch is emphasized.

Many aspects of the developing scour technology are addressed. These include identification of scour-critical bridges, hydrology, hydraulics, scour analysis, countermeasures selection and design, and continued monitoring.

Soil scour studies and remediation are required. It is important to compute scour depths accurately and identify scour-critical bridges and fix the structural deficiency. For hydrologic and hydraulic analysis, flow diagram using HEC-18 and citations from recent literature are presented; for hydraulic assessment, a checklist is included. Effective types of armoring countermeasures are needed. Common types of countermeasures are compared including:

- Riprap, gabions, spurs, dykes, barbs, vanes, concrete slab, and articulated concrete blocks including regular monitoring
- Miscellaneous countermeasures, such as fascine mattress, bio-engineering etc. (and those alternative countermeasures which are listed in HEC-23) which account for 10 percent of the total use of countermeasures
- Rating and vulnerability index of bridges leads to effective maintenance, durability and safety of bridges. Monitoring serves as a preliminary and necessary form of countermeasure.
- Case studies of Guinea Hollow Bridge for providing countermeasures are provided.

European practice for common types of countermeasures is discussed and compared with U.S. practice. For scour-critical bridges, use of the vulnerability index method is recommended to help prevent unexpected failures.

8.12 Bibliography

AASHTO, *LRFD Bridge Design Specifications,* Customary U.S. Units, 7th ed., with 2015 Interim Revisions, Washington, D.C., 2014.

AASHTO, *Guide Specifications for Bridges Vulnerable to Coastal Storms,* Washington, D.C., 2008.

AASHTO *Guide Specifications for LRFD Seismic Bridge Design,* 2nd ed., 2014 Interim Revisions, Washington, D.C., 2013.

AASHTO, *Standard Specifications for Transportation Materials and Methods of Sampling and Testing,* 34th ed., Washington, D.C., 2014.

AASHTO *Manual for Bridge Evaluation,* 2nd ed., with Interim Revisions, 2015.

AASHTO, *Guide Specifications and Commentary for Vessel Collision Design of Highway Bridges,* 2nd ed., Washington, D.C., 2010.

AASHTO, *LRFD Bridge Construction Specifications,* 3rd ed., 2015 Interim Revisions, Washington, D.C., 2015.

AASHTO, *Bridge Security Guidelines,* 1st ed., Washington, D.C., 2011.

American Society of Civil Engineers, *Stream Stability and Scour at Highway Bridges,* Water Resources, Engineering, 1999.

Bridge Management System 2 (BMS2) Coding Manual, Publication 100A, Pennsylvania DOT, July 2009. Web. Sept.-Oct. 2011.

Chow, V. T., *Open-Channel Hydraulics,* Blackburn, Caldwell, NJ, 2009.

Dyhouse, G., Hatchett, J., and Benn, J., *Floodplain Modeling Using HEC-RAS,* Haestad Press, Waterbury, CT, 2003.

"FEMA: Community Rating System," *FEMA | Federal Emergency Management Agency.* Web. 28 Oct. 2011. http://www.fema.gov/business/nfip/crs.shtm

"FEMA FIRM Item ID 42017C0129F." Map. *FEMA Map Service Center.* 1999. Print.

FEMA, *Flood Insurance Study Bucks County Pennsylvania (All Jurisdictions),* 2004.

FHWA, *Evaluating Scour at Bridges,* 4th ed., HEC-18, 2001.

FHWA, *Bridge Scour and Stream Instability Countermeasures,* HEC-23, 1997.

FHWA, *Federal Coding Guide,* 1994.

FHWA, Determination of Unknown Subsurface Bridge Foundations, NCHRP Report no. 21-5, Geotechnical Guideline no. 16, 1998.

FHWA, *Highways in the River Environment,* FHWA-HI-90-016, February 1990, Parker, G., Toro-Escobar, C., and Voigt, R. L., "Countermeasures to Protect Bridge Piers from Scour. User's Guide and Final Report, NCHRP Report 24-7, Highway Research Program, Transportation Research Board, National Research Council, December 1998.

Iranian Journal of Science & Technology, Transaction B, Engineering, Vol. 30, no. B4

"Investigation of Scour Depth at Bridge Piers Using Bri-Stars Mode," Talebbeydokhti, N., and Asadi Aghbolagh, M., Shiraz University, Iran, 2006.

Khan, M. A., Baig, S., and Zarriello, J., "Factors Affecting Stream Stability and Foundation Scour at Two Jersey Bridges," International Conference of Scour and Erosion, Texas A&M, College Station, 2003.

HEC-1, *Flood Hydrograph Package: User's Manual,* Micromedia, Toronto, Ontario, 1990.

Lombard, P. L., and Hodgkins, G. A., "Comparison of Observed and Predicted Abutment Scour at Selected Bridges in Maine," USGS Scientific Investigation Report 2008-5099 (prepared in cooperation with Maine DOT).

Melville, B. W., and Coleman, S. E. *Bridge Scour,* Water Resources Publication, 2000.

NJDOT Bridge Design Manual, 1998.

National Cooperative Highway Research Program, RESEARCH RESULTS DIGEST—Number 241, Hydraulics and Hydrology, "1998 Scanning Review of European Practice for Bridge Scour and Stream Instability Countermeasures, July 1999."

NCHRP "Evaluation of Bridge Scour Research," Research Results Digest 378, August, 2012.

Pennsylvania DOT, *Design Manual Part 2* (Publication 13M). Highway Design, August, 2009). Print. USGS Real-Time Water Data for Pennsylvania." *USGS Water Data for the Nation*. October, 2011. http://waterdata.usgs.gov/PA/nwis/

Pennsylvania DOT, *Scour Critical Bridge—Plan of Action,* Rep. no. 15000103700690, 2010.

"Pennsylvania Code." *The Pennsylvania Code Online Title 25 Environmental Protection*. Web. 27 Oct. 2011. http://www.pacode.com/secure/data/025/025toc.html

Pennsylvania DOT, *Design Manual Part 4,* Vol. 1, 2007.

Pennsylvania DOT, Interpretation of the FHWA code (written communication, with Brian Spangler), 1999.

Richardson, E. V., and Davis, S. R., *Evaluating Scour at Bridges,* 4th ed., HEC-18, NHI 01-004, Federal Highway Administration, Washington, D.C., 2001.

Stankowski, S.J., "Magnitude and frequency of floods in New Jersey with effects of urbanization," New Jersey Special Report 38, 1974.

"Ungauged Sites," *USGS Water Resources of the United States*. Web. 27 Oct. 2011. http://water.usgs.gov/osw/streamstats/ungauged.html

U.S. Army Corps of Engineers, *HEC-RAS River Analysis System Hydraulic Reference Manual Version 4.1,* Hydrological Engineering Center, Jan. 2010. Web. Sept.-Oct. 2011. http://www.hec.usace.army.mil/software/hec-ras/documents/HEC-RAS_4.0

"USGS Pennsylvania Water Science Center." *Water Resources of Pennsylvania*. StreamStats. Web. http://pa.water.usgs.gov/

"USGS Surface-Water Data for the Nation." *USGS Water Data for the Nation*. USGS. Web. 28 Oct. 2011. http://waterdata.usgs.gov/usa/nwis/sw

USGS, Cinotto, P.J., and White, K.E., 2000, Procedures for scour assessments at bridges in Pennsylvania, Open-File Report 00-64, 2000.

CHAPTER 9

Inspections, Monitoring, and Flood Management

9.1 Need for Underwater Inspection, Monitoring, and Vigilance

Bridges on rivers require greater monitoring than those located on land. Every time it rains, the physical conditions in the river change. Moist soil loses its cohesion, and the soil-bearing capacity is reduced. The designed load-carrying capacity of the bridge needs to be maintained at all times. In this final chapter a variety of river-related issues are discussed; namely, structural and environmental conditions, emergency protection measures, and the essentials of the scour monitoring program. In addition, monitoring measures, vulnerable foundations, flood watch, and reconstruction methods are examined. The last part of the chapter deals with scour studies, inspection reports, innovative techniques of monitoring, postflood disaster management in the United States, India, and Pakistan, and the role of environmental agencies.

Monitoring includes structural evaluation and inspection. In addition to inspection and monitoring, vigilance during the rainy season is required. It would require vigilance in terms of flood watch and underwater inspection, forcing immediate action and emergency repairs after flash floods which may occur without warning. However, repeated inspections and monitoring will also identify the extent to which the physical conditions of the bridge and its river environment have changed. An essential part of monitoring is the rating analysis using the new observed and measured data. A series of systematic technical questions about bridge inventory will explain the status of any deficiencies. The computed rating is not likely to improve with the passage of time unless necessary repairs are carried out. Bridges that were designed using innovative methods and materials usually gave good service, with little maintenance and due to unique type of construction they eventually get placed on the national historic register for preservation.

Monitoring is a simple and effective method. When used, monitoring serves as a bridge safety tool. The current procedures used in New Jersey for stage 1 and stage 2 are applicable in many other states. Stage I for preliminary scour evaluation consists of screening all bridges for scour risk.

Stage II in-depth scour analysis leads to detailed design and construction cost evaluation of recommended countermeasures (CMs).

Those bridges that do not have a high sufficiency rating are identified for priority repairs and allocation of emergency funds. The final steps are the detailed design, preparation of construction drawings, technical specifications, approval of the permitting process, and installation of countermeasures. However, based on previous SI&A (bridge inspection) reports, National Weather Service flood warnings can be recorded. Flood watch and monitoring of bridges can be implemented.

Following postflood inspections and documentation of minute bridge details or damage, interim repairs can be carried out so that enough time is available for major repairs or planning for bridge replacement. The decision to replace the damaged bridge may be motivated by other important considerations. These may include the condition of the superstructure, the estimated remaining life of the members and the provision of funds for any pending deck replacement that may be based on future widening requirements.

Risk Classification

Based on structure inventory and assessment (SI&A) by awarding a point system, bridges subjected to flood can be classified as high risk, scour-susceptible, or low-risk and at an earlier stage. The flow diagram procedures may provide guidelines for fixing a wide variety of shallow and deep foundations in the river environment after each flood event.

Those bridges that have a high scour are identified for priority repairs and allocation of emergency funds. Postflood monitoring, followed by design and installation of countermeasures, is widely used for the protection of existing abutment and pier foundations. The standard HEC-23 matrix (Table 2.1) for monitoring is, however, conceptual. It does not provide guidelines for rating or identifying the risk or hydraulic vulnerability of bridges, which should be selected for flood watch and structural safety evaluation.

Recently developed procedures consist of screening all bridges for scour risk and stage I preliminary scour evaluation. Selected bridges with a high-risk rating are investigated for stage II in-depth scour analysis, conceptual design, and construction cost evaluation of recommended countermeasures.

9.2 Floodwater Sources and Rise of Water Level at Bridges

The usual causes for the floods and increase in water level elevations are the increase in the speed of river water resulting from

- Rainfall intensity following dark clouds—monsoons and flash floods
- Snowmelt on mountaintops and glacier melt
- Levees breaking and flooding residential areas
- Dam burst upstream of bridge
- High tides at full moon
- Global warming which raises the sea level in coastal areas

In the hydrologic analysis for a drainage structure, many variables affect a flood. Some of the factors that must be considered on an individual, site-by-site basis include the following:

- Rainfall amount and rate of storm distribution
- Type of precipitation (rain, snow, hail, or combinations thereof)

- Drainage area size, shape, and orientation
- Slopes of terrain and streams
- Moisture conditions
- Storage potential (overbank, pond, wetlands, reservoir, channel, etc.)
- Watershed development potential
- Ground cover and soil type
- Topography, elevation, and mixed-population density
- Obstructions to flow from trees, bridges, and culverts

9.3 Maintaining the River Environment

Floods create any number of environmental issues. Sometimes the volume of water is so large that it washes away homes, inundates streets, causes epidemics, destroys crops and agriculture, etc. In addition, adverse impacts on marine life can be caused by the engineer's choices of remedial actions such as reconstruction in the river or providing countermeasures within the depth of water. There should be adequate justification to obtain an environmental permit, without which no construction activity is allowed. The bridge engineer needs to address the following environmental issues.

Environmental Concerns

The purpose of reconstruction is to benefit transport facility users. The process of short-term construction and long-term impacts should not be adverse to road users or to local residents. A large number of environmental concerns must be addressed related to water, including

- Maintaining water quality
- Providing fish passage
- Avoiding wetland contamination
- Avoiding stream encroachment
- Improving drainage
- Preventing construction impacts on the floodplain
- Minimizing the erosion of native substrate due to sediment transport after the installation

Issues related to ecology include

- Minimization of impacts on natural vegetation by controlling construction access points
- Revegetation of disturbed areas with similar vegetation species possibly required
- Landscape preservation
- Preservation of endangered species

Maintenance of air and water quality:

- Avoidance of air contamination and pollution
- Noise mitigation from construction vehicles, pile driving, concreting, excavation, welding
- Relocation hazards of underground and bridge-supported utilities
- Reactions with acid-producing soils

Additional environmental concerns:

- Historical and cultural aspects, aesthetics, landmarks
- Temporary work and scaffolding
- Traffic disruption
- Socioeconomic aspects, relocation
- Right-of-way issues
- Permitting considerations and implementation of EPA and DEP procedures

Action Required by the Environmental Engineer

The following environmental issues supplement the structural or constructional items.

1. Minimize adverse environmental impacts by addressing the environmental concerns.
   - Develop a baseline survey to define current environmental issues.
   - Develop an assessment of the impact of proposed repairs on air pollution or on the water environment.
   - Develop considerations or measures to avoid or mitigate adverse impacts.
   - Eliminate impacts on wetlands by using top-to-bottom construction and temporary bridges.
2. Consider environment-related items.
   - Requirements for paint removal and containment and disposal of contaminants shall be incorporated per the current State Department policy such as DEP or EPA.
   - The need for bridge-mounted sound walls shall be determined in conjunction with the need for sound walls on the adjoining roadway.

Environmental Issues and Permits

Environmental Documentation The scope of a construction project will qualify it for a categorical exclusion (CE) under National Environmental Policy Act (NEPA) regulations. Documentation will be prepared to confirm the eligibility of the project per categorical exclusion.

Cultural Resources By preliminary research ensure there are no resources presently listed that make the bridge eligible for the National Register. Should eligibility be determined,

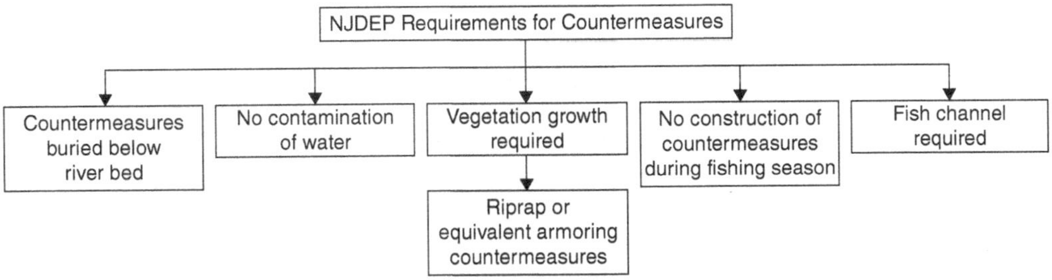

Figure 9.1 DEP general requirements for stream encroachment permit.

consult with the Historic Preservation Office (HPO) on appropriate treatments and include the results of that consultation.

Meeting the Permit Requirements for Reconstruction

1. The proposed construction should neither damage an existing wetland nor adversely affect the historical significance of the bridge itself or its surroundings, except as permitted through the environmental evaluation process. For bridges located on streams, a *flood hazard area general permit* is required.
2. Engineering data and documentation are required for permit approval. Per state DEP and NEPA regulations, the following reports pro forma need to be submitted:
   a. Environmental assessment (EA): An EA is required when the significance of the environmental impact is not clearly established.
   b. Environmental impact statement (EIS): EIS documents need to be prepared when a replacement or new bridge (usually with four or more lanes) has a significant impact on natural, ecological, or cultural resources including endangered species, wetlands, floodplains, groundwater, fauna, and flora. An EIS is required when there are impacts on properties protected by the DOT act or the Historic Preservation Act. Significant impacts on noise and air quality need to be avoided.
   c. Categorical exclusion (CE): An action that does not have a significant effect on the environment falls under CE. Examples provided by FHWA are the reconstruction or modification of two-lane bridges, the addition of pedestrian or bicycle lanes, widening for shoulders, and installation of signs.

9.4 Emergency Bridge Protection Measures

Rainwater and resulting floods can cause settlement of a bridge. Preparedness is required, such as modular bridges and accelerated bridge construction. Other considerations are as follows:

1. Any design and installation of bridge protection measures during high water can be difficult, if not impossible. A planned response for bridge scour is preferable to a reactive response.
2. An emergency installation is typically much more costly than a pre-planned one.

3. There is usually an increased cost of mitigation since damage during an emergency project can be greater and equipment remobilization may be required for postproject mitigation. Project impacts (i.e., damage to trees and vegetation) in carrying out emergency work have to be mitigated in the same way as for projects with normal timing.
4. Impacts of carrying out emergency work should be minimized. Under emergency scenarios, the tendency is to take action to protect a bridge at the expense of existing trees and other vegetation. However, trees and vegetation may be providing protection or may eventually protect a bridge abutment or approach. The trees and vegetation also provide important riparian habitat, and they should be protected even if they don't offer any direct stabilization of bridge countermeasures.

Safety Considerations

The following issues need to be addressed:

1. Working adjacent to fast, unpredictable currents and rapidly rising water levels can be extremely dangerous.
2. Floating or subsurface debris and woody materials contribute to hazard during emergency work.
3. Weather conditions (rain, snow, or darkness) may further endanger safety.

Constructability Review

Addressing constructability issues early in the design process and performing constructability reviews throughout the design process will prove valuable in eliminating changes during construction. A constructability review will include the following to avoid any surprises during construction:

- Site issues, such as transport to site and site accessibility
- Equipment locations
- Maintenance and protection of traffic (MPT)
- Construction work area restriction
- Right-of-way (ROW) encroachment, geometrics
- Access to adjacent properties
- Increased costs
- Project duration
- Material availability
- Fabrication and erection feasibility
- Construction sequencing

A review will be done during the design and performed by construction managers. At the conclusion of the preliminary design, plans are submitted along with the construction cost estimate, outline of specifications, and construction schedule for a constructability review by an independent consultant. The objective is to keep traffic moving while providing a safe project work zone for all. Explore and evaluate methods to minimize construction costs and time as well as the aforementioned risks.

A workshop will be conducted to evaluate constructability. Topics may include, but are not limited to, construction duration, access and ROW, and material availability. Potential construction restrictions are identified:

- The locations of the longitudinal construction joints in the deck slab may cause weakness due to discontinuity in concrete in the transverse directions. The preferred location to form joints is in the middle of the top flange. Staged construction versus detour shall be considered.
- Staging widths and work area widths need to be developed, to avoid locating the construction joint away from steel or prestressed concrete longitudinal girder flanges. Otherwise, under repeated impacts of heavy truck loads, unsupported construction joints can open at a future stage.
- Since shear connectors require field welds, small angles or channels bolted to top flanges will be considered.
- Staged construction will maintain uninterrupted traffic flow and eliminate potential traffic congestion associated with a detour. Some stages can be eliminated if access is available from under the superstructure. Any impacts to daily commuters will be minimized by maintaining two lanes of traffic.

Coordination with any adjacent contracts is essential to the success of this project. The primary goal for the project will be to design improvements that provide a higher level of safety and comfort to vehicular and pedestrian traffic while having the least possible impact on existing surroundings. The initially preferred alternate (IPA) should be able to be constructed in one season, as shown on the schedule. The fewer stages there are, the shorter the time to completion. The structure width from curb to curb can be reconstructed in three stages by maintaining two 11-ft lanes of traffic at all times.

Evaluate the integrity of the existing shoulders to determine if any improvements will be required prior to shifting traffic onto them. Review existing accident information to determine if any specific type of vehicular accidents may affect the proposed staging plans. Traffic control plans will be prepared in accordance with the *Manual on Uniform Traffic Control Devices* (MUTCD) so that they will be comprehensive and clear in delineating how the traffic control is to be accomplished. Maintenance of pedestrian access will be included in the traffic control plans. The use of various multimedia outlets should be considered for this project. Develop a construction schedule to account for work hour restrictions during Christmas and other holidays.

Maintenance and Protection of Traffic

1. Traffic control plans must comply with the state, MUTCD, and AASHTO LRFD regulations. Prior to developing staging plans, the project manager shall contact traffic operations to determine the maximum allowable lane closure hours in each direction and the maximum number of lanes that can be closed at one time.

2. All traffic control schemes and detour plans on local roads must be approved by local authorities. It is important that early in the design a set of applicable traffic control and staging plans be sent to them for approval.
3. A minimum 10-hour night window may be required for the contractor to properly complete the work. Weekend work may be considered in addition.
4. Staging plans shall show a cross section of the bridge for each stage of construction (two stages are preferred). In each of the planned stages, repair areas shall be distinguished from travel lane areas.

9.5 Essentials of Scour Monitoring Program

A monitoring and inspection program should be developed that includes scour depth measurements, an inspection process, and traffic closures.

A monitoring plan may include the following objectives:

1. Development of a database of photographic records of one or more constant points above and below flow depths for scour-critical bridges
2. Measurement of bank and channel cross sections and bed elevations
3. Regular measurement of scour depths
4. Measurement of lateral migration
5. Identification of eroded areas around the footings after major floods.
6. Measurement of plant densities and species composition
7. Estimation of fish use

A project site should be monitored periodically (two or more times a year) during the first few years to ensure the success of the countermeasures and any compensatory mitigation. During this period, vegetation should be reestablished since bridge protection measures may rely on plants to provide long-term stabilization. After vegetation has been established, monitoring once a year or every other year may be adequate to ensure that the selected countermeasure functions as designed, to identify any "obvious or apparent" channel response impact, and to address any potential risk.

Minimizing Repairs Upstream development should be monitored to ensure that bridge protection measures do not fail. Bridge scour problems should be managed through a planned and integrated approach. This will avoid or minimize the need for "reactive" bridge repair projects.

A monitoring and inspection program should be developed that includes obtaining scour depth measurements, an inspection process, and traffic closures.

A monitoring plan may include the following objectives:

1. Develop a database of photographic records of one or more constant points above and below the flow depths for scour-critical bridges.
2. Measure bank and channel cross sections and bed elevations.
3. Measure scour depths regularly.
4. Measure lateral migration.

5. Identify eroded areas around the footings after major floods.
6. Measure plant densities and species composition.
7. Estimate fish use.

Frequency of monitoring: A project site should be monitored periodically (two or more times a year if necessary) during the first few years to ensure the success of the countermeasures and any compensatory mitigation. During this period, vegetation growth should be reestablished since bridge protection measures may rely on plants to provide long-term stabilization. After obstructions from vegetation have been established, monitoring once a year may be adequate to ensure that the selected countermeasure functions as designed, to identify any "obvious or apparent" channel response impact, and to address any potential risk.

9.6 Description of Monitoring Measures

For scour-critical bridges, monitoring can be used to identify potential bridge scour problems before they develop. Regular monitoring of all bridges is now mandatory by NBIS. The inspection cycle is every 2 years for bridges and 4 years for culverts. However, bridges classified as structurally deficient need to be inspected every year or more frequently, depending on the condition of the bridge. It is considered to be a more economic countermeasure, since it prevents the use of expensive countermeasure by implementing regular maintenance and timely repairs.

Scour-critical bridges typically require maintenance. A well-designed scour monitoring program offers the following advantages:

1. *Remedial action*: It can be used as an effective tool for early identification of potential scour problems and can provide a continuous survey of the progression of scour around bridge foundations. Remedial action can be taken to offset a scour event.
2. *Adaptive management*: If monitoring indicates that a bridge protection countermeasure is no longer functioning, then adjustments can be made to ensure continued long-term function of the countermeasure.
3. *Ensuring long-term performance*: Monitoring of dumped riprap, gabions, grout bags, etc., at regular intervals ensures that they are functioning as designed and facilitates the development of detailed field knowledge of their long-term performance.

Details of Monitoring

Instrumentation: Monitoring can be accomplished by using instrumentation or visual inspection. A well-designed instrumentation program for scour monitoring can be very cost-effective. Two types of instrumentation can be used to monitor bridge scour: fixed instrumentation and portable instrumentation.

Objectives: For scour-critical bridges, monitoring can be used to identify potential bridge scour problems before they develop.

Frequency: Regular monitoring of all bridges is now mandatory by the National Bridge Inspections Standards (NBIS). The inspection cycle is every 2 years for bridges and 4 years for culverts. However, bridges classified as *structurally deficient* need to be inspected every year or more frequently, depending on the condition of the bridge. Such a frequency is considered as a more economic countermeasure, since it prevents the use of expensive countermeasures resulting from negligence and by implementing regular maintenance and timely repairs.

Advantages: Scour-critical bridges typically require maintenance. Changes or deterioration in a scour problem over time and through flood events can only be evaluated through an adaptive management program based on monitoring. A well-designed scour monitoring program offers the following advantages:

1. It can be used as an effective tool for early identification of potential scour problems and can provide a continuous survey of the progression of scour around bridge foundations.
2. Remedial action can be taken to offset a scour event.
3. If monitoring indicates that a bridge protection countermeasure is no longer functioning, then adjustments can be made to ensure continued long-term function of the countermeasure. Such maintenance is called *adaptive management* because it helps in identifying, over time, functionally effective countermeasures while minimizing impacts to the natural habitat.
4. Monitoring of dumped riprap, gabions, grout bags, etc., at regular intervals ensures that they are functioning as designed and facilitates the development of detailed field knowledge of their long-term performance.

U.S. Department of Agriculture Forest Service Inspection Procedures

A code 6 is used for the scour-critical bridge (scour calculation or evaluation has not been made). The five categories and corresponding scour-critical bridge (Item 113) codes are as follows:

1. Low-scour-risk Item 113: corresponding codes 4, 5, 7, 8, 9
2. Scour-susceptible Item 113: corresponding codes 6
3. Scour-critical Item 113: corresponding codes 0, 1, 2, 3
4. Unknown foundations Item 113: corresponding code U
5. Tidal Item 113 corresponding code T

The above categories are the same as recommended in HEC-18, Chapter 5. According to the U.S. Forest Service, Watershed Management and Restoration, "forested lands provide 50 percent of U.S. water supply" (http://www.fs.fed.us/research/water-management/).

Miscellaneous Countermeasures in Use (Less Than 1 Percent of Total)

Neglecting the use of suitable countermeasures is seen to be infinitely expensive in terms of collapse of the bridge and shutdown of the highway carrying the bridge. Table 9.1 shows that monitoring combined with dumped riprap accounts for 92 percent of the countermeasures in practice, leaving only 8 percent for alternatives. In recent years there has been an increase in the use of gabion baskets.

Countermeasure	% Use with Monitoring	% Use without Monitoring
1. Monitoring use as an alternative countermeasure	76—monitoring use by itself	N/A
2. Dumped riprap	16	67
3. Extended footings	2	8
4. Rock gabions	1.5	6
5. Spurs	1.1	4
6. Pavements	0.7	3
7. Check dams	0.25	1
8. Rock bank protection	0.25	1

TABLE 9.1 Importance of Monitoring and Inspection with Countermeasures

Monitoring decreases the risk of failure. Miscellaneous countermeasures were described in Chapter 8. The estimated total miscellaneous countermeasures in use are less than 10 percent of all types of countermeasures.

9.7 Methods for Investigating Scour Conditions for Vulnerable Foundations

The two inspection forms are the Waterway/Channel Form and the Substructure Scour Form. Inspection of piers and abutments is carried out with special emphasis on piles and pile bent inspection. The levels of inspection are

1. Visual inspection
2. Detailed inspection with partial cleaning
3. Detailed inspection with nondestructive testing (special equipment such as fathometers possibly required)

Visual Inspection Visual scour monitoring is a standard monitoring practice for inspecting bridges at regular intervals during low-flow events and on a smaller interval (i.e., increased monitoring efforts) during high-flow events. Typically, bridges are inspected on a biennial schedule where channel bed elevations at each pier location are measured. Periodic inspections, especially after major floods or coastal storm surges, should be carried out in addition to biennial inspections.

A normal 2-year inspection cycle with soundings (where required) for all scour-critical bridges should be followed. Channel elevations should also be taken during and after high-flow events. If measurements cannot be safely collected during a high-flow event, the bridge owner should determine whether the bridge is at risk and closure is necessary. The following guidelines given below in New Jersey DOT *Underwater Inspection and Evaluation of New Jersey Bridges Guidelines Manual*, 1994, may be followed for underwater inspections.

Direct Methods These include majority of countermeasures in flowing water. Probes, augers, rotary drilling, or digging test pits adjacent to footings may be used to locate the top of existing footings and unknown foundations, for which construction drawings do not exist. Digging of test pits is not employed owing to the possibility of increased undermining.

Probing of Soil in the Riverbed To identify the areas of scour holes, probing is carried out adjacent to existing footings.

Nondestructive Tests Ground-penetrating radar or acoustical emission techniques are considered to determine footing depths. However, because of the unknown type of soil and stone masonry supported on stone substrata, it may not always be possible to determine an accurate depth of footing.

Drilling Another method of inspection is drilling and obtaining test boring logs and then analyzing the results to determine the elevation of footing and bedrock.

Standard Procedures of Underwater Inspection Used in New Jersey Bridges over waterways shall be inspected in accordance with the 1994 New Jersey DOT *Underwater Inspection and Evaluation Guidelines Manual* and OSHA Guidelines CFR 1910, Subpart T.

- *Frequency of inspections*: All required underwater inspections shall be completed on a 48-month inspection cycle or on a 24-month inspection cycle for scour-critical bridges and bridges with low freeboard. Scour-critical (Item 113 coded 3 or less) bridges that have a depth of water not less than 4 ft at the substructure units shall have underwater inspections performed every cycle (24-month frequency). Additionally, bridges with low freeboard must have an underwater diving inspection performed every cycle (24-month frequency).
- *Minimum depth of water*: The depth of water in the channel near the substructure units must be verified by the team leader that an underwater inspection is indeed required.
- *Qualifications and training*: Diving shall be performed by a fully qualified diver experienced in the underwater inspection of concrete, steel, and timber structures. As per NBIS, all underwater divers must have completed comprehensive, federally approved bridge inspection training.

FHWA Coding Guide Standards Please refer to Figure 9.2 below for further action in each case of relative scour depth. Scour monitoring of bridge foundations and actions required when the computed scour depth elevation is:

1. Shallow—Above the top of footing base
2. Medium depth—Within the footing or limits of piles
3. Very deep—Below the footing base or pile tips.

Application of Recording and Coding Guide

In the absence of scour analysis, the Coding Guide of the NBIS shall be used to classify the bridge if it is scour-critical. Items 61, 71, and 113 are of concern during underwater inspections. Item 113 is used for scour rating.

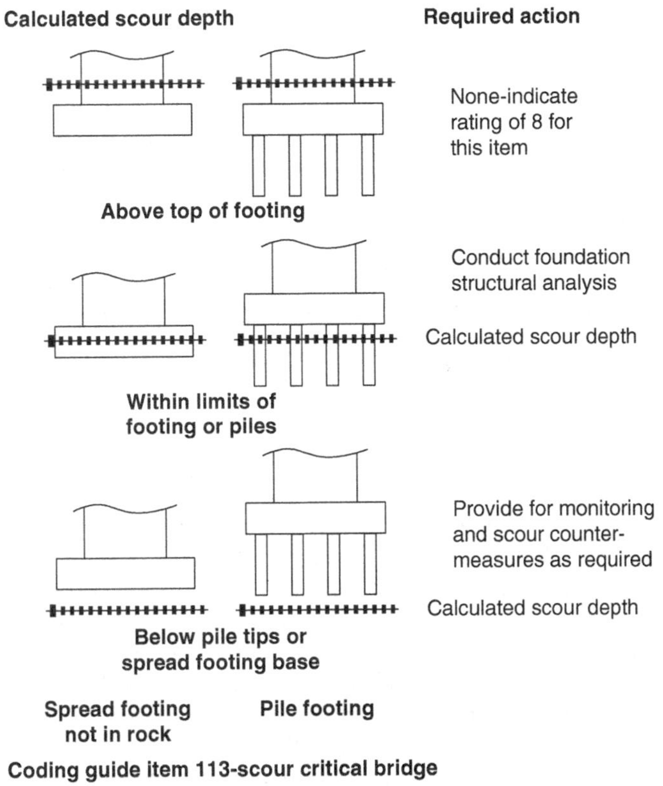

FIGURE 9.2 FHWA Coding Guide Item 113 for Spread and Pile Footings.

Table 9.2 shows that no action is to be taken if the observed level of scour depth is above the footing. Scour analysis is required if the scour depth is within the thickness of footing. In such case, further investigation is required. If the scour depth is below the footing, scour countermeasures need to be provided as an interim measure.

Calculated Scour Depth	Action Required
Above top of footing	No action required except assigning rating = 8
Within limits of footing	Conduct structural analysis for foundation and provide nominal countermeasures
Below spread footing base	Provide scour countermeasures and regular monitoring—Recording and Coding Guide
Within limits of piles	Conduct structural analysis for foundation and provide nominal countermeasures
Below pile tip elevation	Provide scour countermeasures and regular monitoring

TABLE 9.2 FHWA Coding Guide Item 113

National Bridge Inventory Item 67: Structural Evaluation

This important item for structural evaluation is calculated by the Edit/Update program based on Table 1 (page 47 of FHWA *Recording and Coding Guide*) and need not be coded by the bridge inspector. The structural evaluation rating code (from 0 to 9) in the table is selected by comparison of three categories of average daily traffic (0 to 500, 501 to 5000, and greater than 5000). FHWA Memorandum dated April 27, 2001 revised Item 60 (substructure) and Item 113 (scour-critical bridges). Item 67 was updated for rating factors in FHWA Attachment 5, dated August 2013.

9.8 Adopting a Flood Watch List for Scour-Critical Rivers

1. *Establishment of a scour-critical state watch list*: Establish a watch list for scour-critical bridges, and monitor these bridges during flood events until corrective action is completed.
2. *Correlation of water management areas, stream gauging stations, and watch list bridge locations*: Develop a methodology to evaluate real-time data from streams and watersheds to determine threshold values when site monitoring will be required.
3. *Establishment of a procedure to monitor watch list bridges during floods*: Establish a procedure for real-time monitoring of scour-critical bridges during flooding events. The first indication of a potential monitoring situation will be a flood warning posted by the National Weather Service (NWS). This will prompt the structural evaluation group to begin tracking the USGS gauges for the affected storm area on the Internet. The following sequence of actions are suggested but the guidelines of structural evaluation department of the relevant highway agency will be followed:
    - The operations group will dispatch field crews to perform inspection monitoring of the scour-critical bridges within the flood-prone area.
    - The dispatched file crew will observe and assess the affected bridges by completing a standard inspection monitoring form that addresses both critical and noncritical items. The field crew will utilize pre-prepared sheets for each bridge that describe the key data related to scour including projected scour depth, substructure and foundation type, and history of scour problems and debris.
    - Before any closed bridge can be reopened, a post-event inspection of the structure following the normal NBIS procedures will be required.
    - The structural evaluation staff will be responsible for determining if any repairs or scour countermeasures are required as well as when a bridge can be safely reopened to traffic.

Simultaneous Monitoring of Twin Bridges

When twin bridges, such as the northbound and southbound, are located close to each other on the same stream or river, both bridges are likely to suffer from similar scour problems. Hence, monitoring of the scour problem at one bridge should be accompanied by a similar monitoring and rating of the other, irrespective of the bridge prioritized for inspection.

Frequency of Monitoring

Scour-critical bridges should be inspected after every flood and after any installation of a countermeasure. Frequency of monitoring is a site-specific issue which depends on following:

1. Erosion to bridge footings
2. Performance assessment of installed countermeasures
3. Migrating meander forms
4. Adjustments to water and/or sediment supply from upstream
5. Impacts to vegetation survival from on-site land use
6. Head cuts from downstream activities

Developing Flow Diagrams

Flow diagrams are useful tools for scour management, and they can ensure bridge safety against scour. The flow diagrams presented cover a wide variety of shallow and deep foundations. The flow diagrams developed utilize physical features and geotechnical and hydraulic data, including the latest scour conditions of the abutments and piers observed in the field. The diagrams can be modified for use for bridges located on rivers having different geometric, hydraulic, and geotechnical conditions elsewhere. Selection of countermeasures is linked to the scour components and the parameters on which the scour equations are based. Both physical hydraulic modeling in a laboratory and numerical computer modeling are among the standard techniques available to analyze the scour problem and design countermeasures.

9.9 Plans and Preliminary Estimates

Use of Cofferdams and Temporary Works

Reconstruction in rivers in postflood conditions requires river diversion or cofferdams using sheet piles. All temporary works should be in keeping with the guidelines given in AASHTO Guide Design Specifications for bridge temporary Works for constructability.

Plans and specifications will be developed in accordance with the policy and procedures of the agency and the applicable AASHTO design standards for load resistance factor design (LRFD).

- Members will be checked for the effective stress range for fatigue (load- and displacement-induced) and seismic and thermal analyses.
- State- or AASHTO-recommended permit loads will be applied.
- Impact on traffic from staged construction will be addressed.
- Structural stability of substructure members will be considered in design.
- All constructability issues will be investigated prior to reconstruction.
- Superstructure jacking will be required to remove and replace the bearings.
- Cracking of fracture-critical members (FCMs) and diaphragms will be investigated.

- In the case of expansion dams, if a field visit has shown nonfunctioning deck expansion joints, adequate expansion joints need to be provided.
- If existing bearings need to be replaced, the type will be investigated. Consideration of seismic retrofit for bearings will be coordinated with the systemwide seismic study. The width of the bearing seat must meet current AASHTO LRFD requirements. Based on detailed field verification, the repair requirements for the steel bearings should be identified to preclude the need for major repairs. These repairs are likely to be designed as "replacements" in kind for the keeper angles, shoulder bolts, anchor bolts, etc.

The scope of the project is to complete the work within the existing footprint. Therefore, the first task for the design team is to identify all substandard geometric features. This will establish the baseline conditions as a checklist that will be used to help minimize significant impacts that could slow down the implementation of the proposed improvements.

A design exception report will be prepared by utilizing the latest design exception manual for substandard controlling design elements that cannot be easily corrected. The existing deck cross-slopes will be evaluated, and a spread analysis will be performed and compared with the allowable spread. If it is found to be substandard, action will be taken to correct this deficiency in the design of the replacement deck and additional scuppers will be added, as necessary. Otherwise, a design exception will be requested from the highway agency for approval. Existing lighting will be evaluated and upgraded as necessary.

Final Report Preparation and Specifications

After the in-depth scour analysis, the final steps to construction are the detailed design, preparation of construction drawings, technical specifications, approval of permitting process, and installation of countermeasures. National Weather Service flood warnings can be recorded. Flood watch and monitoring of bridges can be implemented. Based on a point system, bridges subjected to flood can be classified as high risk, scour-susceptible, or low-risk at an earlier stage.

Following postflood inspections and documentation of bridge damage, interim repairs can be carried out, so that enough time remains for major repairs or for bridge replacement.

Requirements of a Good Construction Drawing

Contract documents should provide complete design plans and details using standard scales, which should show the work to be performed. The details should show the size, section, material grades, layout of foundations, substructure, bearings, girders, deck slab, parapets, railings, variable-message sign structures, and location of utilities. Construction notes, plan, elevation and sections, dimensions, elevations, and cambers with clarity shall be shown, for preparing shop drawings and for fabrication. Notes will show reactions, codes of practice, estimate of quantities, and precautions. Use of a Request for Information (RFI) from the contractor to the owner and the consultant is encouraged.

Superstructure and Substructure Repairs and Superstructure Replacement

A field inspection will be required to determine whether conditions to the bridge have changed since the last inspection. During peak floods if there is overtopping of the deck, superstructure repairs may also be required in addition to the substructure. Typical repairs encompass the following:

- Structural steel painting
- Drainage system repairs and retrofits of deck scuppers
- Any specialty repairs to reinforced concrete slab and beams
- Sign structures, bridge-mounted signs, and asphalt resurfacing of deck
- Bridge railing and barrier repairs
- Roadway approach resurfacing, curb addition or repair

The following actions apply to superstructure repairs and replacement:

1. Superstructure repairs will be made to the bridge to restore serviceability and original functionality and extend the bridge life. Repairs may include the corroded sectional areas of flanges, deteriorated asphalt overlay, curbs, sidewalks, underdeck, exposed rebars, encasement deterioration, severe spalls, severe scaling, and spread of cracks.
2. The deck will be evaluated using nondestructive testing such as ground-penetrating radar to determine the need for repairs, prior to milling and resurfacing.
3. Minimum service life of 25 years is targeted. Any cracks in deck and parapet concrete will be repaired by pressure grouting. Upgrading existing simply supported beams to continuous beams will eliminate deck joints over piers. This will also increase both the rating (load-carrying capacity of beams) and the life of the deck slab. Continuity can be established by bolting flange plates at the piers. As-built plans and other records of bridge history will be consulted. Similarly, foundation and substructure repairs will be considered. Regarding soil criteria, because of the difficulty of obtaining borehole information at each site, the national seismic hazard maps developed by the USGS can be used as a basis for considering soil vulnerability characteristics in the selection process. The repair process may also apply to the substructure:
   - Pier heights—range of tall and short piers
   - Foundation types—pile foundations/spread footings
   - Varying abutment heights and types

According to Pennsylvania Code for scour, bearing vulnerability ratings will be based on the following characteristics:

- All major bearing types and number of bearings
- Expansion rocker bearings with varying heights
- Anchor bolt arrangements (various diameters, number per bearing, strength, condition)
- Seat length vulnerability based directly on AASHTO specifications

If major repairs are required, it may be more economical to replace the bridge.

Interference from Underground Utilities

Installation and maintenance of underground utility pipes close to a bridge foundation may cause significant disturbance to the soil. Detailed environmental investigation may be needed if utilities have to be relocated because of interference with a proposed countermeasure. However, provision should be kept for placing the utility pipes suspended below the deck and between girders.

9.10 Pennsylvania DOT Bridge Scour Evaluation Program

Table 9.3 shows the format used in Pennsylvania for documentation of scour-critical bridges prior to performing inspections, and Table 9.4 shows a format for data concerning elevations and computed scour depths.

Bridge no.:	Route no.:	Mile Post Location	
Functional classification:			
Name:			
Year built:	Year reconstruction:	Length:	Width:
Number of spans:	Spans over floodplain:		
Structure type:			
Substructure type: Abutments: Piers:			
Foundation type: Abutments: Piers:			
Stage 1 findings:			
Date of evaluation:		By:	

TABLE 9.3 Documentation of Scour-Critical Bridges

Item	North Abutment	South Abutment
Top of footing elevation		
Approximate bottom of footing elevation		
Approximate riverbed elevation		
Bedrock elevation		
Scour depth to bedrock		
Theoretical maximum scour (discharge event: 50-, 100-, 500-year)		

TABLE 9.4 Data Required for Scour Depth Computation

Item	Purpose	Dates Performed	Data Collection	Methodology for Obtaining Data	Duration of Each Activity
Field surveys					
Underwater surveys					
River cross sections					
Soil parameters					
Foundation details					
Test borings					

Table 9.5 Data Collection Items

Findings The findings of the scour evaluation program will be based on data collection items as listed in Table 9.5 and including the following:

- Field survey and investigation of existing conditions
- Independent hydrologic and hydraulic studies and scour analyses
- Repairs performed prior to implementing countermeasures
- Laboratory test sample results on materials

9.11 Scour Studies Listed in Inspection Reports of Sample Bridges

The following types of damages were encountered in the study of several inspection reports:

1. The inspection report requested to backfill scour holes along the abutments.
2. One of the inspections (as shown in Figure 9.3) also observed a medium vertical crack that was detected in the middle of the near abutment's breast wall.
3. Lateral displacement at the vertical construction joint showed that the abutment wall had rotated.
4. The footing exhibited a wide crack below the construction joint, as seen in Figure 9.3.
5. The conclusion of the report suggested placing riprap, grout-filled bags, or other countermeasures along the near abutment.

It was observed that scour existed along the near abutment. The scour is mainly due to the channel entering the bridge on a sharp skew which directly impacts the near abutment. Up to 3 ft of scour existed at the near abutment which covered a large area (25 ft wide by 50 ft long) and exposed the footing for one-half the length.

These reports also observed a large sediment buildup along the full length of the far abutment. In addition, the measured height from river bottom to the bottom of the fascia beam varied from approximately 13 ft at the far abutment to 16 ft at the near abutment. Both abutments received low observed scour ratings.

Plan of Action

If a bridge receives a minimum Scour-Critical Bridge Indicator (SCBI) rating of 3 and a minimum observed scour rating of 4, the bridge is to be listed as a category B bridge.

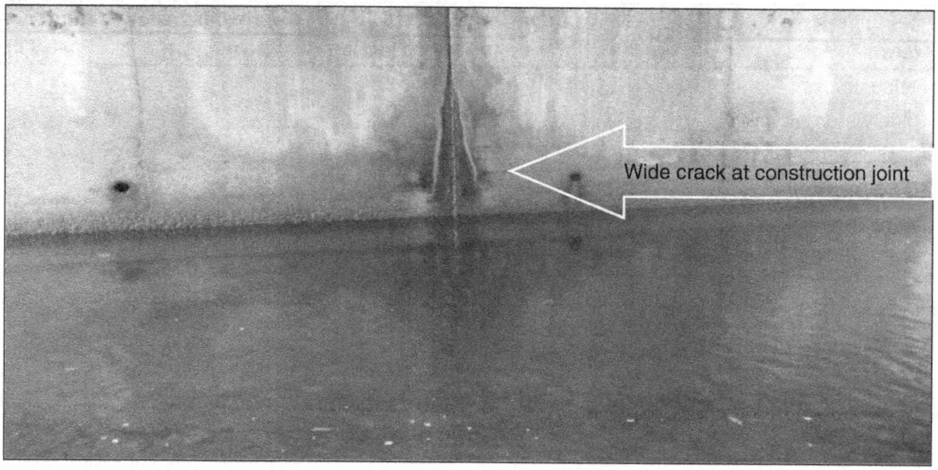

FIGURE 9.3 Rotation of construction joint at abutment and vertical hairline cracks.

Category B bridges are defined as scour-critical bridges, and therefore a scour Plan of Action (POA) must be prepared.

1. Typical scour damages are advanced section loss, deterioration, spalling, or scour.
2. There is an alluvium mixture of fine particles with some larger aggregate that has the potential for scour during ordinary high water; advanced scour may be present or may have occurred.
3. There is highly erodible rock with advanced or serious scour that could advance and threaten substructures during high flows.

The most recent POA states that the bridges are in a Structure Deficient Status due to scour at the near abutment. Backfilling of the scour hole and providing rock protection at the near abutment are required.

A scour plan of action (POA) must be prepared. The most recent documented POA also focused on the removal of debris along the upstream section of some of the bridges inspected.

9.12 Innovative Techniques for Monitoring

University of Michigan Recommends Use of Nanotechnology to Reveal Cracks and Corrosion

A new "skin" for bridges could be a sixth sense for inspectors looking for cracks and corrosion that could lead to a catastrophic failure, as in the recent Minneapolis bridge collapse.[1]

[1]Details of the bridge collapse, which occurred in 2007, are listed in Minnesota DOT, I-35W Bridge Collapse.

Bridges are scrutinized every 2 years, and inspectors rely heavily on their visual observations to find weak points. If they see red flags, they do more tests. According to a paper published in the journal *Nanotechnology*, researchers at the University of Michigan have recently developed a coating that could be painted or sprayed on structures. The research on nanotube composite thin films was performed by a team headed by Jerome Lynch at College of Engineering at University of Michigan in 22007.

This coating would allow inspectors to check for damage without physically examining a structure. When the time comes to examine the health of the structure, an inspector could push a button and in minutes the skin would generate an electrical resistance map and wirelessly send it to the inspector. This sensing skin is an opaque, black material made of layers of polymers. Networks of carbon nanotubes run through the polymers. Carbon nanotubes are a fundamental building block of the nanotechnology revolution.

The microprocessor then creates a two-dimensional visual map of that resistance. The map shows inspectors any corrosion or fracturing too small for the naked eye to detect. This sensing skin would revolutionize the way current bridge health assessment is conducted, resulting in safer structures and lower cost of inspection. The skin could be a permanent veneer over strain- and corrosion-prone hot spots of joints in bridges.

University of Miami Develops a Self-Powered Monitor System

The University of Miami has developed a self-powered monitoring system for bridges using wireless sensors. With a scarcity of inspectors and tens of thousands of bridges, the visual inspection process can be long and laborious. Thousands of bridges erected during the 1960s and 1970s, when much of the nation's infrastructure was built, do not have sensors installed.

A team of University of Miami College of Engineering researchers is implementing a self-powered monitoring system for bridges that can continuously check their condition by using wireless sensors. Sensors can harvest power from structural vibration and wind energy. They plan to place newly developed wireless sensors—some as small as a postage stamp, others no longer than ballpoint pen—along strategic points inside 27-year-old Long Key Bridge in the Florida Keys and on the Northwest 103rd Street quarter-mile steel overpass that leads into Hialeah, Florida. The sensors, developed by project collaborators Virginia Tech University and Physical Acoustics Corporation ([PAC] Princeton Junction, NJ), record many types of data, from vibrations and stretching to acoustic waves and echoes emitted by flaws such as cracks. Even the alkaline levels in the concrete of bridge supports are being measured.

The work is part of the National Institute of Standards and Technology (NIST) Innovation Program and is aimed at developing a more effective system to monitor the health and predict the longevity of bridges. The joint venture is led by PAC.

Innovative Approach for Bridge Inspections

A new imaging program has been developed that automatically detects irregularities in bridges. Research scientists at the Fraunhofer Institute for Industrial Mathematics (ITWM) in Kaiserslautern, Germany have developed the specialized software jointly with fellow scientists from Infracom Italia in Verona, Italy. The engineers have been using the new software successfully to inspect bridges in Italy. The software automatically examines the photographs of a bridge for certain characteristics and irregularities, for marked discoloration. Even minor damage is identified and signaled.

No two bridges are alike. They differ in terms of their shape, construction material, and surface structure. The color of the images found in the software depends on the material, dirt, or degree of humidity.

The changing effects of weather and temperature, road salt, and the increasing volume of traffic all quickly cause damage such as hairline cracks, flaking concrete, and rust penetration. The software can handle these discrepancies. The researchers have extracted metrics from photographs that include the characteristically elongated shape of a hairline crack, the typical discoloration in damp places, and the structures of the material, which are different for a concrete bridge than for a steel bridge.

All this information is stored in a database. When the researchers load a photograph into the program, the software compares the features of the new image with those of the saved images. If it detects any discrepancies, it marks the respective area on the photograph. The bridge inspector can decide how serious the damage is and whether something needs to be done. The earlier damage is identified and clearly categorized, the simpler and less expensive it is to repair.

The Virtual Design Tool for Visual Management

The virtual design (VD) tool for visual management of large, complex projects is an extraordinary tool that construction professionals can employ to collaboratively manage all stages in the design and construction process.

A walk-through software can read directly from CAD platforms and provide a compact and less hardware-intensive file size by reducing redundant tasks in the preconstruction phase. The usual project phases on a large project are as follows:

1. Conceptual design
2. Preliminary design
3. Detailed design
4. Final design
5. Construction phase

The objectives of VD are to offer a compressed schedule, alternative design and delivery methods and reduced costs. For the successful planning of a project, bridging the gap between design and construction is necessary. Engineering decisions should be based on visualization, communication, and collaboration.

Use of Innovative Materials

University of Maine Develops Lightweight Carbon Arches

University of Maine Composites Center researchers have the following developed an experimental bridge and estimated their bridge's carbon footprint to be about one-third smaller than that of a standard concrete bridge and one-fourth smaller than that of a standard steel bridge. The "bridge in a backpack" innovative technology uses carbon-fiber tubes that are inflated, shaped into arches, and infused with resin before being moved into place. The tubes are then filled with concrete, producing arches that are harder than steel yet resistant to corrosion. Finally, the arches are overlaid with a fiber-reinforced decking. The technology has a potential for future use due to its light weight and the portability of its components.

9.13 Essentials of Postflood Disaster Management

Floods cause casualties and victims requiring medical treatment, food, and clean water. After the floods, agricultural produce may be destroyed, houses washed away, and roads inundated. Relief measures are described here.

Although a flood warning may have been issued by the weather bureau, the performance of a bridge during floods cannot be predicted in advance. If a riverbank is breached and highways and bridges are inundated, transportation is suspended. Sometimes an emergency must be declared, and the entire population of the surrounding area has to be relocated. Roads in the floodplain must be shut down due to erosion of the soil. Providing food, medical treatment, and temporary shelter to hundreds of people on very short notice may be necessary. An administrative strategy should be in place other than relying on voluntary organizations such as FEMA and the Red Cross. Huge funding may be required and should be made available without delay for short-term relief and for long-term bridge and highway repairs.

Floodplain Caused by Hurricanes

The East Coast and Northeast of the United States have been subjected to hurricanes such as Floyd, Irene, and Sandy in the recent past. As a result, floodplains around riverbanks were formed and homes in the vicinity were flooded. During Hurricane Floyd in 1999, Route 46 in New Jersey was submerged in water and no traffic could pass. Peckman's River flooding caused the collapse of the Route 46 bridge. The author was responsible for the redesign of the bridge and used wider span and less skew to prevent overtopping from any future floods.

Earthquakes, on the other hand, may not directly cause floods due to significantly low water table and water elevation in rivers at the time of the earthquake. Dams and water-retaining structures are normally designed for seismic activity.

The author was a member of a UNESCO-sponsored team that visited Erzurum in Eastern Turkey in 1987 following a devastating earthquake. In addition, as a member of the National Academy of Sciences (NAS) team, the author visited north Pakistan after a magnitude 7.6 earthquake in October, 2005, in which some highways and bridges were destroyed. In the postearthquake scenario, although bridges were damaged there was no evidence of overtopping of riverbanks or flood conditions.

Based on the findings of the postflood condition assessment, potential repair and remediation recommendations will be developed. Fluctuating river elevations from intermittent rain will be taken into account when developing repair recommendations and reviewing the construction feasibility. Evaluation shall include, but not be limited to evaluation of constructability issues. No construction staging would be required for substructure rehabilitation work.

The Most Frequent Natural Disasters

In the United States and many other countries, the most frequent natural disasters are tornados, hurricanes, high winds, and cyclones which need to, and these should always be fully documented and addressed. Many technical issues accompany the social issues that result from these. Avoiding billions of dollars in recurring damages and loss of life is the objective of current study and research.

Effective mitigation by means such as advance precautions, preparation, disaster management, and provision of reasonable engineering solutions is key. Practical recommendations will be made in the light of advancements in science and technology.

9.14 Role of Environmental Agencies Such as FEMA and EPA to Combat Floods

Plan, prepare, and mitigate—these are the actions that should be taken before, during, and after an event, and they are unique to each situation. Identify the hazards that have occurred or could occur in your area, and plan for the unique actions for each type. Local emergency management offices can help identify the possible hazards in your area and outline the local plans and recommendations for each.

Prepare for Flooding Precautions

1. FHWA HEC-18 recommends river training measures such as river diversion, building auxiliary openings, and strengthening foundations against erosion and settlement by providing armoring and hydraulic countermeasures, as described earlier.
2. FEMA has available flood-elated publications and videos which are useful sources for both riverine floods and coastal floods. (FEMA P-259; Engineering principles and practices for retrofitting flood-prone residential structures. http://www.fema.gov/about-fema). The recent edition describes technical parameters such as these:
    - Flooding characteristics (depth, velocity, frequency, duration, and debris impact)
    - Site characteristics (location, erosion vulnerability, and soil type)
    - Flood forces and loads (flood depth, hydrostatic forces, saturated soil forces, hydrodynamic forces and impact loads, wind, and earthquake)
    - Site drainage
    - Land subsidence and bearing capacity
    - Safe design of levees and flood walls
    - Case studies of practical disasters

Because of the importance of understanding the nation's coastal flood risk, FEMA has initiated coastal flood risk studies for 100 percent of the populated coastline. Information is available about the efforts to scientifically study and understand coastal risks effort in partnership with other federal agencies, states, regional authorities, tribal nations, communities, and nonprofit organizations (https://www.fema.gov/coastal-flood-risk-study-process). Through the Risk MAP effort, FEMA is updating the nation's coastal flood insurance studies (FISs) and flood insurance rate maps (FIRMs), where appropriate, and publishing new FIRMs in densely populated areas that were not previously mapped.

FEMA National Flood Insurance Program: As part of its National Flood Insurance program, FEMA will continue to identify and publish special flood hazards and flood risk zones as authorized and required by Congress. Bridge engineers should utilize and apply experience gained from the past storms causing major floods such as Katrina, Floyd, and Sandy.

Water Supply and Wastewater Facilities For flood preparedness, the concerned highway agencies shall plan activities to help facilities prepare. Although the linked information is written by FEMA for hurricane preparedness, much of it will still apply to flooding preparedness activities.

Necessary Precautions to Be Taken by Affected People

During Flooding State and local response agencies are the primary responders for people who are concerned about or were affected by flooding.

Avoid Contact with Floodwater due to Potentially Elevated Levels of Contamination The contamination that occurs in flooding is associated with raw sewage and other hazardous or toxic substances that may be in the floodwater. EPA and the Department of Health and Human Services urge everyone in contact with floodwaters to follow these guidelines:

- Wash your hands before drinking and eating. Wash frequently, using soap—especially disinfecting soap.
- Limit direct contact with contaminated floodwater.
- Report cuts or open wounds as well as all symptoms of illness. Vaccinations may be necessary.
- Always call 911 if you are in immediate danger and need emergency help.

Drinking Water and Food Clean Water is required for daily drinking, cooking food, and for agriculture.

- Boil water or disinfect water
- Keep food safe during an emergency

Recovery after Flooding Generator exhaust is toxic. Always put generators outside well away from doors, windows, and vents. Never use a generator inside homes, garages, crawl spaces, sheds, or similar areas. Carbon monoxide (CO) is deadly, impossible to detect, can build up quickly, and can linger for hours. Never try to heat your home using a "combustion appliance" such as a gas stove, oven, barbeque grill, or dryer. Never operate any gas-burning heater or other appliance in a poorly vented or closed room, or where you are sleeping.

Limit Contact with Floodwater Floodwater may contain high levels of raw sewage or other hazardous substances. Early symptoms from exposure to contaminated floodwater may include upset stomach, intestinal problems, headache, and other flu-like discomfort. Anyone experiencing these and any other health problems should immediately seek medical attention.

- Do not attempt to drive through floodwater. It is the leading cause of flood-related injury and death.
- Take precautions when reentering a flooded home.
- Mold caused by flooding can cause serious health problems. The key to mold control is moisture control. After the flood, remove standing water and dry indoor areas. Remove and discard anything that has been wet for more than 24 to 48 hours.

Private Wells and Septic Systems

For home septic systems, do not use the sewage system until water in the soil absorption field is lower than the water level around the house. If you have a home-based or small business and your septic system has received chemicals, take extra precautions to prevent contact with water or inhaling the fumes. Proper cleanup depends on the kinds of chemicals in the wastewater.

For home or private drinking water wells, *do not turn on the pump*. There is danger of electrical shock and damage to your well or pump, if flooded. Do not drink or wash with well water. Water from a private well that has been flooded may be contaminated.

Disaster Debris

Communities that have had a large flood should plan to handle exceptionally large amounts of disaster debris from damaged or destroyed buildings, supplies, trees or other green waste, carcasses, or other materials. Burning large amounts of debris to reduce volume may not be an option.

Cleanup and Renovation

Natural disasters such as tornadoes, hurricanes, or floods often result in the need for emergency renovations to damaged homes and other structures. When common renovation activities such as sanding, cutting, and demolition occur in structures that contain *lead-based paint,* they can release lead-based paint hazards, including lead-contaminated dust.

Underground Storage Tanks

During a flood, underground storage tank (UST) systems may become displaced or damaged and release their contents into the environment, causing soil, surface water, and groundwater contamination.

Flooding in U.S. Coastal Areas

The seawater level rise and the tides have caused greater damage than riverine floods. Six of the top 10 most expensive natural disasters in our nation were caused by coastal storms. Historically, the cities, towns, and villages that settled around ports along the nation's coastline areas are extremely important to the nation's economy. Studies have been done on coastal cities such as Boston and New Orleans and the impacts on the transportation infrastructure. Recent climate changes and glacier melt are causing sea levels to rise. See Suarez et al. (2005). Both FEMA and the National Flood Insurance Program (NFIP) (Floodsmart.gov) have training workshops and publications. Other U.S. coastal flood and weather agencies include the National Oceanic and Atmospheric Administration (NOAA), the U.S. Army Corps of Engineers (USACE), and the U.S. Coast Guard. They all have useful publications on this important subject.

Although the coastal areas of the United States comprise only one-fifth of the land area of the contiguous 48 states, they account for more than one-half of the nation's population and housing supply. In 1990, more than 133 million Americans lived in the 673 counties along the Atlantic and Pacific oceans, the Gulf of Mexico, and the

Great Lakes. Coastal hazards such as storm surge, hurricane force winds, and flooding place this population at risk; the risk is greater if they are not aware of the natural hazards that surround them and steps taken to mitigate them.

FEMA is working on the following related issues for coastal areas:

- Producing additional data and hazard mitigation tools
- Developing strategic emergency preparedness and mitigation planning in coastal communities through the Risk MAP effort (see https://www.fema.gov/coastal-flood-risk-resources/)

9.15 Flood Disaster Management in India and Pakistan

India and Pakistan have many flood-prone rivers such as the Ganges, Indus, Kabul, and Brahmaputra which are subject to recurring floods from monsoons and snowmelt at the foot of the Himalayas. Another reason for fast floods is the sharp drop in longitudinal slopes (from the elevated foothills of Himalayas to the low-lying Arabian Sea or Bay of Bengal). Both these flood-affected countries can ill afford huge losses of lives, properties, crops and cattle every few years. They have set up national disaster management authorities (NDMAs) for the purpose of reducing these losses.

The Disaster Management Act in India

By enactment of the Disaster Management Act 2005 (DMA), the government of India ordered the creation of the National Disaster Management Authority (NDMA) as opposed to creating a separate ministry. The importance that the government gave to this newly created body can be judged by the fact that the NDMA is chaired by the Prime Minister himself, with a Vice Chairman of the status of Cabinet Minister and eight members of the status of Ministers of State. The salient features of the official policy is that DMA have the legal authority to respond and take action as demanded by the natural disasters and are backed by an institutional framework. The DMA has financial support through the creation of a Response Fund and a Mitigation Fund. The disaster management setup was structured in three levels: national, state, and district. State Disaster Management Authorities (SDMAs) were created for each state. The SDMAs were headed by the Chief Ministers. At the district level, District Disaster Management Authorities (DDMAs) were set up. These were headed by the District Collectors and cochaired by elected representatives of the local authorities. All these authorities were charged with the responsibility of formulating holistic and integrated plans for disaster relief. Damage to infrastructure or its restoration is not addressed in postflood conditions. The following simple and precautionary directives from the NDMA should be useful to citizens facing flood situations.

Precautions If a Flood Is Likely to Hit Your Area

- Listen to the radio or television for information.
- Be aware that flash flooding can occur. If there is any possibility of a flash flood, move immediately to higher ground. Do not wait for instructions to move.
- Be aware of streams, drainage channels, canyons, and other areas known to get flooded suddenly. Flash floods can occur in these areas with or without such typical warnings as rain clouds or heavy rain.

Remember these evacuation tips:

- Do not walk through moving water. Six inches of moving water can cause you to fall. If you have to walk in water, walk where the water is not moving. Use a stick to check the firmness of the ground in front of you.
- Do not drive into flooded areas. If floodwaters rise around your car, abandon the car and move to higher ground if you can do so safely. You and the vehicle can be quickly swept away.

Pakistan's Monsoon Contingency Plan of 2008

Figure 9.4 shows an investigation into seismic damage to dams. The dam here, Rawal Dam, was designed to resist earthquakes and remained undamaged.

In the National Disaster Management Authority (NDMA), Pakistan, the contingency process is emphasized as a preparedness measure for response to natural hazards. Following contingency planning, this exercise focuses on (1) planning for the upcoming monsoon season to identify and analyze related risks for their humanitarian impact and (2) defining the roles and responsibilities of diverse national stakeholders for preparedness and response. Key anticipated outcomes are the creation of awareness for building institutional capacities for response, creation of political will for earmarking required resources, building of integrated planning capacities, and most importantly, the definition of preparatory measures for monsoon hazards. The functions and duties are defined

FIGURE 9.4 Author's visit to inspect Rawal Dam in Pakistan with the U.S. AID Team for damage after a 7.6 magnitude earthquake in October 2005.

and set by the Constitution of Pakistan in Article 239I. The National Commission is charged with the following duties:

1. To act as the implementing, coordinating, and monitoring body for disaster management
2. To prepare the National Plan to be approved and to implement, coordinate, and monitor the implementation of the national policy
3. To provide necessary technical assistance to the provincial governments and the provincial authorities for preparing their disaster management plans in accordance with the guidelines laid down by the National Commission
4. To coordinate response in the event of any threatening situation or disaster

There have been a number of major floods in the last 10 years in Pakistani rivers. Figure 9.5 shows the failure of Miskin Bridge in northern Pakistan due to severe scour. The volume of flood discharge exceeded the design value and resulted in severe scour of the foundations.

FIGURE 9.5 Foundation scour at Miskin Bridge in northern Pakistan in 2010.

A joint paper on the method of computation of floods was presented by the author and professors at the National University of Science and Technology in Islamabad. The use of regression methods and a software similar to USGS-developed StreamStats was recommended, using the topography and floodplain of the site.

NOTE: *If any website link (given as reference in the book) changes, the source or agency needs to be contacted for latest change.*

9.16 Conclusions

The chapter addressed important issues related to bridge inspections, monitoring and flood management. Precautions need to be taken prior to the occurrence of floods. These include structural planning as discussed above. Similarly, adequate precautions need to be taken during and after floods. These include short- and long-term engineering measures as discussed. Recommendations by FEMA, EPA, and FHWA for flood disasters should be considered.

Inspection procedures are governed by NBI Standards. NBI appraisal ratings and sufficiency ratings help in establishing priorities and allocating funding. The national weather service can provide predictions and flood warnings to establish flood watch in addition to NBI Standards.

Over the years inspections have become specialized. Inspection reports evaluate the structural conditions of the bridge. This procedure helps to prioritize scour-critical bridges for emergency and interim repairs.

In the countermeasures matrix, HEC-23 has classified monitoring as an essential countermeasure. Rather than rely entirely on scour analysis, the quick and practical approach is visual inspection. Monitoring of scour-critical bridges should be promoted using modern techniques such as remote sensors. There are several types of monitoring countermeasures identified.

Underwater inspection reveals any cracks or damages to foundations and substructure. The case studies presented identify such details. Postflood case studies on sample bridges in the Northeast have shown the formation of deep scour holes adjacent to the footings. This may cause settlement of the bridge, and underpinning of foundation is required.

Summary

Inspections are preferably required both immediately prior to floods and immediately after the floods as postflood disaster management. The chapter addresses the following issues:

- Monitoring measures, vulnerable foundations, flood watch, and reconstruction methods are presented. In addition to performing scour studies and assessing damage to foundations, inspection reports are the duty of hydraulic and bridge engineers.
- Innovative techniques of monitoring including use of remote sensors, development of a self-powered monitor system, use of nanotechnology to reveal cracks and corrosion, and the use of virtual design and new materials to help in accelerated bridge construction have been developed by universities and other organizations.

- Postflood disaster management in the United States, India, and Pakistan and the role of environmental agencies are discussed.
- Vigilance requires being aware that flash flooding can occur without warning and knowing about the locations of local streams, drainage channels, canyons, and other areas reputed to flood suddenly. If there is any possibility of a flash flood, move immediately to higher ground and do not wait for instructions.
- Agencies such as FEMA, state agencies, and FHWA can implement emergency evacuation and effect short- and long-term repairs. The importance of combating floods should be acknowledged by the highest level of government. Case studies of a reservoir and a bridge from Pakistan are presented. Administrative policies to combat flood disasters from the recently formed NDMA of India and Pakistan are cited. The flood disaster management issues are being discussed in international conferences on hydrology and hydraulics. There is a need for further research and framing of a code of practice for postflood management, rehabilitation, and reconstruction.
- Flooding of coastal towns gives rise to problems of greater magnitude than riverine floods. FEMA publications address the major issues of the causes and the means to deal with them.

9.17 Bibliography

AASHTO, Guide Design Specifications for Bridge Temporary Works, 1st ed., 2008 Interim Revisions, Washington, D.C., 2009.

FEMA P-259, Engineering principles and practices for retrofitting flood-prone residential structures, January, 2012.

FHWA, HEC-11 "Design of Riprap Revetment," FHWA Publication no.: IP-89-016, Washington, D.C., 1989.

FHWA, HEC-23, Bridge Scour and Stream Instability, Countermeasures: Experience, Selection, and Design Guidance-3rd ed., vol. 1, Washington, D.C., 2009.

Khan, M. A., Irfan, M. and Ahmed, S., "Recent Developments in the Regression Methods for Computing Peak Flood Discharges," paper presented at Omics International Conference, Raleigh, NC, August 2013.

Lagasse, P. F., Clopper, P. E., and Arneson, L. A., "Partially Grouted Riprap as a Pier Scour Countermeasure," ASCE World Environmental and Water Resources Congress, Reston, VA, 2008.

Minnesota Department of Transportation, I-35 W Bridge Collapse, Mississippi River in Downtown Minneapolis, (www.dot.state.mn.us/i35wbridge/index.html), 2015.

National Disaster Management Authority, "Floods Do's and Don'ts," Government of India, New Delhi, February 2014.

Parliament of India (23 December 2005). Disaster Management Act, 2005.

Flooding Hazards in Coastal Areas

FEMA National Flood Insurance Program (NFIP), (Floodsmart.gov), Washington, D.C., January, 2015.

Link, L. E., "The anatomy of a disaster, an overview of Hurricane Katrina and New Orleans." *Coastal Engineering, International Journal for Coastal, Harbour and Offshore Engineers*, Elsevier, New York, 2010.

Nicholls, R. J., Wong, P. P., Burkett, V. R., Codignotto, J. O., Hay, J. E., McLean, R. F., Ragoonaden, S., and Woodroffe, C. D., "Coastal systems and low-lying areas. Climate Change 2007: Impacts, Adaptation and Vulnerability," Contribution of Working Group II to the *Fourth Assessment Report of the Intergovernmental Panel on Climate Change*, Cambridge University Press, Cambridge, UK, 2007.

Suarez, P., Anderson, W., Mahal, V., Lakshmanan, T. R., "Impacts of flooding and climate change on urban transportation: A systemwide performance assessment of the Boston Metro Area," *Transportation Research* Part D, Elsevier, New York, 2005.

U.S. Department of Commerce, National Oceanic and Atmospheric Administration, *Education Strategic Plan, 2009-2029,* Washington, D.C., 2009.

New Techniques

Agrawal, A., Khan, M. A., and Zhihua, Y. *Handbook of Scour Countermeasures*, New Jersey Department of Transportation, Trenton, 2007.

Automated Composite Column Wrapping—Process Overview Manual, XXSys Technologies, Inc., San Diego, CA, October 1997.

Coletti, D., Krueger, M., and Renton, R., "Design Innovation, a Triple Effect," *Roads and Bridges*, August 2007.

"Evaluation of Reinforcing Bars in Old Reinforced Concrete Structures," Concrete Reinforcing Steel Institute, Engineering Data Report no. 48, Schaumburg, IL, 2001.

Fielding, M., "High-Tech Instrumentation Helps Optimize Project Schedules and Manage Risks," *Journal of Bridge Engineering*, American Society of Civil Engineers, March/April 2007.

Jerome Lynch et al "Spatial Conductivity Mapping of Carbon Nanotube Composite Thin Films by Electrical Impedance Tomography for Sensing Applications," *Journal of Nanotechnology*, September, 2007.

Khan, M. A., Baig, S., and Zarriello, J., "Factors Affecting Stream Stability at Scour-Critical Bridges in New Jersey," First International Conference on Bridge Scour, Texas A&M University, College Station, 2002.

New Jersey DOT "Underwater Inspection and Evaluation of New Jersey Bridges Guidelines Manual," Trenton, NJ, 1994.

USDOT, FHWA, *Recording and Coding Guide for the Structure Inventory and Appraisal of the Nation's Bridges*, Report No. FHWA-PD-96-001, Washington, D.C., December 1995.

USDA Forest Service, Watershed Management and Restoration (http://www.fs.fed.us/research/water-management and usda.gov), Washington, D.C.

Glossary

Scour Terminology

AASHTO (LRFD) Requirements on Scour Design The extreme-event limit states relate to events with return periods in excess of the design life of the bridge. A flood event exceeding a 100-year flood is generally considered an extreme event for bridges. A 500-year flood is recommended as the check flood for scour. Other conditions relating to scour that the designer may determine significant for a specified watershed may be ice loads, debris logging, operations, etc. The AASHTO LRFD specifications require that scour at bridge foundations be designed for 100-year flood storm surge tide or for the overtopping flood of less frequent recurrence.

Aggradation Progressive buildup of longitudinal profile of channel bed due to sediment deposition.

Apparent ocean level rise The amount of long-term rise in sea levels. The amount of rise is a combination of actual sea level rise or fall and regional subsidence, uplift, or rebound.

Apron Protective material placed on a streambed to resist scour.

Armor (armoring) Surfacing of channel bed, banks, or embankment slope to resist erosion and scour using riprap, gabions, articulated concrete blocks, etc.

Backshore The portion of the beach between the foreshore and the dunes or the landward extent of the storm beach. The backshore is generally only acted upon by waves during storms.

Bank The sides of a channel between which the flow is normally confined.

Barycenter The center of mass of the earth–moon system. The center of rotation of a two-body system.

Base floodplain The floodplain associated with the flood with a 100-year recurrence interval.

Bay A body of water connected to the ocean by an inlet.

Bed The bottom of a channel bounded by banks.

Bed material Material found in and on the bed of a stream.

Bed slope The inclination of the channel bottom.

Bedrock The solid rock exposed at the surface of the earth or overlaid by soils and unconsolidated material.

Blanket Material covering all or a portion of a stream bank to prevent erosion.

Bridge opening The cross-sectional area beneath a bridge that is available for conveyance of water.

Causeway Rock or earth embankment carrying a roadway across water.

Celerity *See* Wave speed.

Cellular block Interconnected concrete blocks with regular cavities placed directly on a stream bank to resist erosion.

Channel The bed and banks that confine the surface flow of a stream.

Channelization Straightening or deepening of a natural channel artificially.

Check dam A low dam or weir across a channel used to control stage or degradation.

Clear-water scour Scour at a pier or abutment (or contraction scour) when there is no movement of the bed material upstream of the bridge crossing at the flow, causing bridge scour.

Coastal zone The line forming the boundary between the land and water, also commonly called the coastline or shoreline. The strip of land of indefinite width that extends inland to the first major change in terrain features is commonly referred to as the coast or coastal zone. The coastal zone may be several kilometers wide.

Concrete revetment Unreinforced or reinforced concrete slabs placed on the channel bed or banks for the protection from erosion.

Confluence The junction of two or more streams.

Constriction A natural or artificial control section, such as a bridge crossing, channel, or dam, with limited flow capacity.

Contraction scour The component of scour that results from a contraction of the flow area at the bridge, which causes an increase in velocity and shear stress on the bed at the bridge. The contraction can be caused by the bridge or from a natural narrowing of the stream channel. Contraction scour, in a natural channel or at a bridge crossing, involves the removal of material from the bed and banks across all or most of the channel width. Contraction scour occurs when the flow area of a stream at flood stage is reduced, by either a natural contraction or bridge. It also occurs when the overbank flow is forced back to the channel by roadway embankments at the approaches to the bridge. For continuity, a decrease in flow area results in an increase in average velocity and bed shear stress through the contraction. Hence, there is an increase in erosive forces in the contracted area, and more bed material is removed from the contracted waterway than is transported into the waterway. This increase in transport of bed material from the waterway lowers the natural bed elevation.

Countermeasure A measure to prevent, delay, or reduce the severity of hydraulic problems.

Crib A frame structure filled with earth or stone ballast, designed to reduce energy.

Critical shear stress Minimum amount of shear stress required to initiate soil particle motion.

Current meter An instrument used to measure flow velocity.

Datum A reference point for vertical (elevation) measurements. *See* Tidal datum and Fixed datum.

Debris Floating or submerged material, such as logs or vegetation transported by stream.

Degradation (bed) A general and progressive (long-term) lowering of the channel bed due to erosion, over a relatively long channel length.

Depth of scour Vertical distance that a streambed is lowered by scour below a reference elevation.

Design flow (design flood) The discharge that is selected as the basis for the design or evaluation of a hydraulic structure.

Diurnal tide Tide with an approximate tidal period of 25 h.

Drift Alternative term for vegetative debris.

Dyke (groin, spur, and jetty) A structure extending from a bank into a channel that is designed to reduce the stream velocity.

Ebb or ebb tide Flow of water from the bay or estuary to the ocean.

Eddy current A vortex-type motion of a fluid flowing contrary to the main current, such as the circular water movement that occurs when the main flow becomes separated from the bank.

Embayment An indentation in the shoreline.

Estuary Tidal reach at the mouth of a river.

External boundary conditions Upstream and downstream (ocean) constraining variables at the model limits. These include upstream flow hydrographs and downstream stage hydrographs.

Fabric mattress Grout-filled mattress used for stream bank protection.

Fascine A matrix of willow or other natural material woven in bundles and used as a filter.

Fetch The area over water where the wind is unobstructed with fairly uniform speed and direction.

Filter Layer of fabric (geotextile) or granular material (sand, gravel, or graded rock) placed between the bank revetment (or bed protection) and soil filter blanket. For example, a layer of graded sand and gravel is laid between fine-grained material and riprap to serve as a filter.

Fixed datum A reference level surface that has a constant elevation over a large geographical area.

Flood or flood tide Flow of water from the ocean to the bay or estuary.

Floodplain Nearly flat, alluvial lowland bordering a stream that is subject to frequent inundation by floods.

Flow control structure A structure either within or outside a channel that acts as a countermeasure by controlling the direction, depth, or velocity of flowing water.

Foreshore The portion of the beach between the backshore and the low-tide elevation. The foreshore is the active beach where wave uprush and backwash occurs over the range of the tide.

Froude number A dimensionless number that represents the ratio of inertial to gravitational forces in open-channel flow.

Gabion A basket or compartmented rectangular container made of wire mesh. When filled with cobbles or other rock of suitable size, the gabion becomes a flexible and permeable unit with which flow and erosion control structures can be built.

General scour A lowering of the streambed across the stream or waterway at the bridge. This lowering may be uniform across the bed or nonuniform. That is, the depth of scour may be deeper in some parts of the cross section. General scour may result from contraction of the flow or other general scour conditions such as flow around a bend.

Grout A fluid mixture of cement, sand, and water used to fill joints and voids.

Guide bank A dyke extending upstream from the approach embankment at either side or both sides of the bridge opening to direct the flow through the opening.

Hard point A stream bank protection structure whereby "soft" or erodible materials are removed from a bank and replaced by stone or compacted clay. Some hard points protrude a short distance into the channel to direct erosive currents away from the bank.

Hurricane An intense type of tropical cyclone with well-defined circulation and maximum sustained winds of 74 mi/h or higher.

Hydrograph The graph of the stage or discharge versus time.

Hydrology Science concerned with occurrence, distribution, and circulation of water on earth.

Ineffective flow areas Portions of cross sections that provide storage but do not convey flow.

Internal boundary conditions Flow controls at interior portions of a numerical model. These include structures (culverts, weirs, and bridges) and lateral inflow hydrographs.

Intracoastal waterway A series of natural and dredged channels within the U.S. coastline intended for small craft.

Jetty An obstruction built of piles, rock, or other material extending from a bank into a stream, so placed as to induce bank building or to protect against erosion.

Junction confluence A location where channel reaches (main channel and tributaries) connect.

Lateral erosion Erosion in which the removal of material is extended horizontally as contrasted with degradation and scour in a vertical direction.

Launching Release of undercut material (stone riprap, rubble, slag, etc.) downslope or into a scoured area.

Levee An embankment, generally landward of the top bank, that confines flow during high-water periods, thus preventing overflow into lowlands.

Littoral transport or drift Transport of beach material along a shoreline by wave action. Also called longshore sediment transport.

Littoral zone The region that extends seaward from the coastline to just beyond the beginning of the breaking waves. Within this zone, waves and currents transport sediments. A current is generated by the incident waves within the littoral zone.

Local scour Removal of material from around piers, abutments, spurs, and embankments caused by an acceleration of flow and by vortices induced by obstructions to flow.

Mattress A blanket or revetment of materials interwoven or otherwise lashed together and placed to cover an area subject to scour.

Mean high water (MHW) The average of all high tides over a tidal epoch.

Mean higher high water (MHHW) The average of all daily higher high tides (for semidiurnal tides) over a tidal epoch.

Mean low water (MLW) The average of all low tides over a tidal epoch.

Mean lower low water (MLLW) The average of all daily lower low tides (for semidiurnal tides) over a tidal epoch.

Mean sea level The average of hourly tide heights over a tidal epoch.

Mean tide level The average of all high and low tides.

Meander or full meander A meander in a river consists of two consecutive loops, one flowing clockwise and the other counterclockwise.

Migration Change in position of a channel by lateral erosion of one bank.

Mixed tide Semidiurnal tides that exhibit significant differences between the two high and two low tides.

National Tidal Datum Epoch (NTDE) The specific 19-year period adopted by the National Ocean Service (NOS) as the official time segment over which tide observations are taken and reduced to obtain mean values for tidal data.

Neap tide Smaller-than-normal tides that occur approximately twice per month at the first- and third-quarter moon phase when the sun and moon are at right angles to the earth and the tidal forces counteract each other.

Network An assembly of channel reaches, junctions, and storage areas that make up the numerical model geometric description of the waterway.

Numerical stability The ability of the program to converge to a solution at a time step.

Ocean level rise *See* Apparent ocean level rise.

Passage A tidal waterway between two islands or between the mainland and an island.

Pile An elongated member usually made of timber, concrete, or steel that serves as a structural component of a river training structure.

Reaches Segments of the waterway between tributary confluences and between confluences and the upper and lower model limits.

Reinforced earth A retaining structure consisting of vertical panels and bulkhead, attached to reinforcing elements embedded in compacted backfill for supporting a stream bank.

Relief bridge An opening in an embankment on a floodplain to permit passage of bank overflow.

Retard (retarder) A permeable or impermeable linear structure in a channel structure parallel to the bank, intended to reduce flow velocity, induce deposition, or deflect flow from the bank.

Revetment Rigid or flexible armor placed to inhibit scour and lateral erosion.

Riprap Layer of rock or broken concrete dumped or placed to protect footing of bridge piers and abutments, or embankment, from erosion.

River training Measures taken to direct or to lead a stream flow into a prescribed channel.

Rock-and-wire mattress A flat wire cage or basket filled with stone or other suitable material and placed as protection against erosion.

Rubble Rough, irregular fragments of materials of random size used to retard erosion. The fragments may consist of broken concrete slabs, masonry, or other suitable refuse.

Run-up wave Height to which water rises above still-water elevation when waves meet a beach, wall, etc.

Runoff Part of precipitation that appears in surface streams of either perennial or intermittent form.

Sack revetment Sacks (e.g., burlap, paper, or nylon) filled with mortar, concrete, sand, stone, or other available material used as protection against erosion.

Saffir-Simpson hurricane scale A scale of 1 through 5 categories based on the hurricane intensity.

Scour Localized erosion of streambed or stream bank material due to flowing water.

Seepage The slow movement of water through small cracks and pores of the bank material.

Semidiurnal tide Tide with an approximate tidal period of 12.5 h.

Setup wave Height to which water rises above still-water elevation as a result of storm wind effects.

Significant wave height The average height of the one-third largest waves for a specific set of wind, fetch, and water depth conditions.

Sill A structure built underwater with the aim of raising the depth and slope of the streambed.

Slope protection Any measure such as riprap, paving, vegetation, revetment, brush, or other material intended to protect a slope from erosion, slipping, or caving.

Soil-cement A designed mixture of soil and portland cement compacted at a proper water content to form a blanket or structure that can resist erosion.

Spill-through abutment A bridge abutment having a fill slope on the stream side.

Spring tides Larger-than-normal tides that occur approximately twice per month at new and full moon when the sun and moon are aligned and the tidal forces are reinforced. *See* Syzygy.

Spur A permeable or impermeable linear structure that projects into a channel from the bank to alter flow direction, induce deposition, or reduce flow velocity along the bank.

Still-water elevation Flood height to which water rises as a result of barometric pressure and wind occurring during a storm event without including wave heights. The water level without the effects of waves.

Stone riprap Natural cobbles, boulders, or rock dumped or placed as protection against erosion.

Storage areas Network features where the model performs simple storage routing, assuming a level water surface.

Storm surge Coastal flooding phenomenon resulting from wind and barometric changes. The storm surge is measured by subtracting the astronomical tide elevation from the total flood elevation (hurricane surge).

Storm tide Coastal flooding resulting from combination of storm surge and astronomical tide (often referred to as storm surge).

Stream A body of water ranging in size from a large to a small river flowing in a channel.

Stream bank protection Any technique used to prevent erosion or failure of a stream bank.

Stream instability Changes in river cross section due to horizontal and vertical erosion of riverbanks from hydraulic and geomorphic factors.

Stream stability No appreciable changes in cross section of the river from year to year due to conditions of flow.

Sustained winds Wind speeds that persist for duration of 1 min.

Syzygy The time when the moon and sun are aligned such that the tide forces of the bodies act to reinforce each other. This occurs at new and full moon.

Tetrapod Bank protection component of precast concrete consisting of four legs joined at a central joint, with each leg making an angle of 109.5° with the other three.

Thalweg The line extending down a channel that follows the lowest elevation of the bed.

Tidal amplitude Generally, one-half of tidal range.

Tidal current Flow in a tidal waterway that is caused by the rise and fall of tides. *See* Ebb tide and Flood tide.

Tidal cycle One complete rise and fall of the tide.

Tidal datum A local reference elevation relative to a tidal level, such as mean lower low water.

Tidal day Time of rotation of the earth with respect to the moon, assumed to equal approximately 24.84 solar hours.

Tidal epoch A cycle of approximately 18.6 years of the principal tide-producing forces.

Tidal inlet A channel connecting a bay or estuary to the ocean.

Tidal passage A tidal channel connected with the ocean at both ends.

Tidal period Duration of one complete tidal cycle. When the tidal period equals the tidal day (24.84 h), the tide exhibits diurnal behavior. Should two complete tidal periods occur during the tidal day, the tide exhibits semidiurnal behavior.

Tidal prism Volume of water contained in a tidal bay, inlet, or estuary between low and high tide levels.

Tidal range Vertical distance between specified low and high tide levels.

Tidal waterways A generic term that includes tidal inlets, estuaries, bridge crossings to islands or between islands, inlets to bays, crossings between bays, tidally affected streams, etc.

Tides, astronomical Periodic diurnal or semidiurnal variations in sea level that result from centrifugal and gravitational forces between the earth, moon, sun, and other astronomical bodies acting on the rotating earth.

Timber or brush mattress A revetment made of brush, poles, logs, or lumber interwoven or otherwise lashed together.

Toe protection Loose stones laid or dumped at the toe of an embankment, groin; masonry or concrete wall built at the junction of the bank and the channel bed or at extremities of hydraulic structures to counteract erosion.

Total scour The sum of long-term degradation, general (contraction) scour, and local scour.

Tropical depression An organized system of clouds and thunderstorms with a defined circulation and maximum sustained winds of 38 mi/h.

Tropical storm An organized system of strong thunderstorms with a defined circulation and maximum sustained winds of 39 to 73 mi/h.

Tsunami Long-period ocean wave resulting from earthquake, other seismic disturbances, or submarine landslides.

Turbulence Motion of fluids in which local velocities and pressures fluctuate irregularly in a random manner as opposed to laminar flow, where particles of fluid move in distinct and separate lines.

Ultimate scour Maximum depth of scour attained for a given flow condition. Development of ultimate scour may require multiple flow events. In cemented or cohesive soils, ultimate scour may be achieved over a long period.

Uniform flow Flow of constant cross section and velocity through a channel at a given time.

Vortex Turbulent eddy in the flow caused by an obstruction such as a bridge pier or abutment.

Waterway opening Width or area of bridge opening at a specific elevation, measured normal to principal direction of flow.

Wave height The vertical difference between successive wave crests and troughs.

Wavelength The horizontal difference between two successive wave crests or two successive wave troughs.

Wave period Time interval between arrivals of successive wave crests at a point.

Wave speed or celerity The travel speed of a wave equal to the wavelength divided by the wave period.

Weep hole A hole in an impermeable wall or revetment to relieve the neutral stress or pore pressure in soil.

Wire mesh Wire woven to form a mesh.

Bridge Terminology[1]

Abutment Abutment has a dual purpose, with earth retaining and supporting all or part of the bridge. The two abutments define the beginning and end of the bridge and serve as anchors.

[1]The following bridge terminology is widely used in bridge engineering literature. For more information about these terms, please refer to the author's textbook *Bridge and Highway Structure Rehabilitation and Repair* (McGraw-Hill, 2010).

Approach The approach slab interfaces with the abutment to provide continuity to the roadway. The roadway is capped by a slab resting on unyielding soil to prevent differential settlement after a heavy downpour.

Back wall The back wall or the stem separates the approach slab from the bridge. Back walls interface with wing walls.

Barrier A barrier is a railing or reinforced concrete parapet that serves as a protective dwarf wall placed over the deck to guide vehicles and prevent collisions.

Beam A beam or girder is a very important primary member that supports the deck. It spans the river width or the interchange. It is supported by the pier and abutments and is held in position by bearings resting over the substructure. It is made of timber, aluminum, steel, or reinforced or prestressed concrete.

Bearing The bearing is a mechanical device that permits the expansion, contraction, and rotation of beams. Bearings also transfer heavy reactions over the superstructure over a wider area.

Bracing Bracing is a secondary member placed between beams to resist wind load in the horizontal direction.

Column A column comprises vertical load-bearing members for spill-through abutment and pier bents.

Deck The deck is the most important part of the superstructure. It supports the moving loads and can be built in reinforced concrete, timber, or open-grid steel. It is usually made composite with the supporting beams.

Deck joint This joint in the transverse or longitudinal direction helps to release thermal stress in the deck slab. These joints are placed before the abutment back wall. The deck joint consists of steel angles with vertical leg anchored into the deck concrete to shield the corners of concrete edges from any damage resulting from the impact of wheels.

Diaphragm The diaphragm is a secondary member that interconnects the beam in the transverse direction and enables lateral distribution of vertical and horizontal loads. Diaphragms are usually spaced at 20 ft or less.

Embankment An embankment is a body of earth that transitions from deck elevation to the lower ground and is given a gentle slope, using fill material for that purpose.

Flood A flood is an extreme load case for soil erosion, other than an earthquake. A probability of simultaneous flood and earthquake is not assumed.

Footing Footing serves as the fixed feet of the bridge. It is the most important part of the bridge components. Shallow footings transfer the loads from the bridge over the wider area of compacted subsoil. Deep footings are bearing or frictional piles or caissons driven to the required depth deep into soils.

Haunch A haunch is a concrete transition member placed over the flange of the beam and under the deck slab. Similar to a pedestal, its height is adjusted to achieve the required deck elevation. Haunches over 4 in deep require reinforcement.

Integral abutment This abutment wall rests on piles in which the top stub is made integral with the deck slab and the approach slab. Bearings or deck joints are not required. Special boundaries of beam serve as a longitudinal frame. The structural performance of the bridge under seismic conditions is also improved.

Maintenance Maintenance comprises routine or regular activities that are intended to preserve and maintain a structure's original serviceability and functionality.

Parapet A parapet is a barrier placed at the edges of the deck slab. It also supports lighting poles or sign panels.

Pedestal The pedestal act as short columns and support the bearings. They maintain the required bridge seat elevations, and their adjustable heights in concrete are varied to suit.

Pier A pier is a wide column that supports the intermediate parts of the longer-span bridge and shares the loads with abutments. It may be placed in the middle of a river.

Pile A pile acts as a stiff nail driven into a hard soil medium. Acting in a group, they provide stability to the bridge.

Rehabilitation Rehabilitation is the comprehensive repair of a bridge structure's most deteriorated elements which is intended to restore and significantly extend the bridge's original serviceability and functionality.

Repair Repair activities, usually isolated to a portion of one element of a structure, are necessary to restore serviceability or functionality due to distress from vehicle impact damage, observed scour, severe localized deterioration, and other stressors.

Scupper A scupper is used for rapid deck drainage. It is a vertical pipe located at the deck inlet at the face of the parapet and connected under the bridge substructure into a manhole of a drainage system.

Sheeting Sheeting is timber or steel planks driven in the soil to allow construction of a footing. It serves as a temporary retaining wall and is usually left in place.

Sidewalk The sidewalk is an elevated deck for use by pedestrians, and it is placed at the side of a bridge deck. It is usually protected by railing from the traffic.

Sign structure The sign structure supports the sign panels that provide directions to the drivers about the exact location. Overhead, cantilever, and butterfly types are used. It is usually placed well before the exit ramp or intersection.

Underdrain The underdrain is a water drainage pipe usually located behind the abutment. It prevents the water table from rising to the deck elevation.

Wearing surface The wearing surface protects the deck from abrasion and cracking from friction and impact of fast-moving trucks and vehicles.

Wing wall A wing wall is located on the sides of the bridge width. The two splayed or return walls provide stability to the abutment by confining the earth behind the abutment and preventing it from moving.

APPENDIX A

Sample Scour Calculations (HEC-18)

A.1 Determine Scour Variables

100-year discharge at scour location:	72,630 cfs
Is 500-year discharge available?	Yes
Enter 500-year discharge:	123,394 cfs
Enter storm frequency to be calculated:	100-year

A.2 Aggradation / Degradation in Stream Bed

Typically aggradation and degradation are assumed to offset each other: **$Scour_{Agg/Deg}$ = 0.0 ft**

A.3 Evaluate Scour Analysis Method (Live Bed or Clear Water)

Live-bed scour occurs when the flow of water upstream of the bridge is transporting bed material. Conversely, **clear-water** scour occurs when there is no bed material being transported by the upstream flow. Use the following calculation to determine the type of scour:

River station # of upstream cross-section:	6.3	(from HEC-RAS)
River station # at upstream face of bridge:	4.5U	(from HEC-RAS)
Distance between these sections:	235 ft	(should be nearly equal)
Span of bridge:	190 ft	

Median diameter of bed material, D_{50} =	0.1667 ft	(from site observations or geotech)
Bed material D_{95} =	0.1667 ft	(from site observations or geotech)

Evaluated at RS 6.3	100-year	
Average flow depth upstream of the bridge, y_1 =	18.64 ft	(taken from HEC-RAS output)
Critical channel velocity, V_c =	10.0 fps	(see note below)
Average flow velocity upstream of the bridge, V =	10.6 fps	(taken from HEC-RAS output)

If $V > V_c$, bed material will be transported making a Live Bed condition. If $V_c > V$, it will be clear water condition. ⇨ **Live water**

Note: V_c is the velocity needed to transport D_{50} sized particles. From HEC-18 Equation 5.1. $V_c = 11.17\, y^{0.167} D_{50}^{0.333}$

A.4 Live Water Contraction Scour

Input values at Upstream Section RS = 6.3:	100-year	
Flow through channel upstream of the bridge, Q_1 =	15,061.4 cfs	(taken from HEC-RAS output)
Channel width upstream of the bridge, W_1 =	76.0 ft	(taken from HEC-RAS output)
Input values at Bridge Face (u/s side) RS = 4.5U:		
Flow through bridge opening, Q_2 =	26,892.28 cfs	(taken from HEC-RAS output)
Bridge opening width (perpendicular to abutments) =	141 ft	(taken from HEC-RAS output)
Summation of pier widths =	3.5 ft	
Abutment wall skew angle, θ =	0 degrees	
Effective width of bridge opening less pier width W_2 =	137.5 ft	
Calculations for Live Water Scour:		
Fall velocity of D_{50} particle, ω =	1.47 fps	(from HEC-18 Figure 5.8)
Slope of EGL in main channel upstream, S_1 =	0.0015 ft/ft	(taken from HEC-RAS output)
Shear velocity upstream, V^* =	0.95 fps	(see note below)
V^* / ω =	0.65	
Mode of transport exponent, k_1 =	0.64	(from HEC-18 page 5.11)

Note: V^* is the shear velocity in the upstream channel. From HEC-18 Page 5.11. $V^* = (g * y_1 * S_1)^{0.5}$

	100-year	
Average water depth under bridge, y_2 =	21.0 ft	(see note below)
Existing flow depth under bridge before scour, y_0 =	19.46 ft	(taken from HEC-RAS output)
Contraction scour depth, $y_{contraction} = y_2 - y_0$ =	**1.5 ft**	

Note: y_2 is the average depth of flow under the bridge after scour.

Note: Live-water calculation from HEC-18 Equation 5.2. $y_2 = y_1 * (Q_2/Q_1)^{0.857} * (W_1/W_2)^{k1}$

Note: Clear-water calculation from HEC-18 Equation 5.4. $y_2 = [\ (0.0077 * Q_2^2) / (1.16 * D_{50}^{0.666} * W_2^2)\ ]^{0.4286}$

Verification of the live-water and clear-water equations is required by comparing with the equations presented in the latest version of HEC-18.

Comparison with an alternative method such as ABSCOUR or HEC-RAS software is desirable since high scour depths may result and it may not always be practical to provide armoring countermeasures.

A.5 Local Scour: Left Bridge Abutment

Choose Abutment Scour Equation:	100-year	
Length of embankment projected normal to flow, L =	3514.47 ft	(taken from HEC-RAS output)
Width of flow on overbank ineffective or flowing < 2 fps:	0 ft	(taken from HEC-RAS vel. distribution)
Top width of active flow obstructed by embankment, L' =	3514.47 ft	
L' / y_a =	729.1	
If $L'/y_a < 25$, use Froehlich's equation (typical), otherwise, if $L'/y_a > 25$, use use HIRE equation. ⇨	**HIRE equation**	

Note: Fr is the Froude number upstream of the abutment. From HEC-18 page 7.8. $Fr = V_e / (32.2 * y_a)^{0.5}$

Calculate Froude Number:	100-year	
Average depth of flow on left overbank, y_a =	4.82 ft	(from HEC-RAS output or A_e / L)
Flow obstructed by the embankment, Q_e =	55,698.09 cfs	(taken from HEC-RAS output)
Flow area obstructed by the embankment, A_e =	16,956.41 sf	(taken from HEC-RAS output)
Velocity of flow obstructed by the embankment, V_e =	3.28 fps	(Q_e / A_e)
Froude number of flow upstream of the bridge, Fr =	0.26	$Fr = V_e / (32.2 * y_a)^{0.5}$

Calculate Coefficients:		
Vertical abutment walls or spill-through?	*V*	
Wing-walls present?	Yes	
Abutment shape coefficient, K_1 =	0.82	(from HEC-18 Table 7.1)
Angle between embankment & flow, θ =	100 degrees	(from HEC-18 Figure 7.5)
Abutment orientation coefficient, K_2 =	1.01	$(\theta / 90)^{0.13}$

Local scour depth (left abutment), $y_{local(left\ abut.)}$ = 18.8 ft

Note: Froehlich's eq for local scour at abutment. From HEC-18 Eq 7.1. $y_s = y_a * \{ [2.27 * K_1 * K_2 * (L'/y_a)^{0.43} * Fr^{0.61}] + 1 \}$

Note: HIRE equation for local scour at abutment. From HEC-18 Eq 7.2. $y_s = y_a * [4 * Fr^{0.33} * (K_1 / 0.55) * K_2]$

Verification of the complex abutment local scour equations is required by comparing with the equations presented in the latest version of HEC-18.

Comparison with an alternative method such as ABSCOUR or HEC-RAS software is desirable since high scour depths may result and it may not always be practical to provide armoring countermeasures.

A.6 Local Scour: Right Bridge Abutment

Choose Abutment Scour Equation:	100-year	
Length of embankment projected normal to flow, L =	33.25 ft	(taken from HEC-RAS output)
Width of flow on overbank ineffective or flowing < 2 fps:	0 ft	(taken from HEC-RAS vel. distribution)
Top width of active flow obstructed by embankment, L' =	33.25 ft	

L' / y_a =	3.0	
If $L'/y_a < 25$, use Froehlich's equation (typical), otherwise, if $L'/y_a > 25$, use use HIRE equation. ⇨	**Froehlich's eq.**	

Note: Fr is the Froude number upstream of the abutment. From HEC-18 page 7.8. $Fr = V_e / (32.2 * y_a)^{0.5}$

Calculate Froude Number:	100-year	
Average depth of flow on right overbank, y_a =	10.92 ft	(from HEC-RAS output or A_e / L)
Flow obstructed by the embankment, Q_e =	1870.54 cfs	(taken from HEC-RAS output)
Flow area obstructed by the embankment, A_e =	363.14 sf	(taken from HEC-RAS output)
Velocity of flow obstructed by the embankment, V_e =	5.15 fps	(Q_e / A_e)
Froude number of flow upstream of the bridge, Fr =	0.27	$Fr = V_e / (32.2 * y_a)^{0.5}$

Calculate Coefficients:		
Vertical abutment walls or spill-through?	V	
Wing-walls present?	Yes	
Abutment shape coefficient, K_1 =	0.82	(from HEC-18 Table 7.1)
Angle between embankment & flow, θ =	80 degrees	(from HEC-18 Figure 7.5)
Abutment orientation coefficient, K_2 =	0.98	$(\theta / 90)^{0.13}$

Local scour depth (right abutment), $y_{local(right\ abut.)}$ = 25.6 ft

Note: Froehlich's eq for local scour at abutment. From HEC-18 Eq 7.1. $y_s = y_a * \{ [2.27 * K_1 * K_2 * (L'/y_a)^{0.43} * Fr^{0.61}] + 1 \}$

Note: HIRE equation for local scour at abutment. From HEC-18 Eq 7.2. $y_s = y_a * [4 * Fr^{0.33} * (K_1 / 0.55) * K_2]$

Verification of the local scour equations is required by comparing with the equations presented in the latest version of HEC-18. Comparison with an alternative method such as ABSCOUR given in APPENDIX B or HEC-RAS software is desirable since high scour depths may result and it may not always be practical to provide armoring countermeasures.

A.7 Local Scour: Piers

General rule of thumb for scour at piers (for round nose piers aligned with flow):
Scour depth should be less than or equal to 2.4 times the pier width if Froude number is less than 0.8.
Scour depth should be less than or equal to 3.0 times the pier width if Froude number is greater than 0.8.

Does bridge have piers?		
Length of pier, L =	36 ft	
Length of pier subject to angle of attack of flow, L_{eff} =	36 ft	
Pier width, a =	3.5 ft	
Vertical abutment walls or spill-through?	Round nose	
Pier nose shape coefficient, K_1 =	1.0	(from HEC-18 Fig 6.3 and Table 6.1)
L_{eff} / a =	10.3	
Angle between pier & flow, θ =	10 degrees	
Angle of attack of flow coefficient, K_2 =	1.94	HEC-18 Eq 6.4: $(\cos\theta + L/a \sin\theta)^{0.65}$
Bed condition coefficient, K_3 =	1.1	(from HEC-18 Table 6.3)
Median diameter of bed material, D_{50} =	0.167 ft	(if D_{50}< 0.0066, K_4 = 1.0)
bed material D_{95} =	0.167 ft	(if D_{95}< 0.066, K_4 = 1.0)
Calculate K_4:	100-year	
Critical velocity (to move) D_{50} particle, V_{cD50} =	10.08 fps	(from HEC-18 Equation 6.8)
Velocity to initiate scour on D_{50} particle, V_{icD50} =	5.53 fps	(from HEC-18 Equation 6.7)
Critical velocity (to move) D_{95} particle, V_{cD95} =	10.08 fps	(from HEC-18 Equation 6.8)
Velocity to initiate scour on D_{95} particle, V_{icD95} =	5.53 fps	(from HEC-18 Equation 6.7)
V_R =	0.9	(from HEC-18 Equation 6.6)
Bed material armoring coefficient, K_4 =	0.4	(see note below)

Note: If K_4 is not 1.0, $K_4 = 0.4\,(V_R)^{0.15}$. From HEC-18 Equation 6.5. Also refer to HEC-18 Equations 6.6 through 6.8.

Calculate Froude Number:	100-year	
Average depth of flow directly upstream of the pier, y_1 =	19.46 ft	(taken from HEC-RAS output)
Velocity of flow directly upstream of the pier, V_1 =	9.8 fps	(taken from HEC-RAS output)
Froude number of flow directly upstream of the pier, Fr =	0.39	$Fr = V_1 / (32.2 * y_1)^{0.5}$

Note: Fr is the Froude number directly upstream of the pier. From HEC-18 Page 7.8. $Fr = V_e / (32.2 * y_a)^{0.5}$

Local scour depth at pier, y_{pier} = 7.3 ft

Note: From HEC-18 Equation 6.1. $y_{pier} = y_1 * 2 * K_1 * K_2 * K_3 * K_4 * (a/y_1)^{0.65} * Fr^{0.43}$ MAXIMUM = x ft (general note.)

Verification of the pier scour equation is required by comparing with the equation presented in the latest version of HEC-18.

Comparison with an alternative method such as ABSCOUR or HEC-RAS software is desirable since high scour depths may result and it may not always be practical to provide armoring countermeasures.

COMBINED SCOUR DEPTHS for the 100-year storm

Upstream XS used for contraction scour:	RS 6.3
Bridge XS used for contraction scour:	RS 4.5U

Total Scour Left Abutment:

	100-year
Contraction scour depth, $y_{contraction}$ =	1.5 ft.
Local scour depth (left abutment), $y_{local(left\ abut.)}$ =	18.8 ft.
	20.3 ft.

Total Scour Channel incl. Piers:

Contraction scour depth, $y_{contraction}$ =	1.5 ft.
Local scour depth at pier, y_{pier} =	7.3 ft.
Aggradation / degradation scour, $y_{Agg/Deg}$ =	0.0 ft.
	8.8 ft.

Total Scour Right Abutment:

Contraction scour depth, $y_{contraction}$ =	1.5 ft.
Local scour depth (right abutment), $y_{local(right\ abut.)}$ =	25.6 ft.
	27.1 ft.

APPENDIX B

Sample ABSCOUR Solved Example

```
----------------------------------------
 Maryland State Highway Administration
        Pier Local Scour Program
 Version 8 Build 1.03, August 16, 2006
----------------------------------------

Time stamp: 9/23/2009 10:31:10 AM

Input data:

Project information:

Project name:
Project number
Description: with Contracted Bed Elev = -29.0 and WSEL = -0.55
                 Elevations from Existing 11th Street Piers 2 and 3

Pier scour condition: Option 3 Pier with pile cap and pile group exposed
Units used: English Units
Flow depth upstream face of the pier (ft/m): 28.45 ft
Flow velocity upstream face of the pier right at nose (fps/mps): 4.44 fps
Width of pier stem (ft/m): 6.75 ft
Length of pier stem [ft/m): 66.75 ft
Flow attack angle (degree): 0 degree
Median grain size D50 (ft/m): 0.00025 ft
95% finer grain size D95 (ft/m): 0.00026 ft
Pier stem nose shape correction factor (K1): 0.9
K2 calculated by the program
Stream bed condition correction factor (K3): 1.1
K4 calculated by the program
Footing/pile cap width (ft/m): 22 ft
Footing/pile cap length (ft/m): 78.75 ft
Pile cap thickness (ft/m): 10 ft
```

Footing/pile cap shape factor K1f: 1.1
Distance from stream bed to top of footing/pile cap (ft/m): 12 ft
Distance between front edge of footing/pile cap and pier stem (ft/m): 7 ft
Number of pile columns (in pier width direction): 5
Number of pile rows (in pier length direction) 24
Pile center to center spacing in the pier width direction (ft/m): 3.75 ft
Pile center to center spacing in pier length direction (ft/m): 3 ft
Pile size in pier width direction (ft/m): 1.13 ft
Pile size in pier length direction (ft/m): 1.22 ft
Pile shape factor K1p: 1.1

Output Results:

Scour component for the pier stem in the flow:

Froude number Fr for the pier stem: 0.1467
Correction factor for flow angle of attack K2 for pier stem: 1
Correction factor for armoring K4: 1
Coefficient of shielding effect Kh pier: 0.0023
Scour component for the pier stem in the flow: 0.023 ft

Scour component for the exposed footing/pile cap:

Adjusted depth of flow upstream of pier y2: 28.461 ft
Adjusted velocity for the flow approaching the pier v2: 4.438 fps
Equivalent pier width of the footing/pile cap: 4.407 ft
Correction factor for flow angle of attack K2 for footing/pile cap: 1
Correction factor for armoring K4f: 1
Froude number Fr2 for the footing/pile cap: 0.1466
Velocity of critical flow Vc: 30.273 fps
Correction factor for wide pier in shallow water Kw: 1
Scour component for the exposed footing/pile cap: 8.973 ft

Scour component for the exposed pile group:

Adjusted depth of flow upstream of pier y3: 32.948 ft
Adjusted velocity for the flow approaching the pier v3: 3.834 fps
Sum of overlapping projected width of piles: 5.65 ft
Coefficient of pile spacing Ksp: 0.4527
Coefficient of number of aligned pile rows Km: 1.5532
Effective width of the pile group: 3.973 ft
Correction factor for armoring K4: 1
Height of pile group aboved lowered stream bed h3: 6.498 ft
Pile group height factor Kh (pg): 0.6199
Froude number Fr3 for pile group: 0.1177
Scour component for the exposed pile group: 4.98 ft
Total pier local scour depth: 13.975 ft

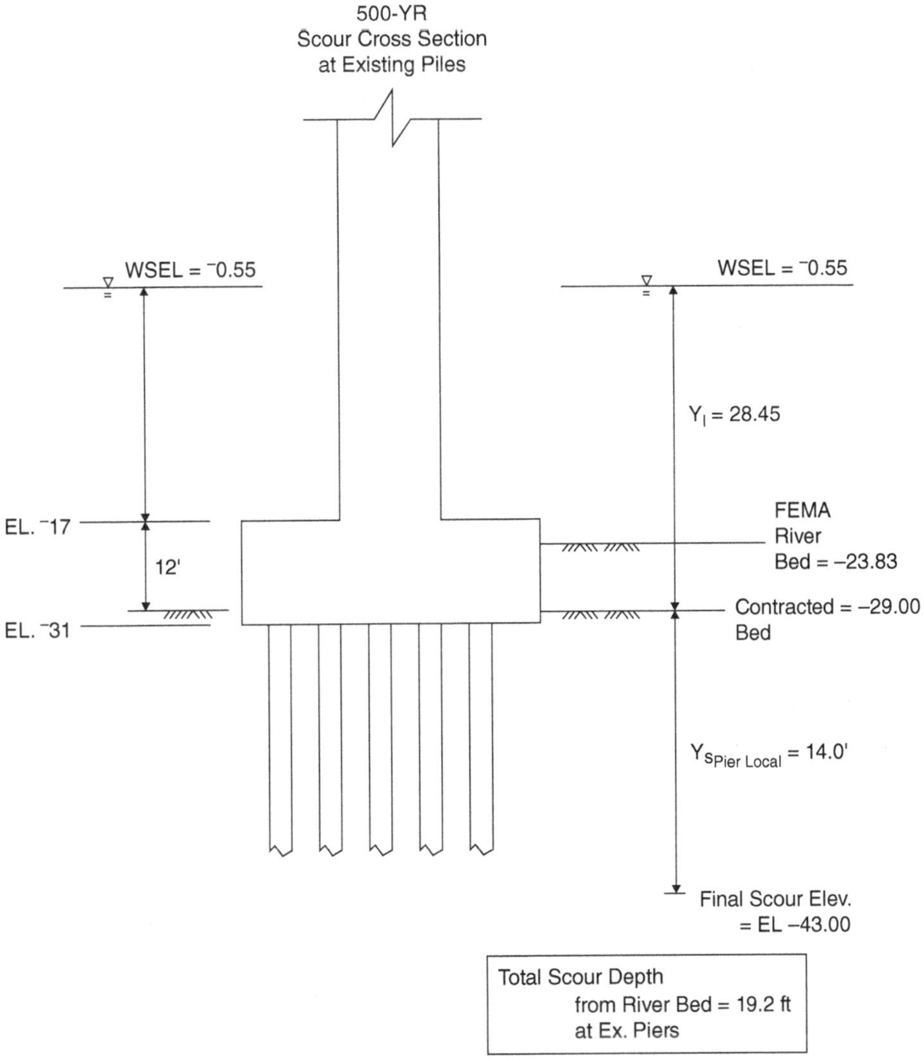

Verification of the equations used by the software is required by comparing with the equations presented in the latest version of HEC-18.

Comparison with an alternative method such as given in Appendix A or HEC-RAS software is desirable since high scour depths may result and it may not always be practical to provide armoring countermeasures.

Index of Proper Names

I

J

K

L

M

N

O

P

Subject Index

W